Simulation and Reliability Assessment of Advanced Packaging

Simulation and Reliability Assessment of Advanced Packaging

Editor

Kuo-Ning Chiang

Basel • Beijing • Wuhan • Barcelona • Belgrade • Novi Sad • Cluj • Manchester

Editor
Kuo-Ning Chiang
Dept. of Power Mechanical Engineering
National Tsing Hua University
Hsinchu
Taiwan

Editorial Office
MDPI
St. Alban-Anlage 66
4052 Basel, Switzerland

This is a reprint of articles from the Special Issue published online in the open access journal *Materials* (ISSN 1996-1944) (available at: www.mdpi.com/journal/materials/special_issues/ Simulation_Reliability_Advanced_Packaging).

For citation purposes, cite each article independently as indicated on the article page online and as indicated below:

Lastname, A.A.; Lastname, B.B. Article Title. *Journal Name* **Year**, *Volume Number*, Page Range.

ISBN 978-3-7258-0972-1 (Hbk)
ISBN 978-3-7258-0971-4 (PDF)
doi.org/10.3390/books978-3-7258-0971-4

Contents

About the Editor

Kuo-Ning Chiang

Professor K.N. Chiang received his Ph.D. from the Georgia Institute of Technology, USA. He is the Chair Professor at the National Tsing Hua University, Taiwan. After graduating from Georgia Tech, he worked for four years as a senior researcher at MSC/NASTRAN, a world-famous finite element system. From 2010 to 2013, he served as Director General of the National High-Performance Computing Center, which is the National Strategic Research Center of Taiwan. He has received outstanding research awards from the National Science and Technology Council of Taiwan three times and has published more than 450 technical papers in international journals and conference proceedings. He has been granted more than 50 invention patents. Among the major awards Professor Chiang received are the Excellence in Mechanics Award from ASME (2021) and the Outstanding Sustained Technical Contribution Award (2020) from IEEE-EPS. He was the Editor-in-Chief of the *Journal of Mechanics* and the Senior Editor of the *IEEE Transactions on Components, Packaging, and Manufacturing Technologies*. Currently, he is the Academic Editor of *Materials* and an Associate Editor of the *Journal of Electronic Packaging*. He is an IEEE, ASME, STAM, and IMAPS Fellow and an International Academy of Engineering academician. He has made significant achievements in simulation-based science and technology.

Preface

For the design and development of packaging structures, simulation technology has been widely used in the electronic packaging (EP) community. Nevertheless, it has faced some challenges in ensuring a trustworthy simulation result. In the electronics packaging community, the design-on-simulation concept has become a trend in the design of advanced packaging, the optimization of its structure, and the evaluation of the reliability life of these advanced packages, which is becoming an increasingly popular design trend within the community. In this reprint, the latest research results of simulation-based technology in advanced packaging are presented in order to provide an overview of the latest developments in this field. Among the topics covered in this reprint are the material characterization of electronic packaging, theoretical or empirical study, modeling, simulation technology, design and validation, AI-assisted design on simulation technology, and reliability life prediction. Moreover, this reprint also covered the topic of artificial intelligence and machine learning methods in depth. The focus of some of the studies in this reprint is on how to combine artificial intelligence/machine learning and finite element simulation to estimate the reliability life of wafer-level packaging and obtain the best structure combination for the packaging.

Kuo-Ning Chiang
Editor

materials

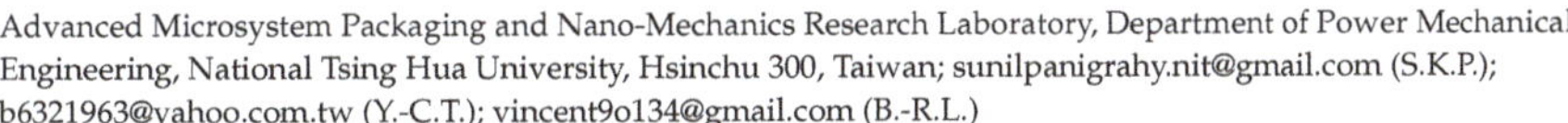

Review

An Overview of AI-Assisted Design-on-Simulation Technology for Reliability Life Prediction of Advanced Packaging

Sunil Kumar Panigrahy, Yi-Chieh Tseng, Bo-Ruei Lai and Kuo-Ning Chiang *

Advanced Microsystem Packaging and Nano-Mechanics Research Laboratory, Department of Power Mechanical Engineering, National Tsing Hua University, Hsinchu 300, Taiwan; sunilpanigrahy.nit@gmail.com (S.K.P.); b6321963@yahoo.com.tw (Y.-C.T.); vincent9o134@gmail.com (B.-R.L.)
* Correspondence: knchiang@pme.nthu.edu.tw; Tel.: +886-3-574-2925; Fax: +886-3-574-5377

Abstract: Several design parameters affect the reliability of wafer-level type advanced packaging, such as upper and lower pad sizes, solder volume, buffer layer thickness, and chip thickness, etc. Conventionally, the accelerated thermal cycling test (ATCT) is used to evaluate the reliability life of electronic packaging; however, optimizing the design parameters through ATCT is time-consuming and expensive, reducing the number of experiments becomes a critical issue. In recent years, many researchers have adopted the finite-element-based design-on-simulation (DoS) technology for the reliability assessment of electronic packaging. DoS technology can effectively shorten the design cycle, reduce costs, and effectively optimize the packaging structure. However, the simulation analysis results are highly dependent on the individual researcher and are usually inconsistent between them. Artificial intelligence (AI) can help researchers avoid the shortcomings of the human factor. This study demonstrates AI-assisted DoS technology by combining artificial intelligence and simulation technologies to predict wafer level package (WLP) reliability. In order to ensure reliability prediction accuracy, the simulation procedure was validated by several experiments prior to creating a large AI training database. This research studies several machine learning models, including artificial neural network (ANN), recurrent neural network (RNN), support vector regression (SVR), kernel ridge regression (KRR), K-nearest neighbor (KNN), and random forest (RF). These models are evaluated in this study based on prediction accuracy and CPU time consumption.

Keywords: FEM simulation; WLP; AI; machine learning; ANN; RNN; SVR; KRR; KNN; RF; regression model

Citation: Panigrahy, S.K.; Tseng, Y.-C.; Lai, B.-R.; Chiang, K.-N. An Overview of AI-Assisted Design-on-Simulation Technology for Reliability Life Prediction of Advanced Packaging. *Materials* **2021**, *14*, 5342. https://doi.org/10.3390/ma14185342

Academic Editors: Gábor Harsányi and Balázs Illés

Received: 18 August 2021
Accepted: 14 September 2021
Published: 16 September 2021

Publisher's Note: MDPI stays neutral with regard to jurisdictional claims in published maps and institutional affiliations.

1. Introduction

Electronics packaging plays an important role in the semiconductor industry. Currently, the mainstream electronic packaging structures include heterogeneous packaging, 3D packaging, system-in-packaging (SiP), fan-out (FO) packaging, and wafer-level packaging [1–8]. With the increasing complexity of packaging structures, manufacturing reliability test vehicles, and conducting ATCT experiments have become time-consuming and very expensive processes, the design-on-experiment (DoE) methodology for packaging design is becoming infeasible. As a result of the wide adoption of finite element analysis [9–15], accelerated thermal cycling tests are reduced significantly in the semiconductor industry, and package development time and cost are reduced as well. In a 3D WLP model, Liu [16] applied the Coffin–Manson life prediction empirical model to predict the reliability life of a solder joint within an accurate range. However, the results of finite element simulations are highly dependent on the mesh size, and there is no guideline to help researchers address this issue. Therefore, Chiang et al. [17] proposed the concept of "volume-weighted averaging" to determine the local strain, especially in critical areas. Tsou [18] successfully predicted packaging reliability through finite element simulation with a fixed mesh size in the critical area of the WLP structure. However, the results of simulation analysis are highly dependent on the individual researcher, and the results are usually inconsistent between simulations. In order to overcome this problem, the present work comparatively reviews an artificial intelligence (AI) approach in which electronic packaging design using a machine learning algorithm [19,20]. The use of machine learning for the analysis of electronic packaging reliability is the best way to obtain a reliable prediction result and meet the time-to-market demand.

In recent years, AI theory has been widely used in various research domains. Machine learning involves the use of AI theory combined with big data to guide computers for training and learning; eventually, a simple AI model with input and output relationships will be developed to help researchers make design decisions [21–24]. Machine learning [25–28] can be applied for regression or classification models using either supervised or unsupervised learning. In this review, because the input datasets are labeled, the learning algorithm for predicting the reliability life is considered supervised. Several machine learning algorithms exist, such as artificial neural network (ANN), support vector regression (SVR), K-nearest neighbor (KNN), kernel ridge regression (KRR), recurrent neural network (RNN), random forest (RF), and convolutional neural network (CNN).

The ANN is one of the most common machine learning methods. McCulloch and Pitts [29] proposed ANNs that mimic biological neural structures, using different numbers of hidden layers and neurons to construct different neural network structures. Denoeux [30] also explored data classification using a neural network algorithm. In this algorithm, the data are entered at the input layer and then calculated in the hidden layer. Once the calculation is completed, the result is shown in the output layer. RNN is a class of ANN algorithms in which connections between nodes form a directed network along a temporal sequence, making them more suitable for deep learning [31,32] when a large number of nonlinear datasets is available [33,34].

SVR, proposed by Cortes and Vapnik [35], is suitable for high-dimensional features, but it has not suited for a small amount of dataset. The concept of SVR is similar to the support vector machine (SVM), which is used to solve classification problems. SVM finds the best classifier by searching the hyperplane with the largest margin [36]. SVR is widely adopted in many fields, including biological, behavioral research, image analysis, and medical research [37–39]. Along with SVR, KRR is among the most popular kernel-based methods. Kernel-based methods are useful for nonlinear structural datasets [40]. KRR is simpler and faster to train with its closed-form solution, and it can outperform SVR [41]. Non-parametric calibration models eliminate the normality assumption and can represent almost any type of data, whether they are scattered or follow a certain trend. However, the models can exhibit only one type of prediction behavior, i.e., they cannot combine local and general prediction. Local calibration models, such as KNN interpolation, consider the surrounding neighborhood as input to obtain information about the output [42,43]. KNN is suitable for both classification and regression problems. The KNN classification output is decided by the nearest neighborhood, which depends on more number of nearest neighbors belongs to that class, whereas the regression output is decided by the average of the nearest neighbor value [44,45].

KNN and RF are more suitable for classification tasks; these two algorithms also show prominent performance for regression tasks. The RF algorithm was proposed in 2001 by Breiman [46]. An RF is formed by combining multiple classifications and regression trees (CART); it analyses [47,48] the data features and data distribution to generate multiple decision trees with different structures and finally summarizes the prediction results of all decision trees.

The CNN is a machine learning algorithm for image recognition and image classification. In 2012, AlexNet, proposed by Krizhevsky et al. [49], achieved the highest accuracy in the Image Net competition; consequently, the CNN has become a focus in academia and has developed rapidly [50,51].

This study reviews the use of ANN, RNN, SVR, KRR, KNN, and RF for the reliability life cycles prediction of WLP. The aim is to learn and establish a regression model for the relationship between packaging geometries (input) and life cycle (output) results. Before the implementation of the above AI algorithms, an AI training database was generated using a finite element simulation combined with the Coffin–Manson empirical equation for WLP reliability life cycles prediction, and this standard simulation procedure was experimentally verified before the generation of the massive simulation database.

The rest of this paper is presented as follows: Section 2 implemented the finite element method for WLP. Section 3 discusses different types of machine learning models. Section 4 is the result and discussion for the machine learning models, and finally, we ended with some concluding remarks in Section 5.

2. Finite Element Method for WLP

If the simulation consistently predicts the result of the experiment [52], then the simulation is an experiment; the experimental work can be replaced by a validated simulation procedure to create a large database for AI training and obtain a small and accurate AI model for reliability life cycles prediction. Once we obtain the final AI model for a new WLP structure, developers can simply input the WLP geometries, and then the life cycle can be obtained. Figure 1 illustrates this procedure.

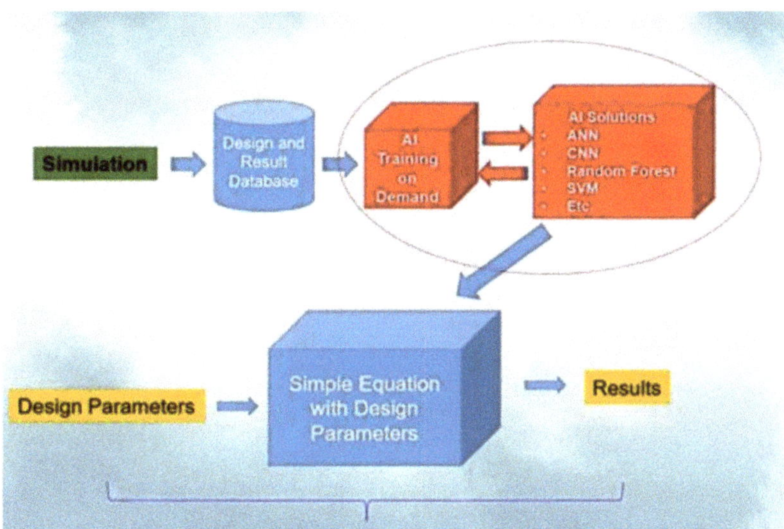

Figure 1. AI-assisted design-on-simulation procedure.

Because a huge amount of data are required to build the AI training model, this work used a two-dimensional finite element method (FEM) model for simulation. Before the database is built, the simulation process must be reliable. This work validated the simulation results with five WLP test vehicles (Tables 1 and 2). All of the sizes and specifications for different materials and the mean times to failure of the test vehicles are presented in the tables (Tables 1 and 2) [53,54]. The simulation method mainly adopted a fixed mesh size at a critical location, determined through appropriate mechanics concepts, and an empirical equation was used to validate the reliability life cycles of all test vehicles. This work used the fixed mesh size of the solder joint at the maximum distance of the neutral point of the WLP to fix the modeling pattern and simulation procedure, as proposed by Tsou et al. [18]. As shown in Figure 2, the width and height of the fixed mesh size were 12.5 μm and 7.5 μm, respectively. The solder joint geometry was generated using Surface Evolver [55].

Table 1. Dimension of WLP TV1 and TV2.

Material	TV1 (mm)	TV2 (mm)
Si Chip	5.3 × 5.3 × 0.33	4 × 4 × 0.33
Cu RDL	0.26 × 0.008	0.26 × 0.008
UBM	–	0.24 × 0.0086
Cu Pad	0.22 × 0.025	0.22 × 0.025
SBL1	5.3 × 5.3 × 0.0075	4 × 4 × 0.0075
SBL2	0.01	0.004
PCB	10.6 × 10.6 × 1	8 × 8 × 1
Low-k	5.3 × 0.005	4 × 0.005
Ball Diameter (mm)	0.25	0.25
Ball Pitch (mm)	0.4	0.4
Ball Counts	121	100
MTTF (cycles)	318	1013

Table 2. Dimension of WLP TV3, TV4 and TV5.

Material	TV3 (mm)	TV4 (mm)	TV5 (mm)
Si Chip	4 × 4 × 0.33	4 × 4 × 0.33	6 × 6 × 0.33
Cu RDL	0.18 × 0.004	0.2 × 0.004	0.25 × 0.0065
UBM	0.17 × 0.0086	0.19 × 0.0086	0.24 × 0.0075
Cu Pad	0.22 × 0.025	0.22 × 0.025	0.22 × 0.04
SBL1	4 × 4 × 0.0075	4 × 4 × 0.0075	6 × 6 × 0.008
SBL2	0.004	0.004	0.0065
PCB	8 × 8 × 1	8 × 8 × 1	12 × 12 × 1
Low-k	4 × 0.005	4 × 0.005	–
Ball Diameter (mm)	0.18	0.2	0.25
Ball Pitch (mm)	0.3	0.3	0.4
Ball Counts	144	144	196
MTTF (cycles)	587	876	904

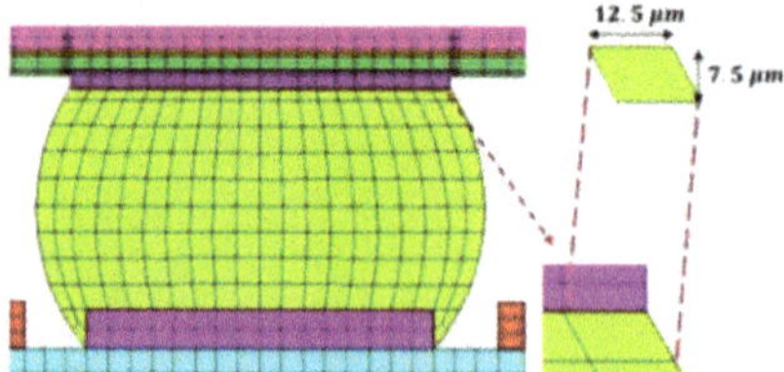

Figure 2. The critical mesh size.

In the simulation process, the solder material was a nonlinear plastic material. Therefore, PLANE182, which has good convergence characteristics and can deal with large deformations, was used as the solder ball element. This work used PLANE42 for other components, which had linear material properties. Table 3 presents the list of individual material properties. The Young's moduli at different temperatures of the solder joint are listed in Table 4. Figure 3 shows the stress–strain curve for an Sn–Ag–Cu (SAC)305 solder joint. The stress–strain curve [56], obtained by tensile testing and the Chabochee kinematic hardening model, was used to describe the tensile curves at different temperatures. Once the model is built, boundary conditions and external thermal loading are required for the WLP simulation.

Table 3. Material properties of WLP.

Material	Young's Modulus (Gpa)	Poisson's Ratio	CTE (ppm/°C)
Solder joint	Temperature-dependent	0.35	25
Silicon chip	150	0.28	2.62
Copper	68.9	0.34	16.7
Low-k	10	0.16	5
Solder mask	6.87	0.35	19
PCB	18.2	0.19	16

Table 4. Temperature-dependent Young's modulus of SAC solder ball.

Temperature	Young's Modulus (GPa)
233 K	45.74
253 K	42.22
313 K	31.66
353 K	24.62
398 K	16.70

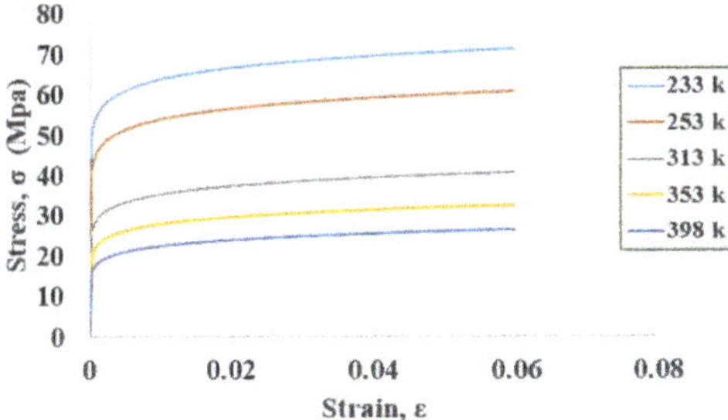

Figure 3. Stress–strain curve for SAC solder.

Electronics packaging geometry is usually symmetrical; therefore, in this study, half of the 2D structure was modeled along the diagonal, as shown in Figure 4. The X-direction displacement on each node was fixed to zero owing to the Y-symmetry. To prevent rigid body motion, the node at the lowest point of the neutral axis, which is at the printed circuit board (PCB), has all degrees of freedom fixed. The complete finite element model and the boundary conditions are shown in Figure 5. The thermal loading condition used in this research was JEDEC JESD22-A104D condition G [57], and the temperature range was −40 °C to 125 °C. The ramp rate was fixed at 16.5 °C/min and the dwell time was 10 min. In a qualified design, its mean-cycle-to-failure (MTTF) should pass 1000 thermal cycles. After the simulation process is completed, the incremental equivalent plastic strain in the critical zone is substituted into the strain-based Coffin–Manson model [58] for reliable life cycle prediction. For a fixed temperature ramp rate, this method is as accurate as the energy-based empirical equation [59,60] but with much less CPU time.

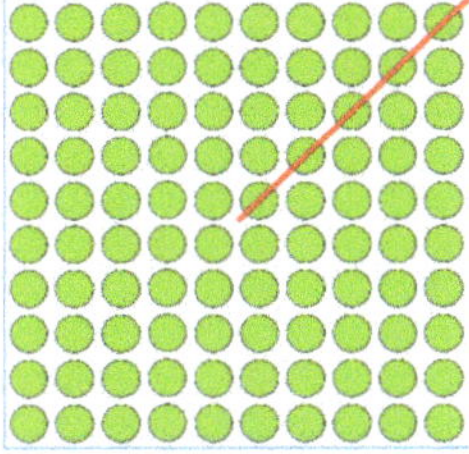

Figure 4. Symmetrical solder ball geometry.

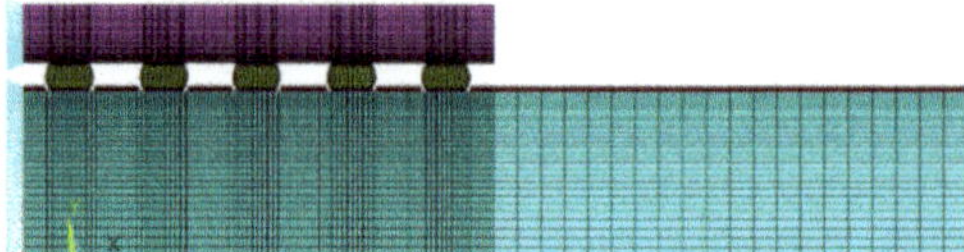

Figure 5. FEM model boundary condition.

The empirical formula for Coffin–Manson equivalent plastic strain model is shown in Equation (1):

$$N_f = C(\Delta\varepsilon_{eq}^{pl})^{-\eta} \tag{1}$$

where N_f is mean cycle to failure, C and η are empirical constants, and $\Delta\varepsilon_{eq}^{pl}$ is the incremental equivalent plastic strain. For SAC solder joint, C and η are 0.235 and 1.75 [61,62].

Table 5 presents the predicted reliability life cycles of the WLP structure. The results show that the difference between the FEM-predicted life cycle and experiment result is within a small range. Therefore, experiments can be replaced by this validated FEM simulation to minimize the cost and time. Compared with the experiment approach, this validated FEM simulation procedure can provide large amounts of data within much less time and can be effectively used to generate a database for AI training.

Table 5. WLP finite element results for five test vehicles.

Test Vehicle	Experimental Reliability (Cycles)	Simulation Reliability (Cycles)	Difference
TV1	318	313	−5
TV2	1013	982	−31
TV3	587	587	0
TV4	876	804	72
TV5	904	885	19

3. Machine Learning

Machine learning is an AI methodology that processes huge datasets to guide a computer for training, learning, and finally, building a simple regression model. In this study, several different machine algorithms were used, including ANN, RNN, SVR, KRR, KNN, and RF, implemented using the Python language. A supervised regression model, e.g., the WLP reliability life cycle prediction model, requires both input data (geometry parameters) and the corresponding output result (life cycles) for machine learning algorithms to build the final AI model of the WLP package. Once the regression model is established for a new WLP structure, the designer can simply input the WLP geometries of each component into the AI regression model to obtain the reliability life of this new WLP. This is a powerful and reliable technique for new packaging design.

3.1. Establishment of Dataset

The WLP structure consists of several components, including the solder mask, solder ball, I/O pad, stress buffer layer, and silicon chip, etc. (Figure 6). For illustration purposes, the four most influential parameters, namely silicon chip thickness, stress buffer layer thickness, upper pad diameters, and lower pad diameters, were selected to build the AI model and predict the reliability life cycles of new WLP structures. These four design parameters were used to generate both training and testing datasets for AI machine learning algorithms. Tables 6 and 7 show the generated training dataset obtained through FEM simulation. First, the number of training features generated in this research was 576 (Table 6), and 1296 (Table 7) datasets. For testing, 54 features were selected randomly from the interpolation of the above training dataset, which can help to build the AI training model. By increasing the training dataset, AI performance improves; however, the CPU/GPU time also increases.

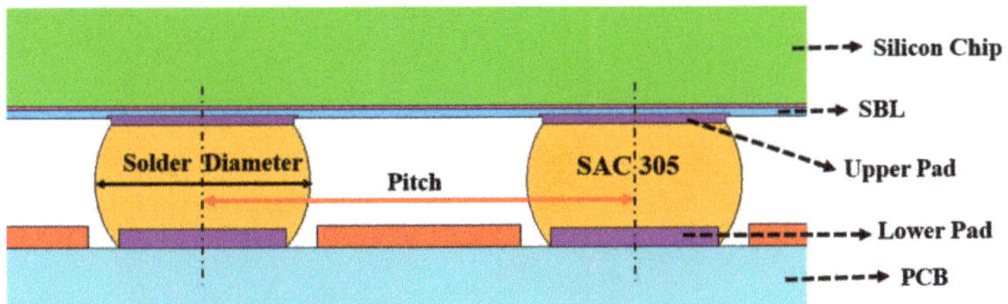

Figure 6. WLP geometry structure.

Table 6. Training data 576 of FEM for four input features.

Feature Name	Level (mm)
Upper Pad Diameter	0.18, 0.20, 0.22, 0.24
Lower Pad Diameter	0.18, 0.20, 0.22, 0.24
Chip Thickness	0.20, 0.25, 0.30, 0.35, 0.40, 0.45
Stress Buffer Layer Thickness	0.0075, 0.0125, 0.0175, 0.0225, 0.0275, 0.0325
Total Number	576

Table 7. Training data 1296 of FEM for four input features.

Feature Name	Level (mm)
Upper Pad Diameter	0.18, 0.19, 0.20, 0.21, 0.22, 0.23
Lower Pad Diameter	0.18, 0.19, 0.20, 0.21 0.22, 0.23
Chip Thickness	0.20, 0.25, 0.30, 0.35, 0.40, 0.45
Stress Buffer Layer Thickness	0.0075, 0.0125, 0.0175, 0.0225, 0.0275, 0.0325
Total Number	1296

3.2. ANN Model

The ANN model is based on the concept of the brain's self-learning ability, mimicking the human nervous system to process information. It is a multilayer neural network, as shown in Figure 7. The model consists of three layers: the input layer, where the data are provided; the hidden layer, where the input data are calculated; and the output layer, where the results are displayed [63]. As the numbers of neurons and hidden layers are increased, the ability to handle nonlinearity improves. However, these conditions may result in high computational complexity, overfitting, and poor predictive performance.

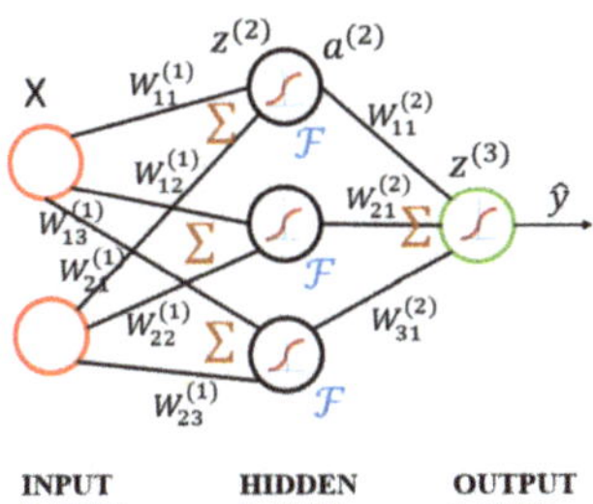

Figure 7. Schematic diagram of artificial neural network.

In the above ANN model, a_i^l is the ith activation element of the lth layer in the hidden layer. b_i^l is bias, a_i^l is equal to input value times weight W_{ji}^l and add the bias in Equation (2):

$$z_i^l = \sum_{i=1}^{n} w_{ji}^l a_i^l + b_i^l \tag{2}$$

From the calculation point of view, at first, the input layer data combined with bias and weight to obtain some value. The calculated input value, i.e., a_{i}^l, is substituted into activation function, e.g., Sigmoid, to be converted into a nonlinear form as an input of a_i^{l+1} for the next layer.

$$a_i^{l+1} = \phi(z_i^l) \tag{3}$$

where the activation function is shown in Figure 8.

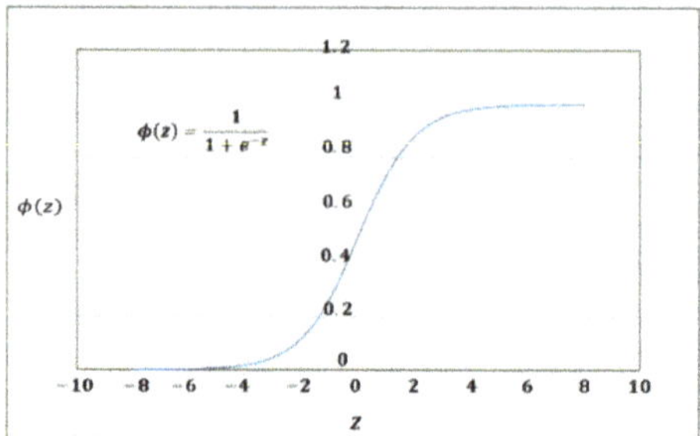

Figure 8. Sigmoid activation function.

3.3. RNN Model

RNN is a type of neural network that can model "time-like"-series data, and it commonly adopts a nonlinear structure in deep learning. RNN [64,65] works on the principle that the output of a particular layer is fed back to the input layer to realize a time-dependent neural network and a dynamic model. Consequently, an ANN with nodes connected in a ring shape is obtained, as shown in the left half of Figure 9. The ring-shaped neural network is expanded along the "time" axis, as shown in the right half of Figure 9, where the "time" step t and the hidden state s_t can be expressed as a function of the output from the previous (s_{t-1}) "time" steps and previous layers (x_t). U, V, and W denote the shared weights in RNN models during different "time" steps. Generally, the RNN series model can be divided into four types according to the number of inputs and outputs in given "time" steps; that is, one to one (O to O), one to many (O to M), many to one (M to O), and many to many (M to M). To synchronize the input features with the output results, RNN models can be subdivided into different series models, as shown in Figure 10 [66].

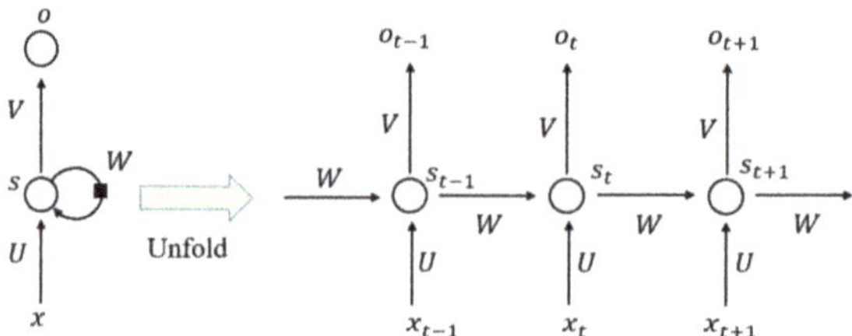

Figure 9. Schematic structure of recurrent neural network.

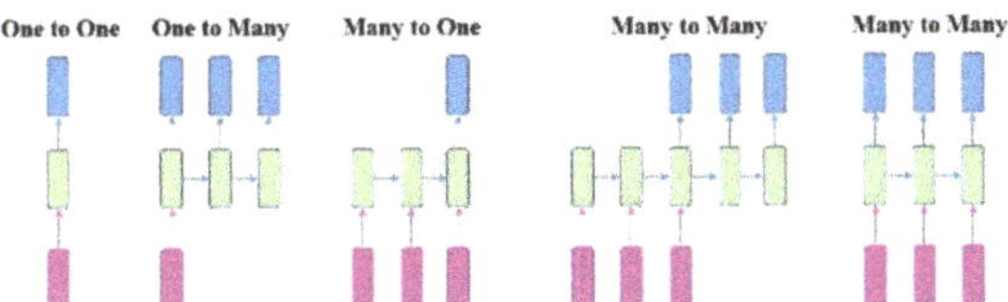

Figure 10. Different series model for RNN.

3.4. SVR Model

This regression method evolved from the support vector machine algorithm. It transforms data to high-dimensional feature space and adapts the ε-insensitive loss function (Equation (4)) to perform the linear regression in feature space (Equation (5)). In this regression method, the norm value of w is also minimized to avoid the overfitting problem. In other words, $f(X, w)$, which is the function of the SVR model, will be as flat as possible. The SVR concept is illustrated in Figure 11. The data points outside the ε-insensitive zone are called support vectors, and two slack variables, ξ_i and ξ_i^*, are used to record the loss of each support vector. Thus, the whole SVR problem can be seen as an optimization problem (Equation (6)).

$$L(y, f(X)) = \left\{ \begin{array}{ll} 0 & if\,|y - f(X)| \leq \varepsilon \\ |y - f(X)| - \varepsilon & otherwise \end{array} \right\} \tag{4}$$

$$f(X) = \langle w, \phi(X) \rangle + b \tag{5}$$

$$\text{minimize } \tfrac{1}{2}\|w\|^2 + C \sum_{i=1}^{1} \xi_i + \xi_i^*$$
$$\text{subject to } \left\{ \begin{array}{l} y_i - \langle w, \phi(X) \rangle - b \leq \varepsilon + \xi_i \\ \langle w, \phi(X) \rangle + b - y_i \leq \varepsilon + \xi_i^* \\ \xi_i, \xi_i^* \geq 0 \end{array} \right. \tag{6}$$

where $L(y, f(X))$ is the LaGrange function of a single out variable y as a function of n input variables X using a function $f(X)$. w is the weight parameter, b represents bias, and $\varnothing(X)$ is the transformation equation. C is a penalty factor that is used to control the accuracy of the regression model; if C is set to infinity, it means you are only concerned about accuracy rather than model complexity.

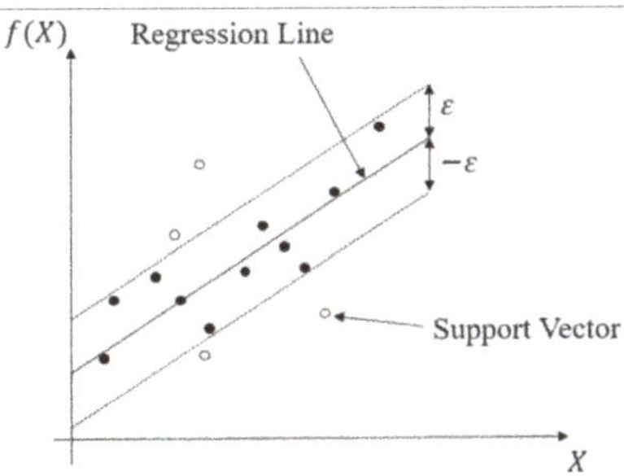

Figure 11. Schematic diagram of SVR.

The SVR problem can be solved easily as a dual problem, and the kernel function $K(x_i, x_j)$, which satisfies Mercer's condition in the objective function, is used. Here, α_i and α_i^* are Lagrange multipliers, and data points with positive and non-zero α_i and α_i^* are support vectors.

In order to solve the optimization problem, the regression model is built as shown in Equation (7), where b is the bias of the SVR model.

$$f(x) = \sum_{i=1}^{l} (\alpha_i^* - \alpha_i) K(X, X_i) + b \tag{7}$$

In Equation (7), the term $K(X, X_i)$ is known as the kernel function, and X_i is the training sample, with X as an input variable. This kernel function should be chosen as a dot product in the high-dimensional feature space [67]. There are numerous types of kernel functions. The commonly used kernel functions for SVR are the linear kernel, polynomial kernel, radial basis function (RBF) kernel, and sigmoid kernel.

All of the kernel functions satisfy Mercer's condition; however, the regression results of the kernels vary. Therefore, it is essential to choose the best kernel function for the SVR algorithm to obtain optimal performance.

3.5. KRR Model

KRR combines ridge regression with the kernel "trick". This model can learn a linear function in the space induced by the respective kernel and the dataset. Nonlinear functions in the original space can be used by the nonlinear kernels. The KRR algorithm also analyzes several kernels such as the RBF kernel, sigmoid kernel, and polynomial kernel to find the suitable kernel function for the WLP nonlinear dataset.

The KRR is possibly the most elementary algorithm that can be kernelized to ridge regression [68]. In this study, a linear function that models the dependencies between the covariate input variable x_i and the response variable y_i is found. The classic method is used to minimize the quadratic cost, as shown in Equation (8). However, for the nonlinear dataset, the lower-dimensional feature space replaces the higher-dimensional feature space; that is, $X_i \rightarrow \Phi(X_i)$. To convert lower-dimensional space to higher-dimensional space, the predictive model undergoes overfitting. Hence, to avoid overfitting, this function requires regularization.

$$C(W) = \frac{1}{2} \sum_i (y_i - W^T X_i)^2 \tag{8}$$

where $C(W)$ is the cost function of the weight-decay W and W^T is the initial weight required for the input samples for the KRR model. A simple and effective way to regularize is to penalize the norm of W. This is called "weight-decay", and it remains to be determined how to choose λ that is known as regularizing factor. Another way, the algorithm can be used cross-validation to avoid over-fitting. Hence, the total cost function becomes

$$C = \frac{1}{2} \sum_i (y_i - W^T X_i)^2 + \frac{1}{2} \lambda \|W^2\| \tag{9}$$

Equation (9) needs to be minimized. Therefore, the derivative of the equation must be obtained and then equated to zero.

To optimize the above cost function C, this study introduces Lagrange multipliers into the problem. Consequently, the derivation step becomes similar to that in the SVR case. After the optimization problem is solved, the resulting KRR regression model is shown in Equation (10).

$$f(x) = \sum_{i=1}^{N} \alpha_i^* K(x, x^i) \tag{10}$$

where $K(x, x^i)$ is the kernel function of the x^i training sample with x as the input variable, and α_i^* is the weight of the KRR model and is equal to

$$\alpha_i^* = (K + \lambda I)^{-1} y \tag{11}$$

Hence, Equation (11) is very simple and more flexible due to introducing kernel function K, λ is the regularize factor with the identity matrix I, and y is the response variable. This model can also avoid both model complexity and computational time.

One big disadvantage of ridge regression is that there is no sparseness in the α vector; that is, there is no concept of support vectors. Sparseness is useful because when a new example is tested, only the support vectors need to be summed, which is much faster than summing the entire training set. In SVR, the only source of sparseness is the inequality constraints because, according to the complementary slackness conditions, if the constraint is inactive, then the multiplier α_i is zero.

3.6. KNN Model

The KNN model is a statistical tool for estimating the value of an unknown point based on its nearest neighbors [69]. The nearest neighbors are usually calculated as the points with the shortest distance to the unknown point [70]. Several techniques are used to measure the distance between the neighbors. Two simple techniques are used in this study: the Euclidean distance function $d(x,y)$, provided in Equation (12), and the Manhattan distance function $d(x,y)$, provided in Equation (13).

$$d(x,y) = \|x - y\| = \sqrt{\sum_{i=1}^{n}(x_i - y_i)^2} \tag{12}$$

$$d(x,y) = \sum_{i=1}^{n}|x_i - y_i| \tag{13}$$

where $x = (x_1,\ldots,x_n)$, $y = (y_1,\ldots,y_n)$, and n is the vector size. The K neighbor point that has the shortest distance to the unknown point is used to estimate its value using Equation (14).

$$\hat{y}_i = \sum_{i=1}^{n} w_i y_i \tag{14}$$

where w_i is the weight of every single neighbor point y_i to the query point $\hat{y}$ [71].

The KNN algorithm defined in Equation (14) is the weighted average of the neighborhood. The simplest KNN model is the mean of the contiguity, which is obtained in the case of uniform weights, where all of the neighbor points have the same effect on the estimation $w_i = \frac{1}{n}$. In contrast, when the neighbor points are assumed to have different effects on the query point estimation, different weights can be applied. The simplest weight function is provided in Equation (15).

$$w_i = \frac{d_i}{\sum\limits_{i=1}^{n} d_i} \tag{15}$$

where d_i is the distance between the unknown point and its neighbor. The weight function must reach its maximum value at zero distance from the interpolated point, and as the distance increases, the function decreases [72].

The KNN estimation shown in Equation (14) depends only on the neighbor points; therefore, it neglects the trend of the whole dataset. However, Equation (15) provides better KNN estimations because the weighted distance considers a lower number of nearest values. Finally, this KNN algorithm is more suitable for regression and classification problems according to the simplest weight function.

3.7. The RF Regression Model

RF is a collection of decision trees. These tree models usually consist of fully grown and unpruned CARTs.

The structure of the RF regression model is shown in Figure 11. This algorithm creates an RF by combining several decision trees built from the training dataset. The CART tree selects one feature from all of the input features as the segmentation condition according to the minimum mean square error method.

The RF algorithm procedure comprises three steps. In step 1, the bagging method is used to create a subset that accounts for approximately 2/3 of the total data volume. In step 2, if the data value is greater than the selected feature value, the data points are separated to the right from the parent node, otherwise to the left of the parent node (Figure 12). Afterward, a set of trained decision trees is created. In step 3, the RF calculates the average value of all decision tree results to obtain the final predicted value.

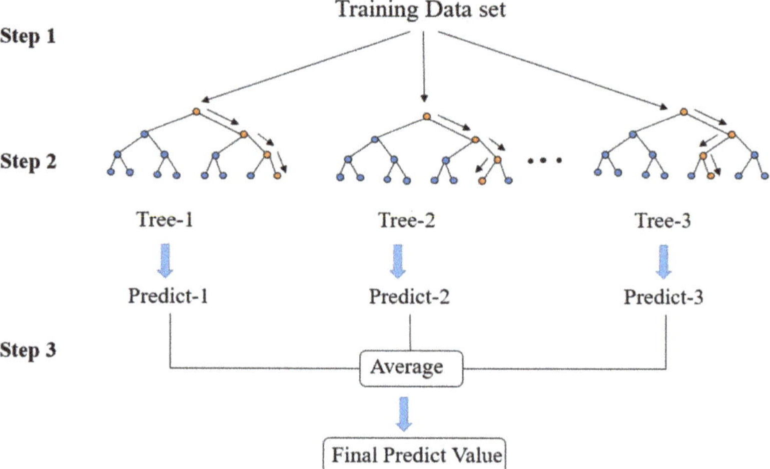

Figure 12. Schematic diagram of random forest structure.

3.8. Training Methodology

To obtain the best performance and avoid overfitting of the final trained regression model, several techniques, including data preprocessing (for standardization), cross-validation (for parameter selection), and grid search (for hyperparameter determination), were applied during training. The AI regression model was estimated using the method in the flowchart given in Figure 13.

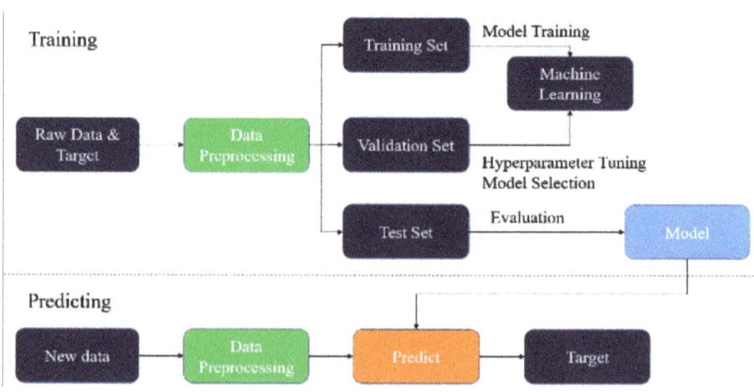

Figure 13. Methodology flow chart.

3.8.1. Data Preprocessing

The dataset values are not in a uniform range. Hence, before the machine learning model is developed, the data need to be preprocessed to standardize all of the input and output datasets and improve the modeling performance. Several data preprocessing methods, including min–max scaling, robust scaling, max absolute scaling, and standard scaling, were used in this study. Hence, of all preprocessing methods, we need to select the one method that provides the most accurately predicted output from the input dataset.

3.8.2. Cross-Validation

Cross-validation is the most frequently used parameter selection method. The basic idea of cross-validation is that not all of the dataset is used for training; a part of it (which does not participate in training) is used to test the parameters generated by the training set. The training data are trained with different model parameters and verified by the validation set to determine the most appropriate model parameters. Cross-validation methods can be divided into three categories: the hold-out method, k-fold method, and leave-one-out method. Owing to the huge calculation burdens of the hold-out method and the leave-one-out method, the k-fold method was chosen for this study (Figure 14). After the choice of data preprocessing method was confirmed, cross-validation was performed to avoid overfitting of the machine learning model, as shown in Figure 14. The dataset was divided into 10 parts, and each part acted as either a validation or training set in different training steps. The validation sets were also used to predict the training results.

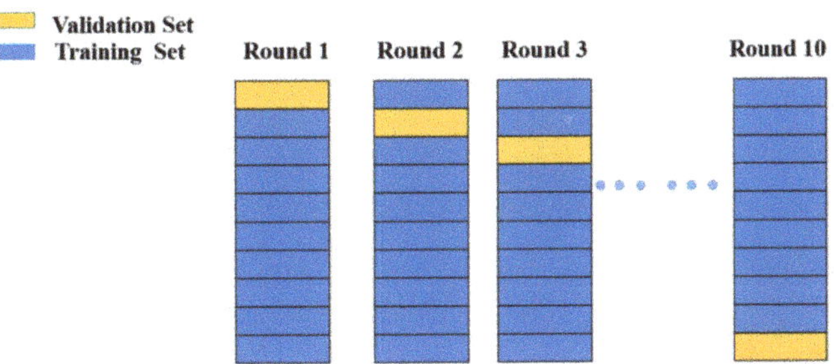

Figure 14. Cross-validation model diagram from Round 1 to Round 10.

3.8.3. Grid Search Technique

Grid search is a large-scale method for finding the best hyperparameter to build the training model. In order to determine the best parameter, the search range value needs to be set by the model builder. Although the method is simple and easy to perform, it is time-consuming. Therefore, to reduce the computation time, this work adopted the grid search technique to find the best hyperparameter as compared to manually searching the hypermeter, and eventually, the training model was fixed with the above hyperparameters to run the best AI model.

4. Results and Discussion

This section discusses the outcomes of the different machine learning algorithms used for the WLP structure design. In this study, we analyzed both the accuracy and CPU time consumption of the algorithms. Regarding accuracy, the mean absolute error (MAE) and the maximum absolute error between the FEM-predicted reliability life cycle and the AI-predicted reliability life cycle of the WLP structure were calculated in this work. We discuss both the training and testing error analysis for different datasets. Similarly, the CPU time required for every regression model to predict the WLP structure reliability life is also discussed. Training datasets comprising 576 and 1296 and a testing dataset comprising 54 WLP geometric combinations were used in this study.

Table 8 presents the ANN regression results for 576 training datasets. As shown in the table, different numbers of neurons, from 10 to 500, and hidden layers, from 2 to 20, were tested. After the numbers of hidden layers and neurons were tuned, the best ANN-predicted number of life cycles was calculated. Validated against the FEM results, the best MAE of AI prediction was five cycles, and the maximum absolute error was 28 cycles, with 10 hidden layers and 200 neurons. Similarly, Table 9 presents the ANN regression results for the training dataset with 1296. From the table, the best MAE was three cycles, and the maximum absolute error was 18 cycles, with five hidden layers and 500 neurons. Therefore, the ANN regression model based on the training dataset with 1296 numbers showed better accuracy than that based on the dataset with 576 numbers. Moreover, the CPU time performances for both datasets were different (Table 10). The CPU time required for 576 datasets was 151 s, whereas 235 s was required for 1296 datasets. Thus, although the dataset with a higher number of features resulted in higher accuracy, the CPU time was higher.

Table 8. Comparison results of FEM and ANN models on 576 training datasets.

Number of Neurons	Hidden Layer (2)		Hidden Layer (5)		Hidden Layer (10)		Hidden Layer (20)	
	Mean Absolute Error in Cycle	Maximum Absolute Error in Cycle	Mean Absolute Error in Cycle	Maximum Absolute Error in Cycle	Mean Absolute Error in Cycle	Maximum Absolute Error in Cycle	Mean Absolute Error in Cycle	Maximum Absolute Error in Cycle
10	57	148	62	157	65	182	51	138
20	60	164	65	179	14	69	10	60
50	63	159	11	51	8	54	11	67
100	66	173	7	51	5	31	9	40
200	64	173	7	47	5	28	5	40
500	10	62	5	43	7	35	7	46

Table 9. Comparison results of FEM and ANN models on 1296 training datasets.

Number of Neurons	Hidden Layer (2)		Hidden Layer (5)		Hidden Layer (10)		Hidden Layer (20)	
	Mean Absolute Error in Cycle	Maximum Absolute Error in Cycle	Mean Absolute Error in Cycle	Maximum Absolute Error in Cycle	Mean Absolute Error in Cycle	Maximum Absolute Error in Cycle	Mean Absolute Error in Cycle	Maximum Absolute Error in Cycle
10	51	142	47	137	47	125	14	57
20	47	140	47	143	12	50	9	57
50	47	127	22	35	7	42	5	43
100	23	104	6	35	4	24	5	27
200	13	70	4	31	3	22	4	27
500	6	43	3	18	3	20	3	19

Table 10. Comparison of ANN regression result for WLP.

Training Data Set (ANN)	Neuron Number	Hidden Layer	Mean Absolute Error in Cycle	Maximum Absolute Error in Cycle	CPU Time in Second
576	200	10	5	28	151
1296	500	5	3	18	235

Table 11 presents the results of the RNN regression models for the training dataset with 576 numbers. The RNN models considered different numbers of neurons and hidden layers. The best MAE was six cycles, and the maximum absolute error was 37 cycles, with 500 neurons and five hidden layers. Table 12 lists the RNN regression results for 1296 training datasets. From the table, the best MAE was three cycles, and the maximum absolute error was 27 cycles. Therefore, the RNN regression model based on the dataset with 1296 numbers exhibited better accuracy. However, the CPU time (698 s) was higher than that of the model based on 576 training datasets (173 s; Table 13). Given the above, the ANN regression model outperformed the RNN regression model in both accuracy and

CPU time. This implies that the ANN model is more flexible than RNN because of the RNN model's complex structure.

Table 11. Comparison results of FEM and RNN models on 576 training datasets.

| | Hidden Layer (2) | | Hidden Layer (5) | | Hidden Layer (10) | | Hidden Layer (20) | |
Number of Neuron	Mean Absolute Error in Cycle	Maximum Absolute Error in Cycle	Mean Absolute Error in Cycle	Maximum Absolute Error in Cycle	Mean Absolute Error in Cycle	Maximum Absolute Error in Cycle	Mean Absolute Error in Cycle	Maximum Absolute Error in Cycle
10	65	166	61	178	61	159	26	101
20	63	163	62	168	13	54	16	64
50	65	157	29	71	8	57	10	43
100	65	159	12	71	7	37	8	65
200	17	79	8	50	9	53	10	65
500	12	69	6	36	6	37	9	50

Table 12. Comparison results of FEM and RNN models on 1296 training datasets.

| | Hidden Layer (2) | | Hidden Layer (5) | | Hidden Layer (10) | | Hidden Layer (20) | |
Number of Neuron	Mean Absolute Error in Cycle	Maximum Absolute Error in Cycle	Mean Absolute Error in Cycle	Maximum Absolute Error in Cycle	Mean Absolute Error in Cycle	Maximum Absolute Error in Cycle	Mean Absolute Error in Cycle	Maximum Absolute Error in Cycle
10	125	355	44	141	11	61	6	44
20	69	215	21	103	10	54	10	57
50	33	154	5	28	4	30	4	30
100	10	51	3	28	4	32	4	32
200	4	36	4	32	6	43	4	32
500	3	27	3	39	159	506	159	506

Table 13. Comparison of RNN regression results for WLP.

Training Data Set (RNN)	Neuron Number	Hidden Layer	Mean Absolute Error in Cycle	Maximum Absolute Error in Cycle	CPU Time in Second
576	500	5	6	36	173
1296	500	2	3	27	698

Table 14 presents the SVR results for 576 datasets. The table presents the accuracy and CPU time analysis results for the SVR model considering different kernel functions and different hyperparameters. The best MAE for the testing data was 13, and the maximum absolute error was 55 for the RBF kernel-based model. The shortest CPU time was 0.093 s. For the training dataset with 1296 numbers, the SVR exhibited better accuracy (Table 15). For this dataset, the best MAE for the testing data was 7.3 cycles, and the maximum absolute error was 30 cycles.

However, the larger dataset had a higher CPU time requirement. Given the above SVR results, we can infer that the RBF kernel plays a more important role in obtaining good SVR performance compared with the sigmoid and polynomial kernel functions.

Table 14. Comparison results of FEM and SVR models on 576 training datasets.

SVR Kernel Function	RBF Kernel	Sigmoid Kernel	Polynomial Kernel Degree 3
Hyperparameter (C)	2250.97	540.31	2563
Hyperparameter (γ)	0.86	1.7	4
Hyperparameter (ε)	10	10	10
Training Score (R^2)	0.996	0.97	0.976
Cross-Validation Score (R^2)	0.964	0.936	0.954
Maximum Absolute Error (cycle) Train Data	68	125	142
Mean Absolute Error (cycle) Train Data	10	30	25
Maximum Absolute Error (cycle) Test Data	55	93	94
Mean Absolute Error (cycle) Test Data	13	29	20
CPU Time In Second	0.093	0.119	0.153

Table 15. SVR result comparison.

SVR Training Dataset	576	1296
Hyperparameter (C)	2250.97	3000.51
Hyperparameter (γ)	0.86	2.37
Hyperparameter (ε)	10	10
Training Score (R^2)	0.996	0.998
Cross-Validation Score (R^2)	0.964	0.981
Maximum Absolute Error (cycle) Train Data	68	46
Mean Absolute Error (cycle) Train Data	10	7.3
Maximum Absolute Error (cycle) Test Data	55	30
Mean Absolute Error (cycle) Test Data	13	8
CPU Time In Second	0.093	6.00

Table 16 presents the KRR results for 576 training datasets. The table presents the accuracy and CPU time analysis for different kernel functions with hyperparameters used in the KRR algorithm. The best MAEs for the training and testing datasets were 8.4 and 12.2 cycles, respectively, for the model with the RBF kernel function. Moreover, the KRR model with the RBF kernel function required a short CPU time (0.093 s). The KRR model with the larger dataset showed better performance (Table 17). The best MAE of the test data was 5.6 cycles, and the maximum absolute error was 24 cycles for the training dataset with 1296. However, the CPU time was higher than that for the training dataset with 576 numbers (Table 18). Similar to the SVR results, the KRR results also show that the RBF kernel exhibited better accuracy and CPU time than the sigmoid and polynomial kernel functions. The KRR algorithm outperformed the SVR model in terms of accuracy and CPU time. Meanwhile, the ANN outperformed the RNN, SVR, and KRR in terms of accuracy. However, KRR and SVR outperformed ANN and RNN in terms of CPU time, owing to the usage of more hidden layers and a greater number of neurons are used in ANN and RNN. The KNN results are shown in Table 19. The table presents the accuracy of KNN in terms of the Euclidean and Manhattan distances versus the number of nearest neighbors (K) used in this algorithm for 576 training datasets. The best MAE and maximum absolute error were 18.2 and 72 cycles with the Manhattan distance when the K value was 9. From Table 19, the algorithm performs better in terms of the Manhattan distance than the Euclidean distance under the same K value. Similarly, Table 20 shows the results of KNN based on the dataset with 1296 numbers. From Table 20, the best MAE and maximum absolute error were 7.5 and 23 cycles, respectively, with the K value as 3. Table 21 compares the cases in which the weight as the uniform is used with the Euclidean distance and the weight as the distance is used with Euclidean distance. From the table, the case of the weight in terms of distance showed better MAE, and maximum absolute error than the case of the uniform weight is used for the KNN model of a WLP structure. With the increase in the number of training datasets from 576 to 1296, the model accuracy improved, whereas the CPU times remained similar, i.e., 0.03 and 0.034 s (Table 22). Table 23 compares the results of RF regression for the two training datasets. From the table, the MAE for the test data was 26.3 cycles for the 1296 dataset and 36 cycles for the 576 datasets. Therefore, the increase in the number of training datasets improved the accuracy. However, the CPU time was higher for the larger training dataset (Table 23).

Table 16. Comparison results of FEM and KRR models on 576 training datasets.

KRR Kernel Function	RBF Kernel	Sigmoid Kernel	Polynomial Kernel Degree 3
Hyperparameter (α)	0.01	8.16	0.1
Hyperparameter (γ)	1	0.09	3.9
Maximum Absolute Error (Cycle) Train Data	57	149	75
Mean Absolute Error (Cycle) Train Data	8.4	38.7	18.9
Maximum Absolute Error (Cycle) Test Data	39	84	57
Mean Absolute Error (Cycle) Test Data	12.2	25.4	17.9
CPU Time In Second	0.093	0.117	0.157

Table 17. Comparison results of FEM and KRR models on 1296 training datasets.

KRR Kernel Function	RBF Kernel	Sigmoid Kernel	Polynomial Kernel Degree 3
Hyperparameter (α)	$1 \times e^{-10}$	$3 \times e^{-9}$	$1 \times e^{-9}$
Hyperparameter (γ)	0.19	0.02	2
Maximum Absolute Error (Cycle) Train Data	40	46	107
Mean Absolute Error (Cycle) Train Data	5.3	7.2	16
Maximum Absolute Error (Cycle) Test Data	24	29	45
Mean Absolute Error (Cycle) Test Data	5.6	7	14.5
CPU Time In Second	0.422	1.495	0.787

Table 18. KRR result comparison.

KRR Training Dataset	576	1296
Hyperparameter (α)	0.01	$1 \times e^{-10}$
Hyperparameter (γ)	1	0.19
Maximum Absolute Error (Cycle) Train Data	57	40
Mean Absolute Error (Cycle) Train Data	8.4	5.3
Maximum Absolute Error (Cycle) Test Data	39	24
Mean Absolute Error (Cycle) Test Data	12.2	5.6
CPU Time In Second	0.093	0.422

Table 19. Comparison results of FEM and KNN models on 576 training datasets.

Number Nearest Neighbor (K)	Euclidean Distance		Manhattan Distance	
	Mean Absolute Error in Cycle	Maximum Absolute Error in Cycle	Mean Absolute Error in Cycle	Maximum Absolute Error in Cycle
1	53.7	172	63	172
2	59.3	128	56.9	124
3	38.2	115	37.9	83
4	28.9	81	28.1	81
5	25.5	74	24.2	66
6	27.6	71	25.6	67
7	22.3	71	21.4	63
8	21.6	77	20.5	77
9	18.9	83	18.2	72
10	21.4	81	18.7	77

Table 20. Comparison results of FEM and KNN models on 1296 training datasets.

Number Nearest Neighbor (K)	Euclidean Distance		Manhattan Distance	
	Mean Absolute Error in Cycle	Maximum Absolute Error in Cycle	Mean Absolute Error in Cycle	Maximum Absolute Error in Cycle
1	2626.37	172	26.3	74
2	18.2	55	14.3	51
3	13	46	7.55	23
4	20.2	54	11.62	41
5	17.8	43	11.15	51
6	14.5	44	11.86	44
7	12.7	37	13.87	52
8	11.5	38	13.78	42
9	11.4	28	11.93	41
10	13.7	36	11.02	42

Table 21. KNN result comparison with different weights.

Number Nearest Neighbor (K)	Weight as Uniform with Euclidean		Weight as Distance with Euclidean	
	Mean Absolute Error in Cycle	Maximum Absolute Error in Cycle	Mean Absolute Error in Cycle	Maximum Absolute Error in Cycle
9	2618.9	8374	18.2	72

Table 22. KNN result comparison with different datasets.

Training Dataset(KNN)	Nearest Neighbor Value (K)	Mean Absolute Error in Cycle	Maximum Absolute Error in Cycle	CPU Time in Second
576	9	18.2	72	0.03
1296	3	7.5	23	0.034

Table 23. RF result comparison with different datasets.

Random Forest Training Dataset	576	1296
Random State	1	1
Number of Tree	81	81
Maximum Absolute Error (Cycle) Train Data	56	28
Mean Absolute Error (Cycle) Train Data	12	6.3
Maximum Absolute Error (Cycle) Test Data	133	103
Mean Absolute Error (Cycle) Test Data	36	26.3
CPU Time In Second	3.5	4

Finally, error analysis and CPU time analysis were conducted for the different machine learning algorithms used to model the WLP structure. Figure 15 shows the MAE analysis with several AI algorithms for the 576-feature training dataset. From the figure, the ANN model exhibited the smallest error, i.e., 4, whereas the RF model exhibited the highest error, i.e., 35.7. The other AI model errors were close to that of the ANN model. Similarly, Figure 16 shows the MAE results for several AI algorithms with 1296 training datasets. From the Figure 16, the ANN model had the lowest MAE, i.e., 2, whereas the RF model had the highest error, i.e., 26.4, and the MAEs of the other AI models were close to that of the ANN model.

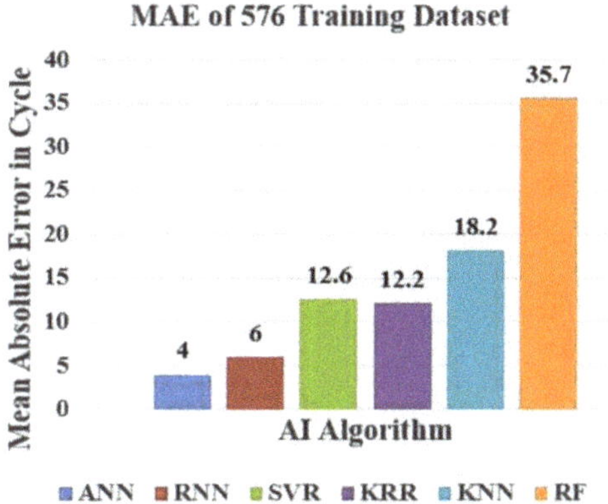

Figure 15. MAE analysis vs. AI model for 576 training datasets.

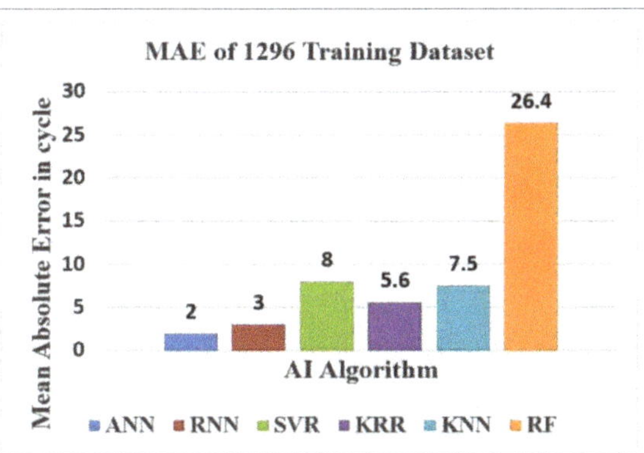

Figure 16. MAE analysis vs. AI model for 1296 training datasets.

Furthermore, the CPU times required by the different AI algorithms for the WLP structure were investigated. Figure 17 presents the CPU time analysis results for different AI models based on the 576- and 1296-feature training datasets.

From Figure 17, for the 576 training dataset, the KNN model required the lowest CPU time (0.03 s), whereas the RNN model required the highest (174 s). The CPU times required by the RF, SVR, and KRR models were close to that of the KNN model. Similarly, for the 1296 training dataset, the KNN model required the least CPU time (0.034 s), whereas the RNN model required the highest CPU time (699 s). Again, the CPU times required by the RF, SVR, and KRR models were close to that of the KNN model. The ANN and RNN models required more CPU times than the KRR, SVR, KNN, and RF models because of the usage of more neurons and hidden layers to estimate the WLP lifetimes.

Eventually, it also makes a comparison between all the above algorithms as per the MAE in cycle and CPU time consumption for different training datasets, which is from 500 to 9000 datasets. Figure 18 shows the MAE in cycle vs. different training datasets for all the above models. Figure 18 demonstrates the increase in the training dataset with the decrease in the testing error for all the above algorithms. From all the algorithms, ANN, RNN, SVR, and KRR testing errors are very close to each other. However, KNN and RF accuracy performance are not as good as other algorithms because these two algorithms are more suitable for classification purposes. Similarly, Figure 19 shows the CPU time analysis with different training datasets for all the above AI models. From Figure 19, it can be seen that CPU time increases with the increase in the training dataset. ANN and RNN required more CPU time due to the usage of more neurons and more hidden layers. The SVR algorithm also required more CPU time as compared to KRR, KNN, and RF algorithms due to the usage of more hyperparameters with the grid search technique to establish the best hyperparameter.

Overall, the use of AI models may have a major impact on the electronic packaging industry. However, the result of the study confirms that a validated FEM solution procedure is crucial for generating reliable training datasets and that increasing the number of design features increases the CPU time needed to build the AI model for the WLP structure.

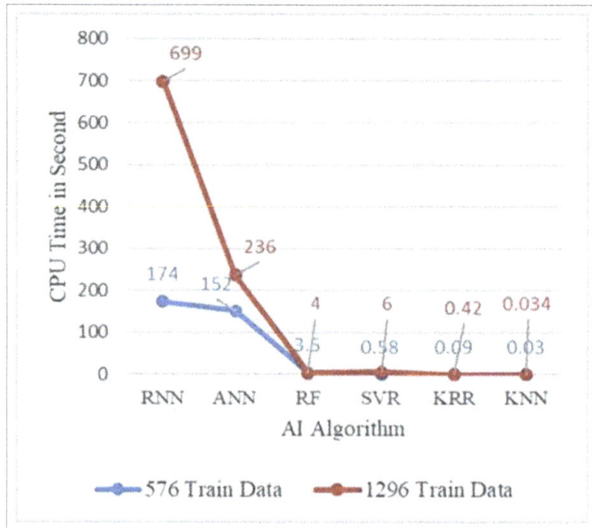

Figure 17. CPU time vs. AI algorithms for 576 and 1296 training datasets.

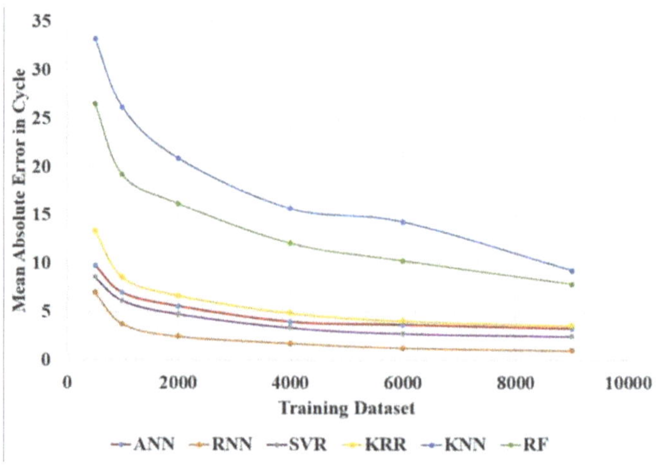

Figure 18. Mean absolute error vs. training dataset for different AI algorithms.

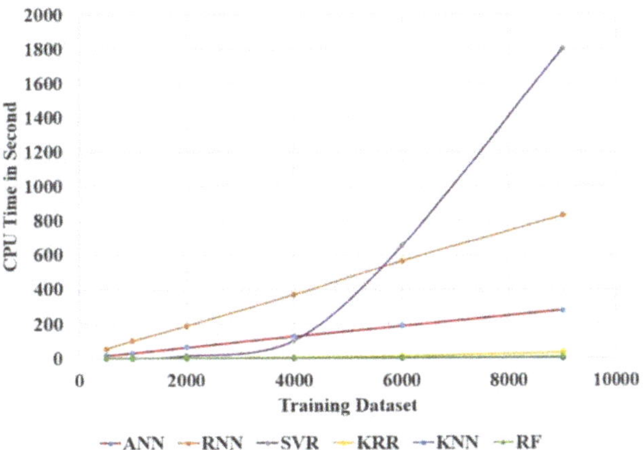

Figure 19. CPU time vs. training dataset for different AI algorithms.

5. Conclusions

The machine learning algorithms used a large database generated by a validated FEM procedure for training analysis and obtained a structural reliability life cycle regression model for WLP. These AI regression models can predict the reliability life cycle of WLP structures when given the WLP geometry as the input. In terms of accuracy, the MAE between FEM and AI was less than 10 life cycles, which is acceptable, and the AI training CPU time consumption of several AI algorithms was small. The ANN algorithm exhibited the best accuracy, whereas the RF regression algorithm exhibited the lowest accuracy, presumably because RF is designed for classification purposes. Other algorithms such as KRR and SVR also showed good accuracy owing to the usage of the RBF kernel function. However, the KRR model slightly outperformed the SVR algorithm in terms of accuracy because of the use of fewer hyperparameters and lower model complexity. KNN is more suitable for classification purposes; nonetheless, the KNN regression model also exhibited good accuracy for the WLP structure database.

The study also investigated the CPU time required for training the above-mentioned AI algorithms to obtain a final regression model for predicting the reliability life of the newly designed WLP structure. The KNN, KRR, and RF regression models required less than 10 s, whereas the ANN, RNN, and SVR regression models required 150 to 1800 s. These times are very low compared with those required by FEM modeling and simulation, which can range from several hours to several days.

It is known by the electronic packaging community that the experiment-based design procedure may take 8 months to a year to complete one run of the WLP test vehicle fabrication and the ATCT test; this has become unacceptable for todays' advanced packaging development. Based on AI/machine learning algorithms and validated finite-element-based simulation technology, this research developed an AI-assisted design-on-simulation technology that can effectively and accurately predict the reliability life cycle of various geometries of WLPs. In addition, after obtaining the AI-trained model of the WLP, developers only need to input geometric data of the new WLP, then the reliability life cycle can be obtained within one second. Therefore, WLP structure optimization becomes feasible because the reliability prediction of any geometric combination of WLP can be completed in a few seconds. The AI-assisted design-on-simulation technology can also be applied to other packaging types such as system-in-packaging, heterogeneous, fan-out, WLP, and 3D packaging.

Materials **2021**, *14*, 5342

Author Contributions: Conceptualization, K.-N.C.; methodology, K.-N.C. and S.K.P.; software, S.K.P., Y.-C.T., B.-R.L.; validation, S.K.P., Y.-C.T., B.-R.L. and K.-N.C.; formal analysis, S.K.P., Y.-C.T., B.-R.L.; investigation, K.-N.C.; resources, K.-N.C.; data curation, S.K.P., Y.-C.T., B.-R.L.; writing—original draft preparation, S.K.P.; writing—review and editing, K.-N.C.; visualization, K.-N.C.; supervision, K.-N.C.; project administration, K.-N.C.; funding acquisition, K.-N.C. All authors have read and agreed to the published version of the manuscript.

Funding: This research was funded by the Ministry of Science and Technology of Taiwan, grant number MOST 108-2221-E-007-007-MY3.

Acknowledgments: The authors would like to thank the Ministry of Science and Technology (Project MOST 108-2221-E-007-007-MY3) of Taiwan for supporting this research.

Conflicts of Interest: The authors declare no conflict of interest.

References

1. Andriani, Y.; Wang, X.; Seng, D.H.L.; Teo, S.L.; Liu, S.; Lau, B.L.; Zhang, X. Effect of Boron Nitride Nanosheets on Properties of a Commercial Epoxy Molding Compound Used in Fan-Out Wafer-Level Packaging. *IEEE Trans. Compon. Packag. Manuf. Technol.* **2020**, *10*, 990–999. [CrossRef]
2. Cheng, H.-C.; Chung, C.-H.; Chen, W.-H. Die shift assessment of reconstituted wafer for fan-out wafer-level packaging. *IEEE Trans. Device Mater. Reliab.* **2020**, *20*, 136–145. [CrossRef]
3. Cheng, H.-C.; Wu, Z.-D.; Liu, Y.-C. Viscoelastic warpage modeling of fan-out wafer-level packaging during wafer-level mold cure process. *IEEE Trans. Compon. Packag. Manuf. Technol.* **2020**, *10*, 1240–1250. [CrossRef]
4. Dong, H.; Chen, J.; Hou, D.; Xiang, Y.; Hong, W. A low-loss fan-out wafer-level package with a novel redistribution layer pattern and its measurement methodology for millimeter-wave application. *IEEE Trans. Compon. Packag. Manuf. Technol.* **2020**, *10*, 1073–1078. [CrossRef]
5. Lau, J.H.; Li, M.; Li, Q.M.; Xu, I.; Chen, T.; Li, Z.; Tan, K.H.; Yong, Q.X.; Cheng, Z.; Wee, K.S. Design, materials, process, fabrication, and reliability of fan-out wafer-level packaging. *IEEE Trans. Compon. Packag. Manuf. Technol.* **2018**, *8*, 991–1002. [CrossRef]
6. Lau, J.H.; Li, M.; Qingqian, M.L.; Chen, T.; Xu, I.; Yong, Q.X.; Cheng, Z.; Fan, N.; Kuah, E.; Li, Z. Fan-out wafer-level packaging for heterogeneous integration. *IEEE Trans. Compon. Packag. Manuf. Technol.* **2018**, *8*, 1544–1560. [CrossRef]
7. Lee, T.-K.; Xie, W.; Tsai, M.; Sheikh, M.D. Impact of Microstructure Evolution on the Long-Term Reliability of Wafer-Level Chip-Scale Package Sn–Ag–Cu Solder Interconnects. *IEEE Trans. Compon. Packag. Manuf. Technol.* **2020**, *10*, 1594–1603. [CrossRef]
8. Zhao, S.; Yu, D.; Zou, Y.; Yang, C.; Yang, X.; Xiao, Z.; Chen, P.; Qin, F. Integration of CMOS image sensor and microwell array using 3-D WLCSP technology for biodetector application. *IEEE Trans. Compon. Packag. Manuf. Technol.* **2019**, *9*, 624–632. [CrossRef]
9. Chen, C.; Yu, D.; Wang, T.; Xiao, Z.; Wan, L. Warpage prediction and optimization for embedded silicon fan-out wafer-level packaging based on an extended theoretical model. *IEEE Trans. Compon. Packag. Manuf. Technol.* **2019**, *9*, 845–853. [CrossRef]
10. Lau, J.H.; Ko, C.-T.; Tseng, T.-J.; Yang, K.-M.; Peng, T.C.-Y.; Xia, T.; Lin, P.B.; Lin, E.; Chang, L.; Liu, H.N. Panel-level chip-scale package with multiple diced wafers. *IEEE Trans. Compon. Packag. Manuf. Technol.* **2020**, *10*, 1110–1124. [CrossRef]
11. Lau, J.H.; Li, M.; Yang, L.; Li, M.; Xu, I.; Chen, T.; Chen, S.; Yong, Q.X.; Madhukumar, J.P.; Kai, W. Warpage measurements and characterizations of fan-out wafer-level packaging with large chips and multiple redistributed layers. *IEEE Trans. Compon. Packag. Manuf. Technol.* **2018**, *8*, 1729–1737. [CrossRef]
12. Qin, C.; Li, Y.; Mao, H. Effect of Different PBO-Based RDL Structures on Chip-Package Interaction Reliability of Wafer Level Package. *IEEE Trans. Device Mater. Reliab.* **2020**, *20*, 524–529. [CrossRef]
13. Qin, F.; Zhao, S.; Dai, Y.; Yang, M.; Xiang, M.; Yu, D. Study of warpage evolution and control for six-side molded WLCSP in different packaging processes. *IEEE Trans. Compon. Packag. Manuf. Technol.* **2020**, *10*, 730–738. [CrossRef]
14. Wang, P.-H.; Huang, Y.-W.; Chiang, K.-N. Reliability Evaluation of Fan-Out Type 3D Packaging-On-Packaging. *Micromachines* **2021**, *12*, 295. [CrossRef]
15. Yang, C.; Su, Y.; Liang, S.Y.; Chiang, K. Simulation of wire bonding process using explicit FEM with ALE remeshing technology. *J. Mech.* **2020**, *36*, 47–54. [CrossRef]
16. Liu, C.-M.; Lee, C.-C.; Chiang, K.-N. Enhancing the reliability of wafer level packaging by using solder joints layout design. *IEEE Trans. Compon. Packag. Technol.* **2006**, *29*, 877–885. [CrossRef]
17. Chiang, K.-N.; Chen, W.-H.; Cheng, H.-C. Large-scale three-dimensional area array electronic packaging analysis. *J. Comput. Model. Simul. Eng.* **1999**, *4*, 4–11.
18. Tsou, C.; Chang, T.; Wu, K.; Wu, P.; Chiang, K. Reliability assessment using modified energy based model for WLCSP solder joints. In Proceedings of the International Conference on Electronics Packaging (ICEP), Yamagata, Japan, 19–22 April 2017; pp. 7–15.
19. Liu, S.; Panigrahy, S.; Chiang, K. Prediction of fan-out panel level warpage using neural network model with edge detection enhancement. In Proceedings of the IEEE 70th Electronic Components and Technology Conference (ECTC), Orlando, FL, USA, 3–30 June 2020; pp. 1626–1631.
20. Yuan, C.C.; Lee, C.-C. Solder joint reliability modeling by sequential artificial neural network for glass wafer level chip scale package. *IEEE Access* **2020**, *8*, 143494–143501. [CrossRef]

21. Gupta, S.; Al-Obaidi, S.; Ferrara, L. Meta-Analysis and Machine Learning Models to Optimize the Efficiency of Self-Healing Capacity of Cementitious Material. *Materials* **2021**, *14*, 4437. [CrossRef]
22. Jianliang, S.; Mengqian, S.; Hesong, G.; Yan, P.; Jiang, J.; Lipu, X. Research on Edge Surface Warping Defect Diagnosis Based on Fusion Dimension Reduction Layer DBN and Contribution Plot Method. *J. Mech.* **2020**, *36*, 889–899. [CrossRef]
23. Kuschmitz, S.; Ring, T.P.; Watschke, H.; Langer, S.C.; Vietor, T. Design and Additive Manufacturing of Porous Sound Absorbers—A Machine-Learning Approach. *Materials* **2021**, *14*, 1747. [CrossRef]
24. Song, H.; Ahmad, A.; Ostrowski, K.A.; Dudek, M. Analyzing the Compressive Strength of Ceramic Waste-Based Concrete Using Experiment and Artificial Neural Network (ANN) Approach. *Materials* **2021**, *14*, 4518. [CrossRef]
25. Farooq, F.; Czarnecki, S.; Niewiadomski, P.; Aslam, F.; Alabduljabbar, H.; Ostrowski, K.A.; Śliwa-Wieczorek, K.; Nowobilski, T.; Malazdrewicz, S. A Comparative Study for the Prediction of the Compressive Strength of Self-Compacting Concrete Modified with Fly Ash. *Materials* **2021**, *14*, 4934. [CrossRef] [PubMed]
26. Huang, X.; Wasouf, M.; Sresakoolchai, J.; Kaewunruen, S. Prediction of Healing Performance of Autogenous Healing Concrete Using Machine Learning. *Materials* **2021**, *14*, 4068. [CrossRef] [PubMed]
27. Salazar, A.; Xiao, F. Design of Hybrid Reconstruction Scheme for Compressible Flow Using Data-Driven Methods. *J. Mech.* **2020**, *36*, 675–689. [CrossRef]
28. Song, S.-H. A Comparison Study of Constitutive Equation, Neural Networks, and Support Vector Regression for Modeling Hot Deformation of 316L Stainless Steel. *Materials* **2020**, *13*, 3766. [CrossRef]
29. McCulloch, W.S.; Pitts, W. A logical calculus of the ideas immanent in nervous activity. *Bull. Math. Biophys.* **1943**, *5*, 115–133. [CrossRef]
30. Denoeux, T. A neural network classifier based on Dempster-Shafer theory. *IEEE Trans. Syst. Man Cybern. Syst.* **2000**, *30*, 131–150. [CrossRef]
31. Tandel, G.S.; Biswas, M.; Kakde, O.G.; Tiwari, A.; Suri, H.S.; Turk, M.; Laird, J.R.; Asare, C.K.; Ankrah, A.A.; Khanna, N. A review on a deep learning perspective in brain cancer classification. *Cancers* **2019**, *11*, 111. [CrossRef] [PubMed]
32. Schmidhuber, J. Deep learning in neural networks: An overview. *Neural Netw.* **2015**, *61*, 85–117. [CrossRef]
33. Kulawik, A.; Wróbel, J.; Ikonnikov, A.M. Model of the Austenite Decomposition during Cooling of the Medium Carbon Steel Using LSTM Recurrent Neural Network. *Materials* **2021**, *14*, 4492. [CrossRef]
34. Yuan, C.C.; Fan, J.; Fan, X. Deep machine learning of the spectral power distribution of the LED system with multiple degradation mechanisms. *J. Mech.* **2020**, *37*, 172–183. [CrossRef]
35. Cortes, C.; Vapnik, V. Support-vector networks. *Mach. Learn.* **1995**, *20*, 273–297. [CrossRef]
36. Braun, A.C.; Weidner, U.; Hinz, S. Classification in high-dimensional feature spaces—Assessment using SVM, IVM and RVM with focus on simulated EnMAP data. *IEEE J. Sel. Top. Appl. Earth Obs. Remote Sens.* **2012**, *5*, 436–443. [CrossRef]
37. Al-Sodani, K.A.A.; Adewumi, A.A.; Mohd Ariffin, M.A.; Maslehuddin, M.; Ismail, M.; Salami, H.O.; Owolabi, T.O.; Mohamed, H.D. Experimental and Modelling of Alkali-Activated Mortar Compressive Strength Using Hybrid Support Vector Regression and Genetic Algorithm. *Materials* **2021**, *14*, 3049. [CrossRef] [PubMed]
38. Benkedjouh, T.; Medjaher, K.; Zerhouni, N.; Rechak, S. Health assessment and life prediction of cutting tools based on support vector regression. *J. Intell. Manuf.* **2015**, *26*, 213–223. [CrossRef]
39. Dhanalakshmi, P.; Kanimozhi, T. Automatic segmentation of brain tumor using K-Means clustering and its area calculation. *Int. J. Adv. Electr. Electron. Eng.* **2013**, *2*, 130–134.
40. Drineas, P.; Mahoney, M.W.; Cristianini, N. On the Nyström Method for Approximating a Gram Matrix for Improved Kernel-Based Learning. *J. Mach. Learn. Res.* **2005**, *6*, 2153–2175.
41. Panigrahy, S.K.; Chiang, K.-N. Study on an Artificial Intelligence Based Kernel Ridge Regression Algorithm for Wafer Level Package Reliability Prediction. In Proceedings of the IEEE 71st Electronic Components and Technology Conference (ECTC), San Diego, CA, USA, 1 June–4 July 2021; pp. 1435–1441.
42. Hamed, Y.; Alzahrani, A.I.; Mustaffa, Z.; Ismail, M.C.; Eng, K.K. Two steps hybrid calibration algorithm of support vector regression and K-nearest neighbors. *Alex. Eng. J.* **2020**, *59*, 1181–1190. [CrossRef]
43. Tan, S. An effective refinement strategy for KNN text classifier. *Expert Syst. Appl.* **2006**, *30*, 290–298. [CrossRef]
44. Ho, W.; Yu, F. Chiller system optimization using k nearest neighbour regression. *J. Clean. Prod.* **2021**, *303*, 127050. [CrossRef]
45. Xia, J.; Zhang, J.; Wang, Y.; Han, L.; Yan, H. WC-KNNG-PC: Watershed clustering based on k-nearest-neighbor graph and Pauta Criterion. *Pattern Recognit.* **2021**, *121*, 108177. [CrossRef]
46. Breiman, L. Random forests. *Mach. Learn.* **2001**, *45*, 5–32. [CrossRef]
47. Hsiao, H.; Chiang, K. AI-assisted reliability life prediction model for wafer-level packaging using the random forest method. *J. Mech.* **2020**, *37*, 28–36. [CrossRef]
48. Mohana, R.M.; Reddy, C.K.K.; Anisha, P.; Murthy, B.R. Random forest algorithms for the classification of tree-based ensemble. *Mater. Today Proc.* **2021**. [CrossRef]
49. Krizhevsky, A.; Sutskever, I.; Hinton, G.E. Imagenet classification with deep convolutional neural networks. *Adv. Neural Inf. Process. Syst.* **2012**, *25*, 1097–1105. [CrossRef]
50. Ortac, G.; Ozcan, G. Comparative study of hyperspectral image classification by multidimensional Convolutional Neural Network approaches to improve accuracy. *Expert Syst. Appl.* **2021**, *182*, 115280. [CrossRef]
51. Petneházi, G. Quantile convolutional neural networks for Value at Risk forecasting. *Mach. Learn. Appl.* **2021**, *6*, 100096.

52. Ramalho, L.; Belinha, J.; Campilho, R. A new crack propagation algorithm combined with the finite element method. *J. Mech.* **2020**, *36*, 405–422. [CrossRef]
53. Hsieh, M.-C. Modeling correlation for solder joint fatigue life estimation in wafer-level chip scale packages. In Proceedings of the International Microsystems, Packaging, Assembly and Circuits Technology Conference (IMPACT), Taipei, Taiwan, 21–23 October 2015; pp. 65–68.
54. Rogers, B.; Scanlan, C. Improving WLCSP reliability through solder joint geometry optimization. In Proceedings of the International Symposium on Microelectronics (IMAPS), Orlando, FL, USA, 30 September–3 October 2013; pp. 546–550.
55. Chiang, K.-N.; Yuan, C.-A. An overview of solder bump shape prediction algorithms with validations. *IEEE Trans. Adv. Packag.* **2001**, *24*, 158–162. [CrossRef]
56. Chang, J.; Wang, L.; Dirk, J.; Xie, X. Finite element modeling predicts the effects of voids on thermal shock reliability and thermal resistance of power device. *Weld. J.* **2006**, *85*, 63s–70s.
57. JEDEC Solid State Technology Association. JEDEC Standard JESD22-A104D, Temperature Cycling. *Jedec. Org* **2005**, *11*, 2009.
58. Coffin, L.F., Jr. A study of the effects of cyclic thermal stresses on a ductile metal. *Trans. ASME* **1954**, *76*, 931–950.
59. Ramachandran, V.; Wu, K.; Chiang, K. Overview study of solder joint reliablity due to creep deformation. *J. Mech.* **2018**, *34*, 637–643. [CrossRef]
60. Lee, C.-H.; Wu, K.-C.; Chiang, K.-N. A novel acceleration-factor equation for packaging-solder joint reliability assessment at different thermal cyclic loading rates. *J. Mech.* **2017**, *33*, 35–40. [CrossRef]
61. Yanjun, X.; Liquan, W.; Fengshun, W.; Weisheng, X.; Hui, L. Effect of interface structure on fatigue life under thermal cycle with SAC305 solder joints. In Proceedings of the International Conference on Electronic Packaging Technology (ICEPT), Dalian, China, 11–14 August 2013; pp. 959–964.
62. Wang, P.; Lee, Y.; Lee, C.; Chang, H.; Chiang, K. Solder Joint Reliability Assessment and Pad Size Studies of FO-WLP with Glass Substrate. *IEEE Trans. Device Mater. Reliab.* **2021**, *21*, 96–101. [CrossRef]
63. Chou, P.; Chiang, K.; Liang, S.Y. Reliability assessment of wafer level package using artificial neural network regression model. *J. Mech.* **2019**, *35*, 829–837. [CrossRef]
64. Tang, Z.; Fishwick, P.A. Feedforward neural nets as models for time series forecasting. *ORSA J. Comput.* **1993**, *5*, 374–385. [CrossRef]
65. Zhang, D.; Han, X.; Deng, C. Review on the research and practice of deep learning and reinforcement learning in smart grids. *CSEE J. Power Energy Syst.* **2018**, *4*, 362–370. [CrossRef]
66. Gers, F.A.; Schmidhuber, J.; Cummins, F. Learning to forget: Continual prediction with LSTM. *Neural Comput.* **2000**, *12*, 2451–2471. [CrossRef]
67. Rui, J.; Zhang, H.; Zhang, D.; Han, F.; Guo, Q. Total organic carbon content prediction based on support-vector-regression machine with particle swarm optimization. *J. Pet. Sci. Eng.* **2019**, *180*, 699–706. [CrossRef]
68. Welling, M.; Kernel Ridge Regression. Max Welling's Class Notes in Machine Learning. 2013. Available online: https://web2.qatar.cmu.edu/~{}gdicaro/10315/additional/welling-notes-on-kernel-ridge.pdf (accessed on 6 August 2021).
69. Hamed, Y.; Mustaffa, Z.B.; Idris, N.R.B. Comparative Calibration of Corrosion Measurements Using K-Nearest Neighbour Based Techniques. In Proceedings of the MATEC Web of Conferences, Amsterdam, The Netherlands, 23–25 March 2016; Volume 52, p. 02001.
70. Hamed, Y.; Mustaffa, Z.B.; Idris, N.R.B. An application of K-Nearest Neighbor interpolation on calibrating corrosion measurements collected by two non-destructive techniques. In Proceedings of the International Conference on Smart Instrumentation, Measurement and Applications (ICSIMA), Kuala Lumpur, Malaysia, 24–25 November 2015; pp. 1–5.
71. Härdle, W.; Linton, O. Applied nonparametric methods. *Handb. Econom.* **1994**, *4*, 2295–2339.
72. Atkeson, C.G.; Moore, A.W.; Schaal, S. Locally weighted learning for control. *Artif. Intell. Rev.* **1997**, *11*, 75–113. [CrossRef]

materials

Article

Theoretical and Experimental Investigation of Warpage Evolution of Flip Chip Package on Packaging during Fabrication

Hsien-Chie Cheng [1,*], **Ling-Ching Tai** [1] **and Yan-Cheng Liu** [1,2]

[1] Department of Aerospace and Systems Engineering, Feng Chia University, Taichung 407, Taiwan; she850303@yahoo.com.tw (L.-C.T.); P0700100@o365.fcu.edu.tw (Y.-C.L.)

[2] Ph.D. Program of Mechanical and Aeronautical Engineering, Feng Chia University, Taichung 407, Taiwan

* Correspondence: hccheng@fcu.edu.tw

Abstract: This study attempts to investigate the warpage behavior of a flip chip package-on-package (FCPoP) assembly during fabrication process. A process simulation framework that integrates thermal and mechanical finite element analysis (FEA), effective modeling and ANSYS element death-birth technique is introduced for effectively predicting the process-induced warpage. The mechanical FEA takes into account the viscoelastic behavior and cure shrinkage of the epoxy molding compound. In order to enhance the computational and modeling efficiency and retain the prediction accuracy at the same time, this study proposes a novel effective approach that combines the trace mapping method, rule of mixture and FEA to estimate the effective orthotropic elastic properties of the coreless substrate and core interposer. The study begins with experimental measurement of the temperature-dependent elastic and viscoelastic properties of the components in the assembly, followed by the prediction of the effective elastic properties of the orthotropic interposer and substrate. The predicted effective results are compared against the results of the ROM/analytical estimate and the FEA-based effective approach. Moreover, the warpages obtained from the proposed process simulation framework are validated by the in-line measurement data, and good agreement is presented. Finally, key factors that may influence process-induced warpage are examined via parametric analysis.

Keywords: flip chip package on package; finite element analysis; viscoelastic behavior; process-induced warpage; trace mapping; effective modeling

Citation: Cheng, H.-C.; Tai, L.-C.; Liu, Y.-C. Theoretical and Experimental Investigation of Warpage Evolution of Flip Chip Package on Packaging during Fabrication. *Materials* **2021**, *14*, 4816. https://doi.org/10.3390/ma14174816

Academic Editor: Fei Tang

Received: 11 July 2021
Accepted: 23 August 2021
Published: 25 August 2021

Publisher's Note: MDPI stays neutral with regard to jurisdictional claims in published maps and institutional affiliations.

1. Introduction

In recent years, there has been explosive and continuous growth in the consumer market for various smart products and Internet of Things (IoT) products, as well as the developing requirements of 5G communication, artificial intelligence (AI), and autonomous vehicles. Advanced packaging technology like flip chip packaging [1], wafer level packaging [2], and flip chip chip-scale packaging (FCCSP) [3,4] was introduced to achieve high I/O density, excellent electrical performance and miniaturization, and thus is commonly used in high-end smart chips in recent years. However, the physical limitations of Moore's law [5] make it difficult for electronic packaging to continuously shrink and functionally improve, prompting the development of new packaging technologies. Among the many solutions, the system-in-package (SiP) for heterogeneous integration is the current alternative, and is one of the most feasible methods for "More than Moore" or even "Beyond CMOS".

SiP technology may have a two-dimensional (2D) planar configuration, a three-dimensional (3D) vertical stacking configuration, or an integrated (hybrid) configuration. 3D packaging technology can be categorized into package stacking, like package-on-package (PoP) and package-in-package (PiP), wire-bonding [6], and through silicon via (TSV)-based 3D IC stacking [7]. In addition to high I/O quantity and miniaturization, further requirements of multi-functionality have aroused the development of the flip chip package-on-package (FCPoP) technology. This packaging technology has attracted a great deal of attention from the semiconductor packaging industry due to its compelling features including heterogeneous integration capability, high electrical performance, high bandwidth, low power consumption, small form factor, low cost, etc., leading to wide potential applications, such as high-performance application CPUs. To date, research on FCPoP has been exceptionally limited. Among the limited literature, the focus was placed on packaging construction [8,9]. For example, Hsieh et al. [9] proposed a PoP technology with the flip chip structure for mobile device applications, which can be a bare die PoP packaging technology, a molded laser PoP packaging technology, or a silicon interposer substrate PoP technology to achieve thickness and warpage reduction.

Reliability and yield are the two most important issues in microelectronics assembly [2]. Process-induced warpage during fabrication is one of the root causes for the poor assembly reliability and yield [10–12] Although the bottom layer structure of the FCPoP assembly employs a mature flip chip packaging technology, it is still indispensable to stack an interposer as a bridge to connect to the memory, potentially causing a more serious mismatch of the coefficients of thermal expansion (CTE) of the different materials. The excessive CTE mismatch together with high process temperatures may induce considerable residual stress/strain, which will generate not only serious warpage but also throughput loss. Therefore, to effectively master and control the process-induced deformations of the assembly is the key to the success of the technology. Compared with FCPoP, there have been many studies in the past on the warpage behavior of FCCSPs [3,12]. It was found that viscoelastic behavior of the molded underfill material would contribute to their warpage performance. In addition, the substrate or interposer is also a crucial factor dominating the warpage behavior. To suppress the warpage of packages, substrates with a core material have been widely used. As portable devices become thinner, so do the package size and substrate, coreless substrates have recently become increasingly popular in electronic packaging [3]. However, a lack of rigid core material for structural support may cause the warpage to be even more sensitive and pronounced.

This research aimed to establish a process simulation framework for predicting the warpage behavior of an FCPoP assembly during fabrication. The viscoelastic behavior and volumetric shrinkage of the epoxy molding compound (EMC) [11,12] were included in the process modeling. The FCPoP consisted of one bottom orthotropic coreless substrate and one top orthotropic core interposer, each of which comprised several copper (Cu) circuit layers of multi-material and multi-scale structures and complex geometric features. These layers may have a significant influence on process-induced warpage because of the high modulus and CTE of Cu. Thus, in order to offer an accurate prediction, these Cu circuitries need to be accurately modeled in the modeling. However, because of their high geometric and structural complexity, efficiently and thoroughly modeling, these Cu circuit layers presents great challenges. In order to greatly improve the computational and modeling efficiency while accommodating the need for good prediction accuracy, a novel effective approach was proposed to effectively simulate the global thermo-mechanical behavior of the orthotropic coreless substrate and core interposer. The effectiveness of the proposed effective approach was demonstrated by comparing the predicted effective elastic properties with the results of other effective approaches. The proposed process simulation model was validated using in-line warpage measurement data. Finally, parametric analysis is performed to assess the influence of several material and geometry parameters on the process-induced warpage of the FCPoP.

2. Structure and Fabrication Process of FCPoP

The research vehicle was an FCPoP assembly, as shown in Figure 1, that was primarily composed of an FCCSP package, an EMC, a core interposer, and Cu core solder balls (CCSBs). The main structure of the FCCSP package included a silicon chip, an underfill, Cu pillar bumps, and a coreless substrate. The FCCSP package usually is used for high-end processors. In addition, the core interposer was applied to facilitate the connection with the

HBM for heterogeneous integration, the CCSBs were utilized to connect the bottom FCCSP and the core interposer, and the EMC was used to protect the solder balls. To minimize the package profile, a three-layer 100 μm embedded trace substrate (ETS) was applied, as schematically depicted in Figure 2, which mainly included two solder mask (SM) protective layers, two prepreg (PP) dielectric layers, and three metal (Cu) layers, with the circuitries filled with either SM or PP material. The chip was 9.36 mm in length, 8.76 mm in width, and 70 μm thick. The chip was connected to the coreless substrate using 2500 Cu pillar bumps, which were 40 μm in length, 70 μm in width and 58 μm in height. The gap between the chip and coreless substrate was filled with an underfill via the capillary action. The top layer was stacked on a 90 μm thick two-layer core interposer. In the FCPoP assembly, there were a total of 550 CCSBs with a diameter of 190 μm. Finally, the EMC was filled between the substrate and interposer to form an FCPoP assembly with a length and width of 14 mm and a thickness of 500 μm. Figure 3 describes the main fabrication process steps; i.e., the die bonding process (steps 1–3), underfill cure process (steps 3–6), interposer bonding process (steps 6–9), mold cure process (steps 9–12) and temperature elevation process (steps 12–13), and also the corresponding process temperatures.

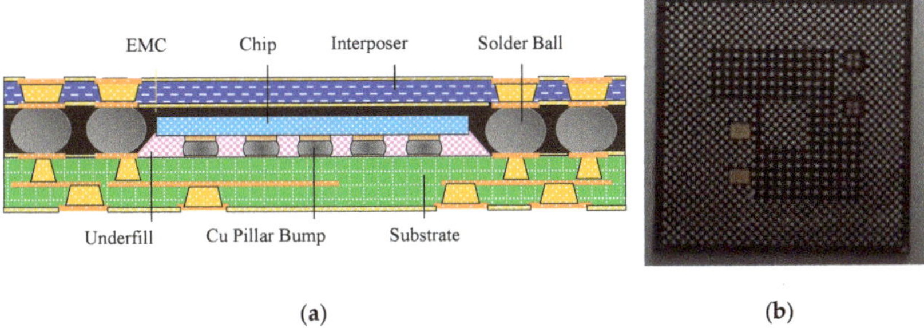

(**a**)

(**b**)

Figure 1. The FCPoP assembly: (**a**) cross-sectional view; (**b**) prototype.

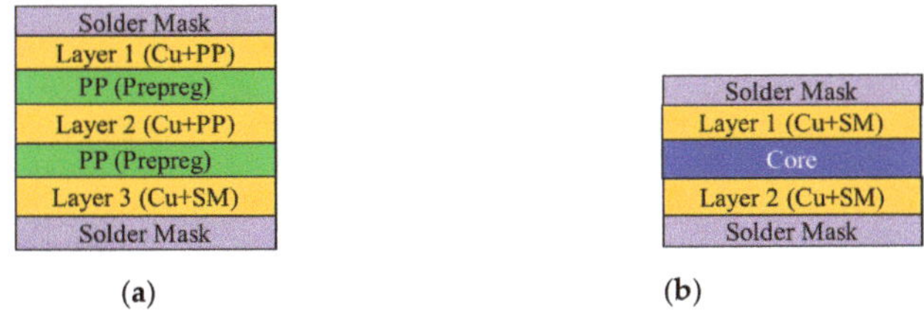

(**a**)

(**b**)

Figure 2. Schematic cross-sectional view: (**a**) top coreless substrate; (**b**) bottom core interposer.

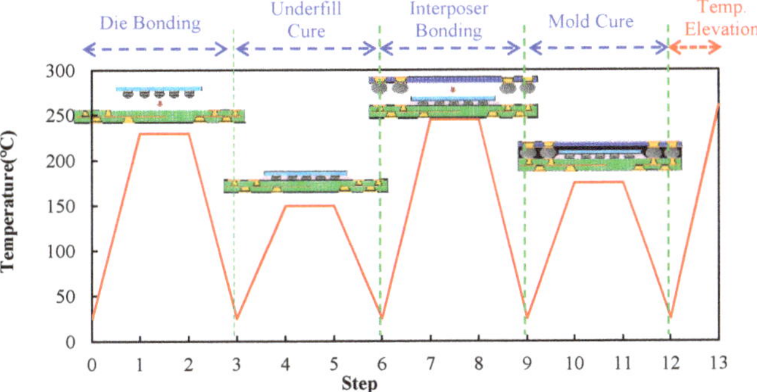

Figure 3. Fabrication process steps with temperature loads.

3. Theoretical Models

3.1. Linear Viscoelasticity

The properties of viscoelastic materials can be divided into elastic and viscoelastic parts. The elastic part can react immediately and obey Hooke's law when subjected to a fixed load. Conversely, the viscoelastic part gradually increases the strain (creep) or reduces the stress (relaxation), and eventually reaches stability. Polymer materials typically show viscoelastic relaxation behavior [13–16]. The stress relaxation effect is usually dominated by chemical phenomena at high temperatures for a long period of time. To describe the viscoelastic relaxation behavior, the generalized Maxwell model is most commonly used. It is composed of several Maxwell elements and an independent spring combined in parallel. The time-dependent viscoelastic stress relaxation modulus can be expressed by a Prony series mathematical representation [11]:

$$E(t) = \sum_{k=1}^{n} E_k \exp\left(-\frac{t}{\varsigma_k}\right) + E_\infty \tag{1}$$

where n is the number of Maxwell elements, E_k denotes the modulus of each Maxwell element, E_∞ represents the relaxed modulus, ς_k stands for the relaxation time, and t is the time. Furthermore, the modulus of each Maxwell element can be described as:

$$E_k = c_k E_0 \tag{2}$$

where c_k denotes the weighting factor and E_0 represents the unrelaxed modulus, expressed as

$$E_0 = \sum_{k=1}^{n} E_k + E_\infty \tag{3}$$

Combining Equations (1)–(3) yields:

$$E(t) = E_0 \left[c_\infty + \sum_{k=1}^{n} c_k \exp\left(-\frac{t}{\varsigma_k}\right) \right] \tag{4}$$

The properties of polymer materials, in either a glass or a rubbery state, show a strong relationship with temperature. The glassy state refers to the polymer material at a temperature higher than the glass transition temperature (T_g). In contrast, the rubbery state refers to the polymer material at a temperature lower than T_g. The Young's modulus and CTE have a large variation during the phase transition state. The time-temperature superposition (TTS) principle is often applied to depict the time-temperature dependence of the linear viscoelastic behavior. The TTS principle illustrates that the relaxation curve of the material's modulus and time (or frequency) at a certain temperature is analogous to the relaxation curve of the adjacent temperature. The relaxation curve at each temperature, except the reference temperature, is translated in the logarithmic time domain to form the relaxation curve of the material at the reference temperature. The value of this translation highly depends on the temperature, the reference temperature and the properties of the polymer materials. The TTS principle can be simply expressed as follows:

$$E(t, T) = E(\tau, T_0) \tag{5}$$

where τ stands for the reduced time when $T_0 < T$, which can be derived below:

$$\tau = \int t \kappa_T(t) dt \tag{6}$$

In Equation (6), κ_T is the shift factor of temperature. This parameter can be approximated by the well-known Williams-Landel-Ferry (WLF) model [17] as:

$$\log_{10} \kappa_T = \frac{-a_1 \left(T - T_{ref} \right)}{a_2 + \left(T - T_{ref} \right)} \tag{7}$$

where T_{ref} is the reference temperature, and a_1 and a_2 represent the coefficients of the curve fit. Basically, they are highly dependent on the materials and the reference temperature.

3.2. Linear Elastic Mechanics

Polymer materials, such as EMCs, will exhibit volumetric changes or chemical shrinkage during the mold cure process due to chemical reactions. In essence, the extent of volumetric change highly depends on the material's cure state in an isothermal isobaric ensemble. Consider that a polymer cube has a length of one-unit side before curing. The volumetric change ($\widetilde{V}$) of the unit cube after full curing is written as:

$$\widetilde{V} \approx 1 - 3\Delta l \tag{8}$$

where Δl is the variation of the unit side length. The corresponding strain due to the volumetric change ε_{vc} can be expressed as:

$$\varepsilon_{vc} = \frac{1}{3} \widetilde{V} \tag{9}$$

The total strain of the polymer materials is the sum of the volumetric change-inducted strain (ε_{vc}), elastic strain (ε_e), and thermal strain ($\varepsilon_{thermal}$). According to Hooke's law, the relationship between the elastic strain and stress can be written as follows:

$$\{\varepsilon_e\} = [B]\{w\} - \{\varepsilon_{vc}\} - \{\varepsilon_{thermal}\} \tag{10}$$

where $\{w\}$ is the nodal displacement vector and $[B]$ represents the strain-displacement matrix.

4. Numerical Modeling and Material Characterization

4.1. Effective Modeling

4.1.1. The Proposed Effective Method

To effectively capture the process-induced warpage behavior of the FCPoP assembly relies on a dependable and accurate thermo-mechanical characterization of the core interposer and coreless substrate, consisting of several Cu circuit layers with multi-material and multi-scale structures and complex geometric features. Accurately and fully modeling them presents great challenges due to the requiring extensive, tedious effort and cost required. To ease the modeling challenges, the core interposer and coreless substrate were approximated as an equivalent homogeneous medium and their effective elastic properties were evaluated by using an effective approach that made use of the powerful electronic computer-aided design (ECAD) trace mapping (TM) method together with the rule-of-mixture (ROM) technique and finite element analysis (FEA).

The ECAD TM method enables a more efficient and accurate representation of tiny, delicate, complex Cu traces, pads and vias surrounded by the PP dialectic material and the SM protective material on the coreless substrate and core interposer [18]. A flowchart of the ECAD TM method is shown in Figure 4. It is noted that the Cu circuit layers consisted of not only Cu traces, pads and vias, but also SM or PP dielectric material. First of all, based on the ECAD model, a uniform regular background mesh with a significant number of very fine first-order brick elements was established on each Cu circuit layer of the coreless substrate and core interposer. Then, the spatially non-uniformly distributed Cu circuitries in the ECAD model were mapped onto the background mesh to obtain a high resolution (HR) Cartesian Cu circuit map and a finite element (FE) model. Based on the volume ratio of Cu and neighboring materials, such as PP or SM on each brick element of the background mesh, the effective isotropic elastic properties were calculated using an ROM technique, by which a complete material property map of the Cu circuit layer was derived (Figure 5). The benefits of the ECAD TM method were a very uniform regular mesh, flexible mesh density control and close match to the geometry of the Cu circuitries.

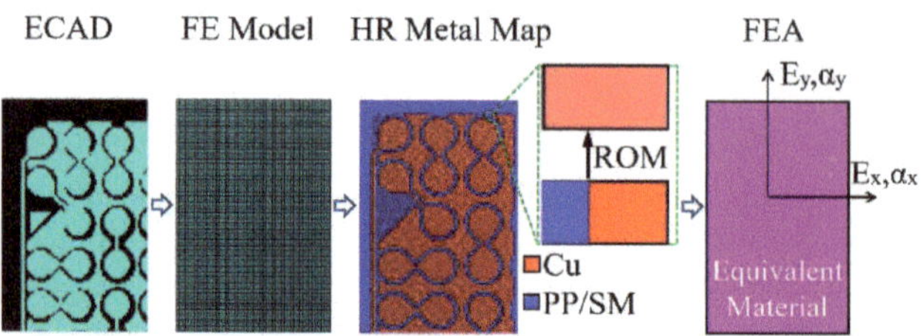

Figure 4. A flowchart of ECAD TM method.

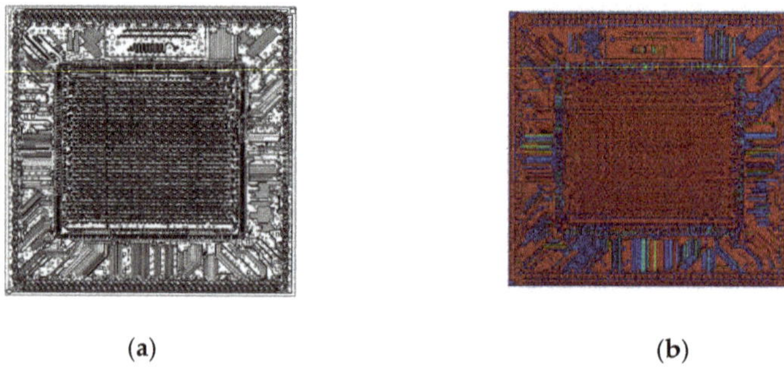

(a)   (b)

Figure 5. A Cu circuit layer in a substrate: (**a**) Cu circuitry pattern; (**b**) Approximate FE model.

As soon as the material property map and 3D FE model of the Cu circuit layer were created, FEA was applied to calculate its effective orthotropic elastic properties. For modeling simplification, the orthotropic Cu circuit layer could be approximated as a transversely isotropic material by averaging the effective in-plane elastic properties and further as an isotropic material by averaging the effective in-plane and out-of-plane elastic properties. The 3D FE models of the mapped Cu circuit layer, PP dielectric and SM layers could be combined together to form an integrated 3D FE model of the substrate and interposer. Finally, the effective orthotropic elastic properties of the substrate and interposer as a whole could be derived using FEA. It is worth mentioning that the final modeling step; i.e., approximation of the substrate and interposer as a homogeneous equivalent continuum, may not be indispensable, since FE modeling of the substrate and interposer could be directly carried out using the integrated 3D FE model. This effective approach is hereinafter termed the TM/FEA effective method.

4.1.2. The ROM/Analytical Estimate

The effective in-plane and out-of-plane CTEs of the Cu circuit layers could be also assessed using an analytical estimate integrated with an ROM method (It is alternatively termed the ROM/analytical estimate). Specifically, the effective in-plane CTE $\alpha_{x,y}$ of the Cu circuit layers was evaluated according to the literature [19] using an energy approach, and the effective out-of-plane CTE α_z also was derived based on the work of [19], and also as presented in [7]:

$$\alpha_z = \frac{E_1 \alpha_1 \zeta_1 + E_2 \alpha_2 \zeta_2}{\zeta_1 E_1 + \zeta_2 E_2} \tag{11}$$

$$\alpha_{x,y} = (1 + v_1)\alpha_1 \zeta_1 + (1 + v_2)\alpha_2 \zeta_2 - \alpha_z \bar{v} \tag{12}$$

where $E_1(E_2)$, $\alpha_1(\alpha_2)$, and $\xi_1(\xi_2)$ are the Young's modulus, CTE and volume fraction of the Cu (SM or PP), respectively. In Equation (12), the effective Poisson's ratio $\bar{v}$ can be simply approximated using ROM as:

$$\bar{v} = v_1\xi_1 + v_2\xi_2 \tag{13}$$

where ξ_1 and ξ_2 are the Poisson's ratio of the Cu and SM or PP respectively. Likewise, the effective in-plane elastic modulus $E_{x,y}$ and out-of-plane elastic modulus E_z of the Cu circuit layers can be also estimated as:

$$E_{x,y} = \frac{E_1 E_2}{E_1\xi_2 + E_2\xi_1} \tag{14}$$

$$E_z = E_1\xi_1 + E_2\xi_2 \tag{15}$$

4.1.3. The FEA-Based Effective Approach

FEA using a detailed fine mesh model could be directly applied to derive the effective orthotropic elastic properties of the Cu circuit layers. This method can be very effective in accurately grasping the crucial parameters affecting the effective properties but require a very tedious, time-consuming and complex procedure to model and simulate the material models [20]. The method is briefly termed the FEA-based effective approach [7]. The underlying idea behind this approach is that the elastic responses of the homogeneous equivalent continuum should be consistent with those of the original continuum.

The effective CTEs of the Cu circuit layers could be simply calculated based on the strength of the materials,

$$\alpha_i = \delta_i / (\Delta T) L_i \ (i = x, y, z) \tag{16}$$

where δ_i is the thermal deformation, $\alpha_i (i = x, y, z)$ stands for the effective CTE in the i-th direction, ΔT denotes the temperature increment, and L_i represents the side length of the Cu circuit layers in the i-th direction.

In accordance with the generalized Hooke's law, the stress-strain relationship of an orthotropic material is expressed as:

$$\varepsilon_{xx} = \frac{\sigma_{xx}}{E_x} + \frac{v_{yx}}{E_y}\sigma_{yy} + \frac{v_{zx}}{E_z}\sigma_{zz} \tag{17}$$

$$\varepsilon_{yy} = \frac{v_{xy}}{E_x}\sigma_{xx} + \frac{\sigma_{yy}}{E_y} + \frac{v_{zy}}{E_z}\sigma_{zz} \tag{18}$$

$$\varepsilon_{zz} = \frac{v_{xz}}{E_x}\sigma_{xx} + \frac{v_{yz}}{E_y}\sigma_{yy} + \frac{\sigma_{zz}}{E_z} \tag{19}$$

$$\gamma_{yz} = \frac{\tau_{yz}}{G_{yz}} \tag{20}$$

$$\gamma_{xz} = \frac{\tau_{xz}}{G_{xz}} \tag{21}$$

$$\gamma_{xy} = \frac{\tau_{xy}}{G_{xy}} \tag{22}$$

where $\varepsilon(\varepsilon_x, \varepsilon_y, \varepsilon_z)$ and $\sigma(\sigma_x, \sigma_y, \sigma_z)$ are the normal strain and stress, respectively, $\gamma(\gamma_{xy}, \gamma_{yz}, \gamma_{xz})$ and $\tau(\tau_{xy}, \tau_{yz}, \tau_{xz})$ represent the shear strain and stress, respectively, and $v(v_{xy}, v_{yz}, v_{xz})$ denotes the Poisson's ratio. In total, there are nine independent effective elastic constants to be determined for an orthotropic elastic material, which are E_x, E_y, E_z, v_{xy}, v_{yz}, v_{xz}, G_{xy}, G_{yz}, and G_{xz}. These constants can be simply derived based on Equations (17)–(22) through FEAs with a set of different loading and boundary conditions. The rest of the effective elastic constants v_{yx}, v_{zy}, v_{zx} can be readily derived from the fact that the compliance matrix is symmetric.

4.2. Process Modeling

In this study, a process simulation framework that incorporated the proposed TM/FEA effective method, thermal and mechanical FEAs and the element death and birth method in ANSYS was introduced. Due to symmetry, a quarter of the FCPoP assembly was simulated via the proposed process simulation framework. Figure 6 reveals the constructed 3D FE model of the FCPoP assembly, which comprised 193,401 nodes and 185,211 solid elements. It consisted of the main components of the FCPoP assembly, including a coreless interposer, a core substrate, an EMC, a silicon chip, solder balls, Cu pillar bumps, and an underfill. The displacement boundary conditions are set to simulate the symmetry boundary condition, where the out-of-plane displacement of the nodes on the symmetry planes were constrained. In addition, the bottom node on the intersecting line of these two symmetry planes was fixed in the z-direction to prevent rigid body motion.

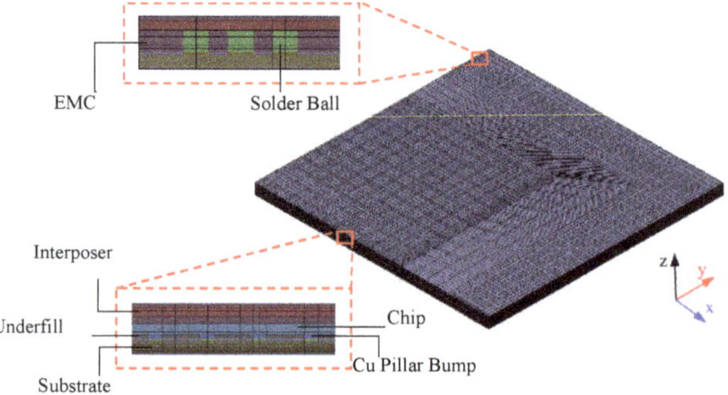

Figure 6. 3D FE model of the FCPoP assembly.

All the materials in the assembly were assumed to be either linearly elastic and isotropic or orthotropic except the EMC, which was assumed to be linearly viscoelastic. It was noteworthy that the temperature dependence of these materials and the effects of curing shrinkage of the EMC were also taken into account in this investigation. The temperature-dependent elastic properties of the components in the assembly are characterized using a thermal-mechanical analyzer (TMA) (TA Instruments, New Castle, DE, USA) and a dynamic mechanical analyzer (DMA) (TA Instruments, New Castle, DE, USA), as shown in Figure 7.

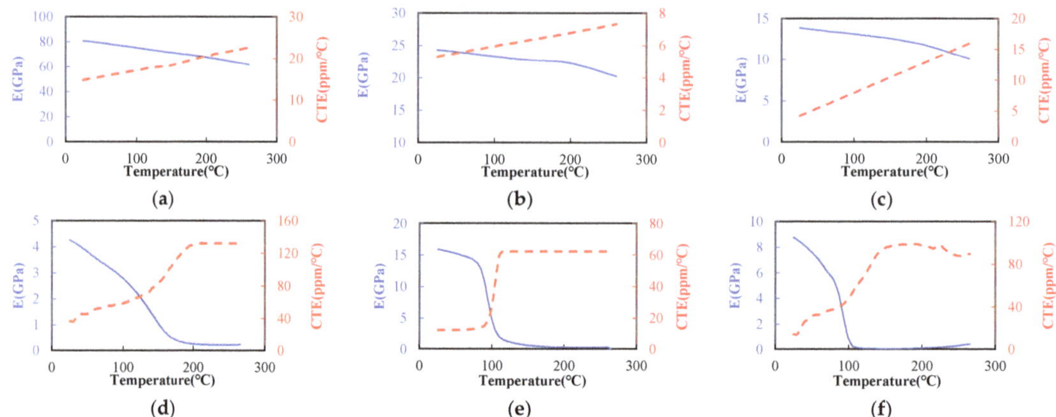

Figure 7. Temperature-dependent Young's modulus (solid line) and CTE (dashed line) of the components: (**a**) Cu; (**b**) core material; (**c**) PP; (**d**) solder mask; (**e**) EMC; (**f**) underfill.

The stress-free temperature of the EMC was defined as its cure temperature. The curing process of the EMC involved two processes: in-mold cure (IMC) and post-mold cure (PMC). Typically, the IMC process applies a lower temperature and a shorter curing time to increase the stiffness of the EMC. Subsequently, a PMC process with a higher temperature and a longer duration was utilized to completely cure the EMC. However, during the curing, the EMC would experience a volumetric (chemical) expansion or contraction. The measured volumetric change data provided by the manufacturer was applied. It was reported that the EMC during the curing process from the gel point (where the stiffness of the EMC is nearly developed) to a full cure state would cause a 0.09% volumetric shrinkage. The process modeling for the warpage prediction of the FCPoP assembly during fabrication closely adhered to the process steps shown in Figure 3.

5. Results and Discussion
5.1. Characterization of EMC Viscoelastic Properties

In this work, a DMA measurement system was applied to conduct the stress relaxation experiments in the frequency domain through a three-point bending mode. The storage moduli of the EMC under a 0.5% applied strain over a wide frequency scan ranging from 0.1 Hz to 100 Hz and a broad isothermal temperature range of 25–260 °C with 5 °C increment were derived, and some of the results are shown in Figure 8a. These stress relaxation storage moduli were further approximated by the Ninomiya–Ferry method [21] as:

$$E(t) = E'(\omega) - 0.4E''(0.4\omega) + 0.014E''(10\omega) \tag{23}$$

where $\omega = 1/t$, E represents the stress relaxation modulus, E' is the storage modulus, and E'' denotes the loss modulus. It is clear that the stress relaxation storage modulus would have a strong temperature correlation at temperatures neighboring the T_g of the EMC, which was around 100 °C. In addition, the stress relaxation storage modulus near the T_g showed a great time dependence, and at temperatures lower than 50 °C and higher than 200 °C exhibited trivial time and temperature correlations.

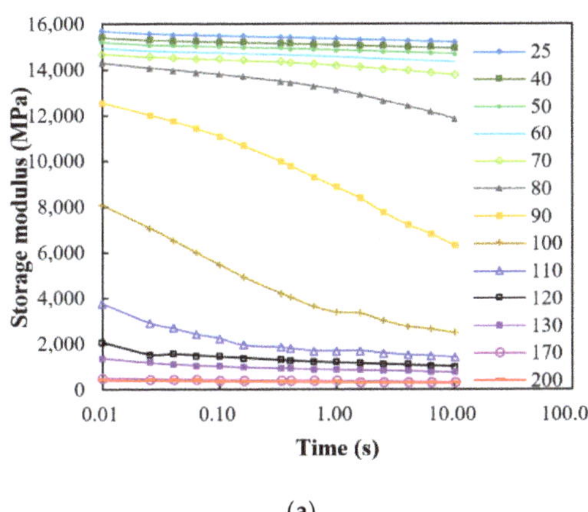

(a)

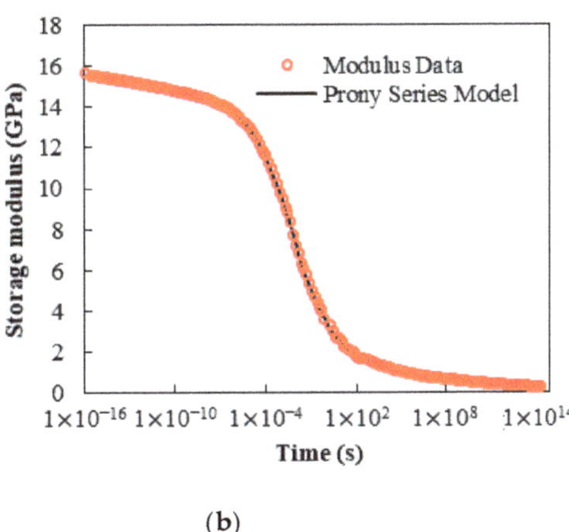

(b)

Figure 8. (**a**) Stress relaxation storage moduli at different isothermal temperatures; (**b**) Construction of single master curve and its Prony series representation.

Based on the TTS principle, a single master curve could be constructed by shifting these frequency-dependent storage moduli at different temperatures along the time axis, as shown in Figure 8b, where the reference temperature was set to the T_g of the EMC. The master curve could be well fitted by a Prony series equation using 22 Prony elements; the fitted weighting coefficients c_k and the relaxation times ς_k are listed in Table 1. The corresponding temperature shift factors in logarithmic scale are shown in Figure 9, as a function of temperature. These shift factors were further fitted to the curve using the WLF model and the curve fitting result is also shown in Figure 9. Clearly, there also was a very good fit to these shift factors with the fitted constant values $a_1 = 208.9$ and $a_2 = 1092.0$.

Table 1. Fitted values of weighting coefficients and relaxation times.

k	ς_k	c_k	k	ς_k	c_k	k	ς_k	c_k	k	ς_k	c_k
1	1.0×10^{-16}	0.0241	7	1.0×10^{-6}	0.0259	13	1.0×10^{0}	0.0743	19	1.0×10^{8}	0.0092
2	1.0×10^{-14}	0.0128	8	1.0×10^{-5}	0.0465	14	1.0×10^{1}	0.0474	20	1.0×10^{10}	0.0073
3	1.0×10^{-12}	0.0139	9	1.0×10^{-4}	0.0796	15	1.0×10^{2}	0.0233	21	1.0×10^{12}	0.0052
4	1.0×10^{-10}	0.0162	10	1.0×10^{-3}	0.0894	16	1.0×10^{3}	0.0089	22	1.0×10^{14}	0.0035
5	1.0×10^{-8}	0.0169	11	1.0×10^{-2}	0.1411	17	1.0×10^{4}	0.0208			
6	1.0×10^{-7}	0.0122	12	1.0×10^{-1}	0.0948	18	1.0×10^{6}	0.0150			

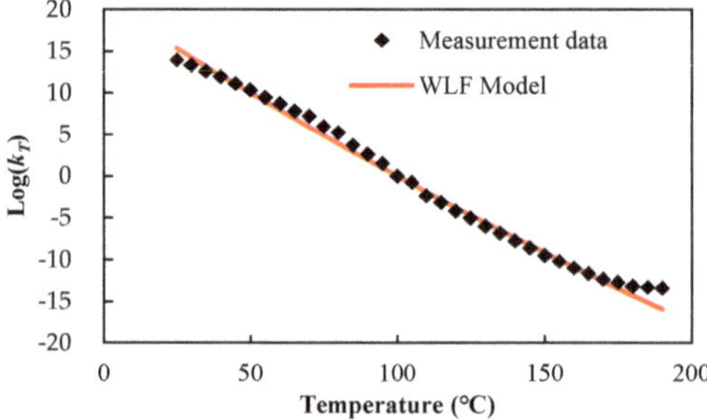

Figure 9. Temperature shift factors and WLF curve fitting.

5.2. Verification of the Effective Models

To demonstrate the feasibility of the TM technique, a fraction of a Cu circuit layer was considered as a test vehicle. The fractional Cu circuit layer was modeled using both detailed FE modeling and the TM technique, and the results are presented in Figure 10. Noticeably, there was a high agreement between them, indicating that the TM technique could not only robustly but also precisely distinguish the Cu circuitries from the PP dielectric or SM material.

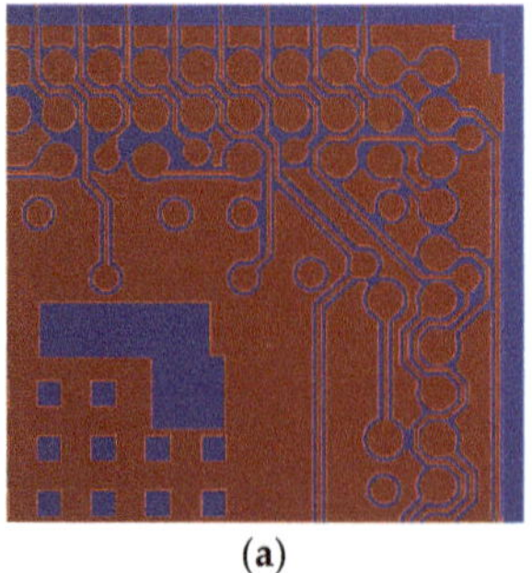
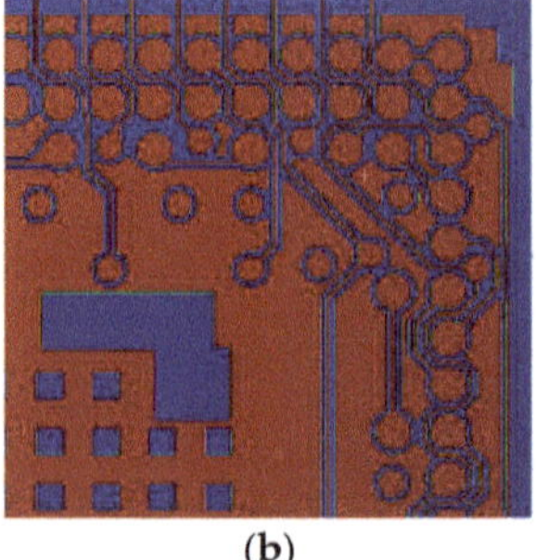

(a)   (b)

Figure 10. Constructed FE models of a fractional Cu circuit layer using: (**a**) detailed FE modeling; (**b**) TM method.

The effectiveness of the proposed TM/FEA effective method was verified by comparing the predicted effective orthotropic elastic properties of the fractional Cu circuit layer displayed in Figure 10 with the ROM/analytical estimate and with the FEA-based effective approach. The latter was considered a benchmark model. It was worth mentioning that for simplification, the orthotropic elastic material is simplified as a transversely isotropic elastic material by averaging the effective in-plane elastic moduli (E_x and E_y) and CTEs (α_x and α_y) to be $E_{\overline{x,y}}$ and $\alpha_{\overline{x,y}}$, respectively. The calculated effective properties are shown in Tables 2 and 3. These two tables illustrate that the effective elastic moduli and CTEs showed a strong temperature dependence, where an elevated temperature would considerably lessen the effective elastic moduli but enlarge the CTEs. Moreover, Table 2 shows that the calculated effective in-plane and out-of-plane elastic moduli by the proposed TM/FEA effective method were much more consistent with those of the FEA-based effective method over the temperature range of 25–260 °C, as compared to the ROM/analytical estimate. On the other hand, the calculated effective in-plane elastic moduli by the ROM/analytical estimate deviated considerably from those of the other two effective approaches across the temperature range. Furthermore, a similar result could be also found for the predicted effective CTEs, as shown in Table 3. Similar to the effective elastic moduli, there is a very pronounced difference in the effective in-plane CTEs between the ROM/analytical estimate and the other two effective approaches.

Table 2. Comparison of calculated effective elastic moduli (MPa) using three different approaches.

T (°C)	FEA-Based		TM/FEA				ROM/Analytical			
	$E_{\overline{x,y}}$	E_z	$E_{\overline{x,y}}$	Diff. (%)	E_z	Diff. (%)	$E_{\overline{x,y}}$	Diff. (%)	E_z	Diff. (%)
25	40,317	56,033	40,690	0.9	55,377	−1.2	30,951	−23.2	58,444	4.3
50	39,536	54,915	39,899	0.9	54,283	−1.2	30,366	−23.2	57,274	4.3
120	37,088	51,125	37,397	0.8	50,581	−1.1	28,650	−22.8	53,272	4.2
150	36,043	49,614	36,337	0.8	49,102	−1.0	27,874	−22.7	51,688	4.2
200	33,914	46,975	34,202	0.8	46,510	−1.0	26,102	−23.0	48,976	4.3
260	30,122	42,614	30,422	1.0	42,219	−0.9	22,817	−24.3	44,546	4.5

Table 3. Comparison of the calculated effective CTEs using three different approaches.

T (°C)	FEA-Based		TM /FEA				ROM/Analytical			
	$\alpha_{\overline{x,y}}$	α_z	$\alpha_{\overline{x,y}}$	Diff. (%)	α_z	Diff. (%)	$\alpha_{\overline{x,y}}$	Diff. (%)	α_z	Diff. (%)
25	11.99	13.76	11.90	−0.8	13.66	−0.7	10.47	−12.6	13.87	0.8
50	12.92	14.62	12.84	−0.6	14.53	−0.6	11.47	−11.2	14.73	0.7
120	15.43	16.90	15.36	−0.5	16.81	−0.5	14.19	−8.1	17.00	0.6
150	16.27	17.58	16.21	−0.4	17.51	−0.4	15.15	−6.9	17.67	0.5
200	18.55	19.82	18.49	−0.3	19.76	−0.3	17.47	−5.8	19.90	0.4
260	20.90	22.03	20.86	−0.2	21.99	−0.2	19.92	−4.7	22.10	0.3

The orthotropic Cu circuit layer was further approximated as a transversely isotropic material by simply averaging the effective in-plane elastic moduli (E_x and E_y) and CTEs (α_x and α_y) as $E_{\overline{x,y}}$ and $\alpha_{\overline{x,y}}$, respectively. The transversely isotropic material could be further simplified as an isotropic material through an average of the effective in-plane and out-of-plane elastic moduli ($E_{\overline{x,y}}, E_z$) and CTEs ($\alpha_{\overline{x,y}}$, α_z) as $E_{\overline{x,y,z}}$ and $\alpha_{\overline{x,y,z}}$, respectively. The feasibility of the three constitutive models, i.e., orthotropic, transversely isotropic and isotropic, was further examined through FEA of the fractional Cu circuit layer shown in Figure 10 when subjected to a temperature load from 25 °C to 260 °C, and the calculated thermal deformations in the x-, y- and z- directions are displayed in Table 4. For comparison, the detailed FEA results, serving as benchmark data, are also listed in the table, which shows that the effective orthotropic model presented the best consistency with the detailed FEA, followed by the transversely isotropic and isotropic models. This result matched with mechanical intuition. Accordingly, the effective orthotropic constitutive model was used in the subsequent warpage process simulation.

Table 4. Comparison of calculated thermal deformations among three different constitutive models.

	U_x (mm)	Diff. (%)	U_y (mm)	Diff. (%)	U_z (mm)	Diff. (%)
Detailed FEA	1.14×10^{-2}	-	1.17×10^{-2}	-	4.56×10^{-4}	-
Orthotropic	1.13×10^{-2}	−0.47	1.17×10^{-2}	−0.44%	4.58×10^{-4}	0.52%
Transversely Isotropic	1.15×10^{-2}	1.04	1.15×10^{-2}	−1.90%	4.22×10^{-4}	−7.52%
Isotropic	1.18×10^{-2}	3.96	1.18×10^{-2}	0.94%	4.34×10^{-4}	−4.84%

5.3. Thermal Analysis of the Interposer Bonding Process

The temperature distribution of the FCPoP assembly during the interposer bonding process may not be uniform across the assembly due to the uneven applied process thermal loading, which would cause a more excessive deformation. Thus, prior to conducting the warpage process simulation, the temperature distribution of the FCPoP assembly in natural convection during the interposer bonding process was characterized using a 3D transient heat conduction FEA. The natural convective heat transfer model proposed in [22] and the standard radiative heat transfer model [23] were applied to depict the natural convective and radiative surface heat transfer, respectively. According to the process condition, a preheat temperature was first set on the top and bottom surfaces of the assembly for 4 s, which were 185 °C and 145 °C, respectively, followed by a temperature increase up to 245 °C in 10 s on the top surface. The ambient temperature is 25 °C. The thermal analysis result at the end of the process is demonstrated in Figure 11. It is important to note that the Cu core solder balls were arranged in two to three rows around the periphery of the substrate and interposer. Evidently, the heat was conducted from the top coreless substrate to the bottom core interposer mainly by way of these Cu core solder balls; as a result, the part of the substrate and interposer adjacent to these periphery Cu core solder balls experienced a higher temperature. In addition, there was a significant temperature non-uniformity and gradient across the assembly. The characterized temperature distribution was imposed as a thermal load in the warpage process simulation for a better prediction accuracy.

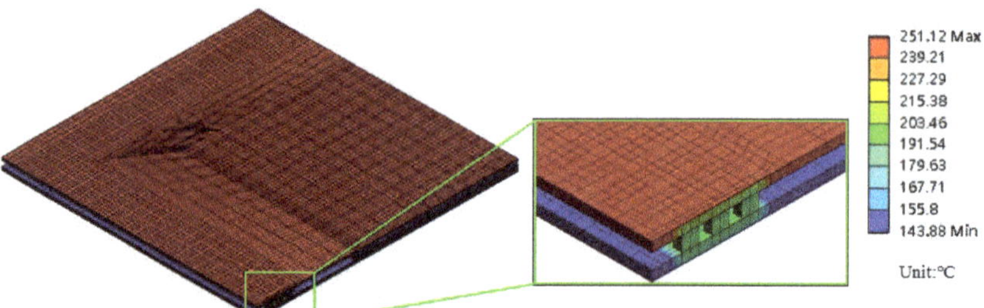

Figure 11. Temperature distribution of the assembly during interposer bonding process.

5.4. Warpage Process Simulation

The calculated temperature-dependent effective orthotropic elastic properties of the coreless substrate and core interposer using the TM/FEA effective approach are demonstrated in Tables 5 and 6, respectively. It was interesting to see that the core interposer was much stiffer than the coreless substrate across the temperature range; On the contrary, the CTEs of the coreless substrate tended to slightly greater than those of the core interposer. The warpage evolution of the FCPoP assembly during the fabrication process is shown in Figure 12a. It can be clearly observed in the figure that the process-induced warpage extensively varied with the process steps, and also showed a significant increase after the die process bonding, underfill curing, and interposer bonding processes. In addition, the maximum warpage occurred after the interposer bonding process; i.e., at around 653.7 µm, rather than after the fabrication process, i.e., at about 82.6 µm. The reason that the interposer bonding process created the maximum warpage was the lack of the EMC helping to resist the shear force caused by the global CTE mismatch between the substrate and interposer.

Table 5. Temperature-dependent effective orthotropic elastic properties of the coreless substrate (unit: MPa, °C/ppm).

T (°C)	E_x	E_y	E_z	v_{xy}	v_{yz}	v_{xz}	G_{xy}	G_{yz}	G_{xz}	α_x	α_y	α_z
25	27,279	25,603	20,012	0.3	0.3	0.3	3195	3196	3244	12.6	12.1	6.2
50	26,709	25,039	19,631	0.3	0.3	0.3	3115	3132	3182	13.7	13.2	7.3
120	24,871	23,212	18,531	0.3	0.3	0.3	2846	2944	3001	16.4	16.0	10.7
150	24,011	22,277	18,003	0.3	0.3	0.3	2669	2841	2912	17.4	16.9	12.2
200	22,463	20,721	16,796	0.3	0.3	0.3	2411	2628	2713	19.9	19.2	14.9
260	19,991	18,347	14,604	0.3	0.3	0.3	2122	2285	2358	22.4	21.8	17.7

Table 6. Temperature-dependent effective orthotropic elastic properties of core interposer (unit: MPa, °C/ppm).

T (°C)	E_x	E_y	E_z	v_{xy}	v_{yz}	v_{xz}	G_{xy}	G_{yz}	G_{xz}	α_x	α_y	α_z
25	37,149	37,205	33,475	0.3	0.3	0.3	4636	6028	6023	12.7	12.4	7.2
50	36,423	36,479	32,967	0.3	0.3	0.3	4533	5934	5929	13.4	13.3	7.7
120	33,884	33,940	31,178	0.3	0.3	0.3	4158	5611	5599	14.9	14.9	9.0
150	32,686	32,745	30,183	0.3	0.3	0.3	3930	5477	5445	15.3	15.2	9.6
200	30,904	30,966	27,634	0.3	0.3	0.3	3533	5295	5168	16.8	16.7	11.0
260	28,160	28,218	25,099	0.3	0.3	0.3	3212	4822	4703	18.3	18.2	12.1

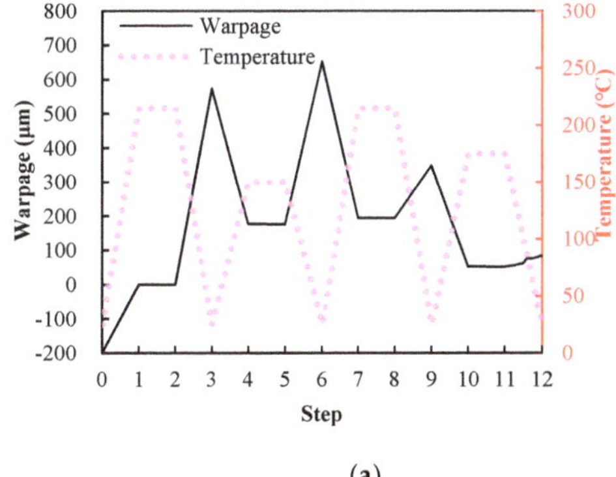

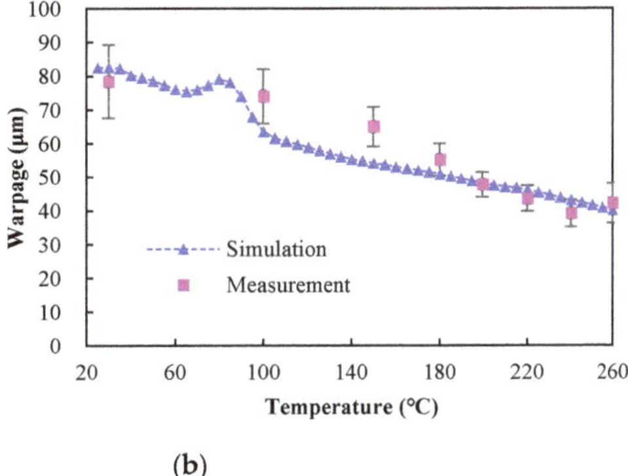

Figure 12. Warpage evolution: (**a**) from die bond to mold cure process; (**b**) during increasing temperature process.

Figure 12b illustrates the simulated warpages during the increasing temperature process. For comparison, the in-line warpage measurement data are also listed in Figure 12, presented as an average value with a standard deviation (SD) (error bar). It was evident that the simulated warpages over the temperature range of 25–260 °C closely followed the measurement data. The fair difference in warpage between the measurement and simulation could be attributed to the uncertainty in the measured temperature-dependent Young's modulus and CTE of the EMC. Additionally, the increased temperature reduced the warpage probably due to the softening of the EMC when exposed to temperatures greater than the T_g. The simulated and measured warpage contour plots of the FCPoP assembly at step 12 (after the mold cure process) are shown in Figure 13. Once again, there were very consistent results between the simulation and measurement. Table 7 illustrates the simulated and average measured residual warpages at 30 °C and 260 °C together with the smallest and largest values of the measured data shown in the bracket. It is clear that these simulation data fell in the respective ranges of the measured data, and in addition, the results of the process simulation were very comparable to those of the measurement, where the maximum warpage difference between them was only around 5%. Moreover, the residual warpage at 30 °C is nearly double that found at 260 °C, and due to this, the residual warpage at 260 °C was not considered in the subsequent parametric analysis.

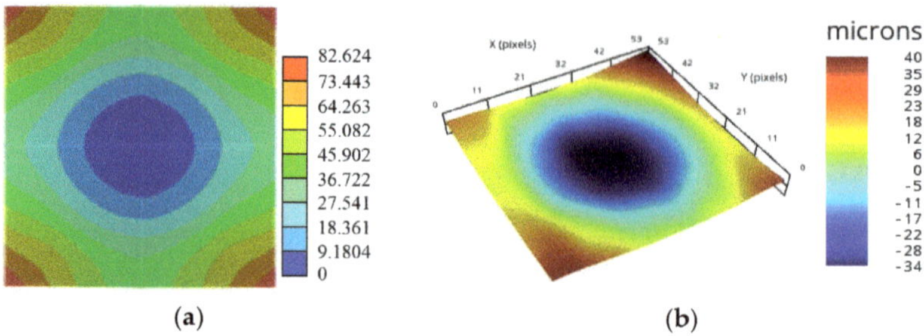

(a)   (b)

Figure 13. Comparison of simulated and experimental warpages after the mold cure process; (**a**) simulation; (**b**) experiment.

Table 7. Warpage comparison between the simulation and measurement.

Method	Warpage (μm)	
	30 °C	**260 °C**
Simulation	82.6	40.2
Measurement	78.4 (60,89)	42.2 (32,49)

5.5. Parametric Study

5.5.1. Effects of Component CTEs

The influences of the effective CTEs of the EMC, core interposer and coreless substrate on the warpage of the FCPoP assembly at 30 °C were addressed. The parametric results of the effect of the EMC CTE are presented in Figure 14a. In the parametric analysis, the effective CTE of the EMC nominally varied from −10% to +10%. The figure shows that the EMC CTE had a minor impact on the residual warpages due to the relatively rigid substrate and interposer. Specifically, an increase in the EMC CTE somewhat decreased the residual warpage. This could be due to an increased EMC CTE reducing its local CTE mismatch with the core interposer and coreless substrate, thereby leading to a lessened residual warpage.

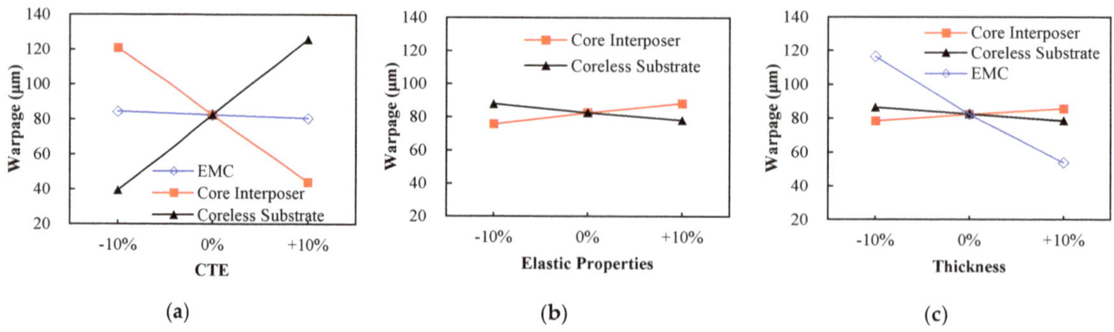

(a)   (b)   (c)

Figure 14. Effects of component material and geometric parameters: (**a**) CTE; (**b**) orthotropic elastic properties; (**c**) thickness.

The effects of the effective CTEs of the core interposer and coreless substrate are investigated, and the parametric results are also displayed in Figure 14a. Note that the three effective CTEs (α_x, α_y, α_z) shown in Tables 5 and 6 simultaneously underwent a

$\pm10\%$ variation from the original value. The figure demonstrates that the effective CTEs had a significant impact on the residual warpage. Specifically, an increase in the effective interposer CTEs dramatically reduced the warpage, whereas there was a totally opposite trend for the effective substrate CTEs. This was principally due to the effective CTEs of the interposer being smaller than those of the substrate. This suggests that the increase in the interposer CTEs reduced the CTE mismatch with the substrate, thereby leading to a reduced residual warpage. Likewise, the result of the effects of the effective CTEs of the coreless substrate can be also explained in the same way.

5.5.2. Effect of Component Orthotropic Elastic Properties

The effects of the effective orthotropic elastic properties of the core interposer and coreless substrate on the warpage at 30 °C were considered. Similar to the parametric analysis of the effective CTEs, there was a $\pm10\%$ variation in these 9 independent effective elastic property data (E_x, E_y, E_z, v_{xy}, v_{yz}, v_{xz}, G_{xy}, G_{yz} and G_{xz}) shown in Tables 5 and 6. The parametric results are presented in Figure 14b. They indicated that increased effective elastic properties of the interposer and decreased effective elastic properties of the substrate would amplify the residual warpage. This was because the core interposer is stiffer than the coreless substrate due to its possessing greater effective elastic and shear moduli. The growth of the effective elastic shear and elastic moduli of the interposer tended to result in a more excessive shear force resulting from the CTE mismatch between the interposer and substrate, thereby causing a greater warpage. On the other hand, the structural rigidity of the FCPoP assembly increased with the increase of the effective elastic and shear moduli of the substrate, which thus led to a reduced warpage. The results totally differed from the effects of the effective CTEs of the interposer and substrate.

5.5.3. Effect of Component Thickness

The dependence of the warpage at 30 °C on the thickness of the core interposer, coreless substrate and EMC was examined. Similarly, the thickness variation is also $\pm10\%$ from their original value. Figure 14c illustrates the parametric results, where the residual warpage would increase both with an increasing interposer thickness and with a decreasing substrate thickness. The results were very consistent with the effects of the elastic properties of the interposer and substrate, and the explanation for this is the same as stated in the pervious section. In the parametric analysis of the thickness effect of the EMC, parametrizing the EMC thickness would, in the meantime, change the height of the CCSBs. Before conducting the parametric study of the influence of the EMC thickness, the height effect of the CCSBs in the assembly without an EMC was first explored. The parametric study, which is not presented here due to limited space, suggested that the CCSBs' height had little impact for the residual warpage. As a result, the parametric results on the thickness effect of the EMC was barely affected by the CCSBs' height. The dependence of the residual warpage on the EMC thickness is also presented in Figure 14c, which shows that the residual warpage significantly decreased with the EMC thickness. This can be attributed to the structural stiffness of the assembly substantially increasing with the EMC thickness, thus resulting in a less residual warpage.

6. Conclusions

This research successfully conducted an effective and robust prediction of the warpage performance of an FCPoP assembly during the fabrication process through the proposed process simulation framework. In this framework, the temperature-dependence of the elastic properties of the components, as well as the viscoelastic behavior and chemical shrinkage of the EMC were taken into account in this investigation. The temperature-dependent elastic properties and viscoelastic properties were experimentally characterized. In order to improve the computational and modeling efficiency while also preserving good prediction accuracy, a novel effective approach; i.e., the TM/FEA effective method, was introduced to assess the effective elastic properties of the orthotropic coreless substrate and core interposer. The effectiveness of the proposed effective method and the proposed process simulation framework were extensively validated. Finally, a parametric analysis was performed to investigate the dependence of the process-induced warpage on some geometric and material parameters.

1. The ECAD TM technique was a very effective and robust way to precisely recognize the Cu circuitries in the PP dielectric or SM material.
2. The DMA results indicated that the storage modulus of the EMC showed great time and temperature dependence particularly at temperatures near its T_g.
3. Both the predicted effective elastic moduli and CTEs of the substrate and interposer turned out to have a negative and a positive temperature coefficient, respectively.
4. The orthotropic constitutive assumption was shown to provide the most accurate prediction of the thermal deformations of the substrate and interposer, as compared to the transversely isotropic and isotropic ones.
5. The thermal analysis results showed that there was a significant temperature non-uniformity across the assembly during the interposer bonding process, which could potentially affect the process-induced warpage.
6. The proposed TM/FEA effective method and proposed process simulation framework were found to be very effective in predicting the effective elastic properties of the substrate and interposer and the process-induced warpage of the FCPoP assembly, respectively.
7. The process-induced warpage of the FCPoP assembly experienced a dramatic change over the process steps, and more importantly, the maximum warpage occurred after the interposer bonding process rather than the end of the fabrication process. In addition, the warpage at 30 °C was roughly twice that of 260 °C.
8. The warpage decreased with temperature during the increasing temperature process, probably because the EMC material became softened at temperatures greater than the T_g.
9. Among the parameters considered in the parametric analysis, the substrate CTE had the greatest influence on the warpage at 30 °C, followed by the interposer CTE and the EMC thickness; moreover, a smaller substrate CTE, a larger interposer CTE and a thicker EMC brought about a reduced warpage.

Author Contributions: Conceptualization, H.-C.C.; methodology, H.-C.C.; software, L.-C.T. and Y.-C.L.; validation, H.-C.C., L.-C.T. and Y.-C.L.; formal analysis, H.-C.C., L.-C.T. and Y.-C.L.; investigation, H.-C.C., L.-C.T. and Y.-C.L.; resources, H.-C.C., L.-C.T. and Y.-C.L.; data curation, H.-C.C., L.-C.T. and Y.-C.L.; writing—original draft preparation, H.-C.C.; writing—review and editing, H.-C.C.; visualization, H.-C.C.; supervision, H.-C.C.; project administration, H.-C.C.; funding acquisition, H.-C.C. All authors have read and agreed to the published version of the manuscript.

Funding: This research received no external funding. The work is partially supported by Ministry of Science and Technology, Taiwan, ROC, under grants 110-2221-E-035-049-MY3.

Institutional Review Board Statement: Not applicable.

Informed Consent Statement: Not applicable.

Data Availability Statement: Data sharing not applicable.

Conflicts of Interest: The authors declare no conflict of interest.

References

1. Tsai, M.-Y.; Wang, Y.-W.; Liu, C.-M. Thermally-induced deformations and warpages of flip-chip and 2.5D IC packages measured by strain gauges. *Materials* **2021**, *14*, 3723. [CrossRef] [PubMed]
2. Chou, P.H.; Chiang, K.N.; Liang, S.Y. Reliability assessment of wafer level package using artificial neural Network regression model. *J. Mech.* **2019**, *35*, 829–837. [CrossRef]
3. Chao, S.H.; Hung, C.P.; Chen, M.; Lee, Y.; Huang, J.; Kao, G.; Luh, D.B. An embedded trace FCCSP substrate without glass cloth. *Microelectron. Reliab.* **2016**, *57*, 101–110. [CrossRef]

4. Liu, W.W.; Weng, B.; Li, J.; Yeh, C.K. FCCSP IMC growth under reliability stress following automotive standards. *J. Microelectron. Electron. Packag.* **2019**, *16*, 21–27.
5. Waldrop, M.M. The chips are down for Moore's law. *Nature* **2016**, *530*, 144–147. [CrossRef] [PubMed]
6. Yang, C.C.; Su, Y.F.; Liang, S.Y.; Chiang, K.N. Simulation of wire bonding process using explicit FEM with ALE remeshing technology. *J. Mech.* **2020**, *36*, 47–54. [CrossRef]
7. Cheng, H.-C.; Li, R.-S.; Lin, S.-C.; Chen, W.-H.; Chiang, K.-N. Macroscopic mechanical constitutive characterization of through-silicon-via-based 3-D integration. *IEEE Trans. Compon. Packag. Manuf. Technol.* **2016**, *6*, 432–446. [CrossRef]
8. Hsieh, M.-C.; Kang, K.; Choi, H.; Kim, Y. Thin profile flip chip package-on-package development. In Proceedings of the 11th IMPACT Conference, Taipei, Taiwan, 26–28 October 2016.
9. Hsieh, M.-C.; Lin, S.; Hsu, I.; Chen, C.Y.; Cho, N. Fine pitch high bandwidth flip chip package-on-package development. In Proceedings of the 21st European Microelectronics and Packaging Conference & Exhibition, Warsaw, Poland, 10–13 September 2017.
10. Cheng, H.-C.; Liu, Y.-C. Warpage characterization of molded wafer for fan-out wafer-level packaging. *ASME J. Electron. Packag.* **2020**, *142*, 011004. [CrossRef]
11. Cheng, H.-C.; Wu, Z.-D.; Liu, Y.-C. Viscoelastic warpage modeling of fan-out wafer level packaging during wafer-level mold cure process. *IEEE Trans. Compon. Packag. Manuf. Technol.* **2020**, *10*, 1240–1250. [CrossRef]
12. Yeh, S.-S.; Lin, P.-Y.; Jeng, S.-P.; Lin, W.-Y.; Yew, M.-C.; Lee, K.-C.; Wang, J.-H.; Lai, P.-C.; Leu, S.-T. Lidded FCCSP warpage evaluation: Process modeling and characterization of the effect of viscoelasticity and cured shrinkage for molded underfill. In Proceedings of the 11th IMPACT Conference, Taipei, Taiwan, 26–28 October 2016.
13. Dacol, V.; Caetano, E.; Correia, J.R. A new viscoelasticity dynamic fitting method applied for polymeric and polymer-based composite materials. *Materials* **2020**, *13*, 5213. [CrossRef] [PubMed]
14. Sadeghinia, M.; Jansen, K.M.B.; Ernst, L.J. Characterization and modeling the thermo-mechanical cure-dependent properties of epoxy molding compound. *Int. J. Adhes. Adhes.* **2012**, *32*, 82–88. [CrossRef]
15. Sangtabi, M.R.; Kiasat, M.S. Long-term viscoelastic properties of an adhesive and molding compound, characterization and modeling. *Polymer* **2017**, *116*, 204–217. [CrossRef]
16. Sadeghinia, M.; Jansen, K.M.B.; Ernst, L.J. Characterization of the viscoelastic properties of an epoxy molding compound during cure. *Microelectron. Reliab.* **2012**, *52*, 1711–1718. [CrossRef]
17. Williams, M.L.; Landel, R.F.; Ferry, J.D. The temperature dependence of relaxation mechanisms in amorphous polymers and other glass-forming liquids. *J. Am. Chem. Soc.* **1995**, *77*, 3701–3707. [CrossRef]
18. Wang, M.; Wells, B. Substrate trace modeling for package warpage simulation. In Proceedings of the IEEE 66th Electronic Components and Technology Conference, Las Vegas, NV, USA, 31 May–3 June 2016; pp. 516–523.
19. Schapery, R. Thermal expansion coefficients of composite materials based on energy principles. *J. Compos. Mater.* **1968**, *2*, 380–404. [CrossRef]
20. Czyzewski, J.; Rybak, A.; Gaska, K.; Sekula, R.; Kapusta, C. Modelling of effective thermal conductivity of composites filled with core-shell fillers. *Materials* **2020**, *13*, 5480. [CrossRef] [PubMed]
21. Ninomiya, K.; Ferry, J.D. Some approximate equations useful in the phenomenological treatment of linear viscoelastic data. *J. Colloid Sci.* **1959**, *14*, 36–48. [CrossRef]
22. Ellison, G.N. *Thermal Computations for Electronic Equipment*; R.E. Krieger Publishing Company: Malabar, FL, USA, 1989.
23. Cheng, H.-C.; Wu, C.-H.; Lin, S.-Y. Thermal and electrical characterization of power MOSFET module using coupled field analysis. *J. Mech.* **2019**, *35*, 641–655. [CrossRef]

materials

Article

Predicting Wafer-Level Package Reliability Life Using Mixed Supervised and Unsupervised Machine Learning Algorithms

Qing-Hua Su and Kuo-Ning Chiang *

Department of Power Mechanical Engineering, National Tsing Hua University, Hsinchu 300, Taiwan;
0967356474shq@gmail.com
* Correspondence: knchiang@pme.nthu.edu.tw

Abstract: With the increasing demand for electronic products, the electronic package gradually developed toward miniaturization and high density. The most significant advantage of the Wafer-Level Package (WLP) is that it can effectively reduce the volume and footprint area of the package. An important issue in the design of WLP is how to quickly and accurately predict the reliability life under the accelerated thermal cycling test (ATCT). If the simulation approach is not adopted, it usually takes several ACTCs to design a WLP, and each ACTC will take several months to get the reliability life results, which increases development time considerably. However, simulation results may differ depending on the designer's domain knowledge, ability, and experience. This shortcoming can be overcome with artificial intelligence (AI). In this study, finite element analysis (FEA) is combined with machine learning algorithms, e.g., Kernel Ridge Regression (KRR), to create an AI model for predicting the reliability life of electronic packaging. Kernel Ridge Regression (KRR) combined with the K-means cluster algorithm provides a highly accurate and efficient way to obtain AI models for large-scale data sets.

Keywords: Wafer-Level Package (WLP); Finite Element Analysis (FEA); machine learning; Kernel Ridge Regression (KRR); Cluster algorithm

Citation: Su, Q.-H.; Chiang, K.-N. Predicting Wafer-Level Package Reliability Life Using Mixed Supervised and Unsupervised Machine Learning Algorithms. *Materials* **2022**, *15*, 3897. https://doi.org/10.3390/ma15113897

Academic Editor: Michele Bacciocchi

Received: 22 April 2022
Accepted: 27 May 2022
Published: 30 May 2022

Publisher's Note: MDPI stays neutral with regard to jurisdictional claims in published maps and institutional affiliations.

1. Introduction

Reliability is an important topic in the field of electronic packaging. Solder ball reliability analysis is the key to measuring the reliability of WLP. One main cause of package failure is thermal-induced CTE (coefficient of thermal expansion) mismatch between different materials. In ATCT, the first solder ball failure usually occurs at the diagonal corner of the package; this is the location with the largest distance from the neutral point (DNP). Although traditional ATCT tests can obtain the reliability life result, they are too time-consuming (usually several months). The long experiment time leads to decreased R&D efficiency, which cannot meet the market demand.

Finite element simulation is a feasible approach to reducing design cycles and time. Lin et al. [1] built a finite element model for WLP. The reliability life can be obtained by substituting incremental equivalent plastic strain into the Coffin–Manson model, and simulation and experiment results are in good agreement. Because the element mesh size in the upper right corner of the solder ball is crucial to the final simulation result, Cheng [2] built a 3D finite element model for an area array type package. In this study, we will build our simulation database for machine learning by following our lab modeling experiences [1–4].

For FEA, different researchers may lead to different results. Moreover, it still takes time (several days or weeks) to get the simulation results. Therefore, it is necessary to introduce machine learning to lower the training threshold of simulation, unify the results, and reduce development time. Machine learning can be divided into supervised and unsupervised learning based on the presence or absence of artificially assigned labels. Among the two algorithms involved in this study, kernel ridge regression (KRR) is supervised learning, and cluster analysis is unsupervised learning. This study uses the K-means algorithm for cluster analysis. Arthur and Robert [5] first proposed the idea of ridge regression (RR) to solve the multicollinearity of the data dimension. With the evolutionary progress of the algorithm, Kernel Ridge Regression (KRR) was proposed in 2000 by Cristianini and Shawe-Taylor [6]. The essence of KRR is the combination of RR and kernel tricks. K-means clustering was proposed by Macqueen [7]. It can perform data partitioning to reduce the workload of KRR; this would significantly reduce the training CPU time of the prediction model.

On the other hand, many researchers have used different types of algorithms to effectively predict the reliability of solder balls, such as Artificial Neural Network (ANN) [8], Support Vector Regression (SVR), Random Forest (RF) [9], and so on. In this study, we compare KRR, and KRR with K-means (K-K) with other algorithms in performance error and training CPU time.

2. Fundamental Theory

2.1. Reliability Life Prediction Model

The reliability life prediction method of the packaging structure can be mainly divided into two types: the energy-based method and the strain-based method. The Coffin–Manson model [10] used in this research belongs to the strain-based method. The incremental equivalent plastic strain is the key to evaluating the solder's reliability life. The expression of the Coffin–Manson model is shown in Equation (1).

$$N_f = \alpha \left(\Delta \varepsilon_{eq}^{pl} \right)^{\phi} \tag{1}$$

N_f is the reliability prediction life (cycle), and $\Delta \varepsilon_{eq}^{pl}$ is the incremental equivalent plastic strain. α and ϕ are empirical constants. In this study, α and ϕ are 0.235 and -1.75.

2.2. Ridge Regression (RR)

We often use the Least Squares Method (LSM) to solve regression problems in statistics. It is a mathematical optimization modeling method that finds the best function match of the data by minimizing the sum of squares of the error. The squared loss function of LSM is shown as Equation (2).

$$L(\beta) = (y - X\beta)^T (y - X\beta) \tag{2}$$

In Equation (2), X is the matrix expression of the data input. The row of the matrix is the number of data samples. The column of the matrix is the data dimension. y is the output. β is the equation coefficient. In LSM, our target value $\hat{\beta}$ is the value that minimizes $L(\beta)$. To do this, we need to solve the partial derivative of $L(\beta)$ to β. The resulting expression of $\hat{\beta}$ is shown as Equation (3).

$$\hat{\beta} = \left(X^T X \right)^{-1} X^T y \tag{3}$$

Training models using LSM enables them to fit known sample points quickly and accurately, but it may cause overfitting. In Equation (3), when there is multicollinearity in the data dimension, $X^T X$ is no longer a full-rank matrix, and it would be challenging to solve its inverse matrix directly. RR is an algorithm proposed to solve this problem. RR is essentially an improved LSM. By giving up the unbiasedness of the LSM, a more realistic mathematical model is obtained at the expense of losing some information and reducing accuracy. Equations (4) and (5) show the new loss function and target value expression.

$$J(\beta) = (y - X\beta)^T (y - X\beta) + k\|\beta\|^2 \tag{4}$$

$$\beta = \left(X^T X + kI \right)^{-1} X^T y \tag{5}$$

In Equations (4) and (5), k is the ridge parameter, and I is a matrix of normal numbers. As k increases, the undetermined coefficient β would stabilize, and we were looking for the smallest k value under the condition that the coefficient is stable.

2.3. Kernel Ridge Regression (KRR)

KRR [11–13] combines RR and kernel tricks. In many cases, it requires mapping data into high-dimensional space to improve machine learning performance. It is found that the same effect can be achieved directly by defining a function K. This function K is called the kernel function. There are three commonly used kernel functions: polynomial kernel, sigmoid kernel, and radial basis function (RBF) kernel, shown as Equations (6)–(8). As we can see, there are three parameters in the polynomial kernel. The sigmoid kernel has two parameters; the RBF kernel only has one parameter, which is its strength. The amount of calculation of KRR is significantly reduced by utilizing the RBF kernel. In this study, we focus on the RBF kernel.

$$K(x_i, x_j) = \left(\gamma \langle x_i, x_j \rangle + b\right)^d \tag{6}$$

$$K(x_i, x_j) = tanh\left(\gamma \langle x_i, x_j \rangle + b\right) \tag{7}$$

$$K(x_i, x_j) = exp\left(-\gamma \|x_i - x_j\|^2\right) \tag{8}$$

In Equations (6)–(8), x_i and x_j represent two data, $\langle \cdot, \cdot \rangle$ represents dot product, and $\|x_i - x_j\|$ is the Euclidean distance between x_i and x_j.

We need to write the RR solution as an inner product to introduce the kernel function. Converting the original formula to a particular form requires the matrix inversion lemma [14].

Consider a general partitioning matrix $M = \begin{pmatrix} E & F \\ G & H \end{pmatrix}$. Assuming that both E and H are invertible, we have:

$$\left(E - FH^{-1}G\right)^{-1} = E^{-1} + E^{-1}F\left(H - GE^{-1}F\right)^{-1}GE^{-1} \tag{9}$$

$$\left(E - FH^{-1}G\right)^{-1}FH^{-1} = E^{-1}F\left(H - GE^{-1}F\right)^{-1} \tag{10}$$

$$\left|E - FH^{-1}G\right| = \left|H - GE^{-1}F\right|\left|H^{-1}\right||E| \tag{11}$$

We use Equation (10) to simplify the optimal solution of β. Let $H^{-1} \triangleq k^{-1}I$, $F \triangleq X^T$, $G \triangleq -X$, $E \triangleq I$, then we obtain a new expression shown as Equation (12).

$$\beta = \left(X^T X + kI\right)^{-1}X^T y = X^T\left(kI + X^T X\right)^{-1}y \tag{12}$$

Now, it is very close to kernelization. Our remaining task is to predict y^* when a new sample point x^* comes in. We write β in the form of vector summation, and let $\alpha \triangleq \left(kI + X^T X\right)^{-1}y$. Then, we rewrite Equation (12) to Equation (13).

$$\beta = X^T \alpha = \sum_{i=1}^{N} \alpha_i x_i \tag{13}$$

We can find that β is just a weighted average of all samples. Thus, the predicted value for a new sample is:

$$y^* = \beta^T x^* = \sum_{i=1}^{N} \alpha_i x_i^T x^* = \sum_{i=1}^{N} \alpha_i K(x^*, x_i) \tag{14}$$

The predicted value is the weighted average of the inner product of the new sample and all the old samples. After converting the original formula to the inner product form, we selected different kernel functions to simplify our calculations.

2.4. K-Means Clustering

K-means [15] is an algorithm that implements cluster analysis based on the principle of minimum distance. The K value must be given in advance, representing the number of cluster centers. For each iteration, we need to calculate the mean of the sample points in the cluster to update the cluster center. K-means clustering divides the n samples into k sets so that the within-cluster sum of squares (WCSS) is the smallest. The formula we use is shown in Equation (15). The updated formula for cluster centers is shown in Equation (16).

$$E = \sum_{i=1}^{k} \sum_{p \in C_i} dist(p, c_i)^2 \tag{15}$$

$$m_i(c_i) = \sum_{j=1}^{n} p_j / n \tag{16}$$

In Equation (15), c_i is the cluster center, p is one sample point, and $dist(p, c_i)$ is the Euclidean distance from p to the cluster center.

3. WLP FEA Model Validation

It is assumed that the CTE difference between the substrate and the wafer is $\Delta\alpha$. The DNP of the solder ball is L. After selecting material parameters, $\Delta\alpha$ is fixed. Solder balls farther from the chip's center have a greater impact on deformation mismatch due to thermal loading. In the case of WLP, the thermal loading failure usually occurs at the outermost diagonal solder ball of the package. This study uses five WLP test vehicles (TV: WLP1-5) [16–18] and one fan-out WLP (FO-WLP, [19]) for FEA model validation. The structure component sizes, materials, and mean-time-to-failure (MTTF) reliability life are shown in Tables 1–3 [17–19]. This research uses these data to verify our simulation results. In order to reduce the computational time cost, this study adopts a simplified two-dimensional finite element model and sets the following basic assumptions: each structure is with homogeneous and isotropic materials; the temperature of the structure is uniform; residual stress is not considered; and all the contact surface between different materials is considered as perfect bonding.

Table 1. Dimension of WLP test vehicles [17,18].

	WLP-1	WLP-2	WLP-3	WLP-4	WLP-5
Silicon Chip	5.3×0.33 (mm)	4.0×0.33 (mm)	4.0×0.33 (mm)	4.0×0.33 (mm)	6.0×0.33 (mm)
Solder Ball Diameter	250 μm	250 μm	180 μm	200 μm	250 μm
Pitch	400 μm	400 μm	300 μm	300 μm	400 μm
Number of Solder Ball	121	100	144	144	196
MTTF (Cycles)	318	1013	587	876	904

Considering that the package body is a symmetric structure, semi-diagonal two-dimensional models were used in this study to simplify the modeling and finite element analysis processes; examples can be seen in Figures 1–4, and PLANE 182 has been selected as the element type in ANSYS (Figure 4). The model boundary condition is fixed for all nodes in the center of the structure in the x-direction. The node at the bottom of the center of the structure is fixed in the x- and y-direction to avoid rigid body motion.

Table 2. Dimensions of fan-out WLP [19].

	FO-WLP
Packaging Size	14 mm × 14 mm × 0.1 mm
Chip Size	10 mm × 10 mm × 0.1 mm
Molding Compound Thickness	190 µm
Die-attach Film Thickness	10 mm × 10 mm × 0.01 mm
Solder Ball Diameter	250 µm
Pitch	400 µm
Number of Solder Ball	540
MTTF (Cycles)	249

Table 3. Material properties for WLP [17,18].

Material	E (GPa)	υ	CTE (ppm/°C)
Solder Ball	Figure 5	0.35	25
Silicon Chip	150	0.28	2.62
Copper	68.9	0.34	16.7
SBL	2	0.33	55
Low-k	10	0.16	5
Solder Mask	6.87	0.35	19

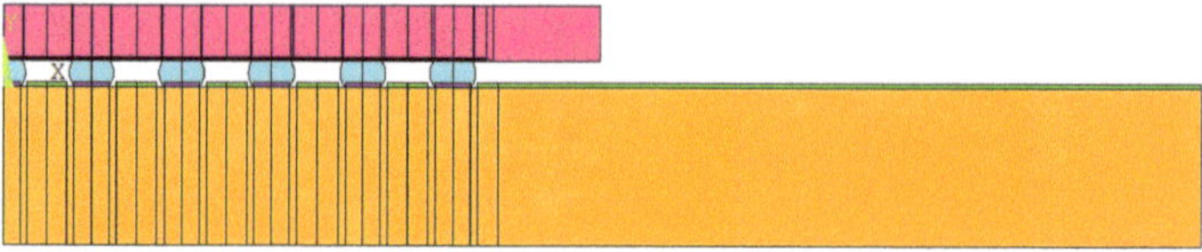

Figure 1. WLP-1 semi-diagonal FEA model.

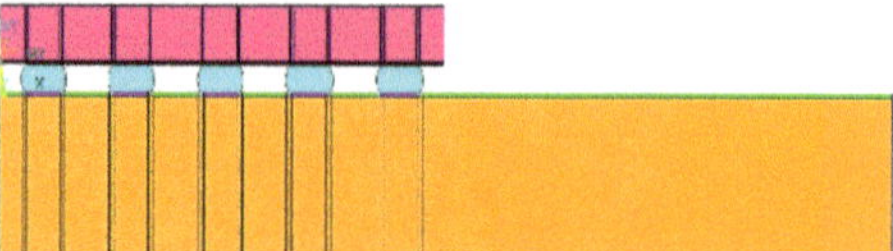

Figure 2. WLP-2 semi-diagonal FEA model.

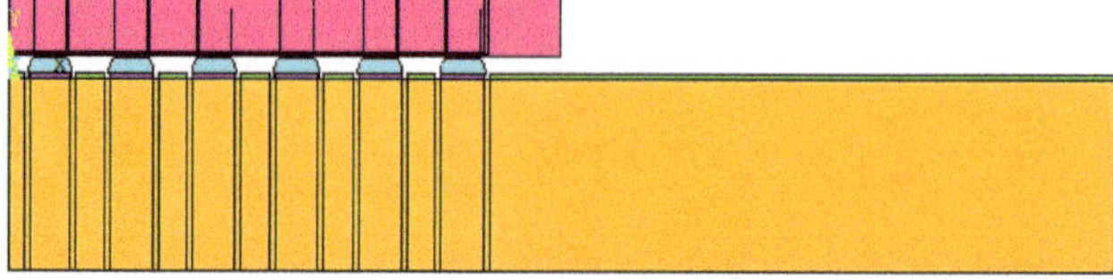

Figure 3. WLP-3 semi-diagonal FEA model.

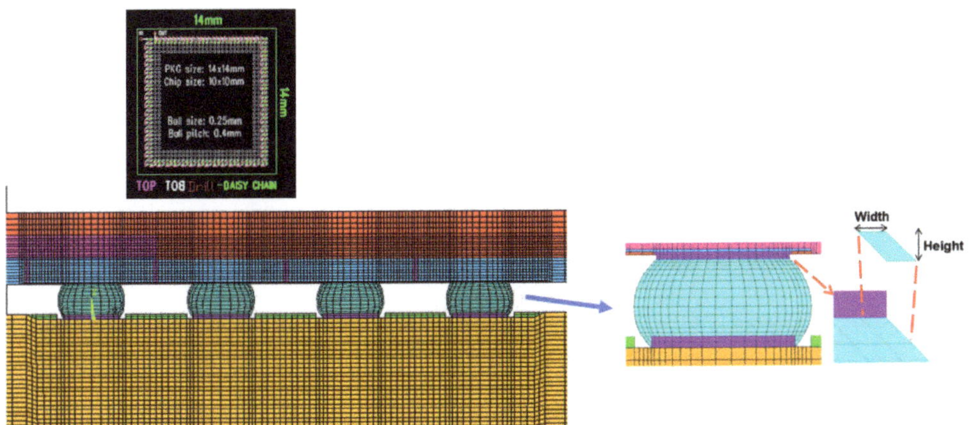

Figure 4. FO-WLP and the schematic of critical mesh size.

The FEA model for WLP 1-5 includes the following materials: silicon chip; low-k layer; stress buffer layer (SBL); redistribution layer (RDL); solder ball; under bump metallurgy (UBM); copper pad; printed circuit board (PCB); and solder mask. In addition, to further simplify the 2D model, the actual model simplifies the connection between UBM and the solder ball. The element mesh size in the upper-right corner of the solder ball would affect the final simulation result. Based on our previous research experience, the mesh size of this key position is fixed; it is located on the upper-right corner of the outmost solder ball. The controlled mesh size in height and width is 7.5 μm and 12.5 μm, and is shown in Figure 4.

Table 3 shows linear elastic material parameters for the WLP model. They are Young's Modulus €, Poisson's Ratio (ν), and CTE. For the solder ball, we use the Chaboche Kinematic Hardening model (Equation 17) to fit the nonlinear behavior of the solder at different temperatures, and the obtained fitting parameter table is shown in Table 4. The stress–strain curve for solder balls in different temperatures in this study is shown in Figure 5 [20].

$$\alpha = \frac{C}{\gamma}\left(1 - e^{-\gamma \varepsilon^{pl}}\right) + \sigma_0 \tag{17}$$

where α is the back stress, σ_0 is initial yield stress, C is constant for proportional to hardening modulus, γ is the rate of decrease of hardening modulus, and $\Delta\varepsilon^{pl}$ is increment plastic strain, individually.

Table 4. Parameters for the Chaboche model.

T (K)	σ_0	C	γ
233	47.64	8894.8	639.2
253	38.87	8573.3	660.0
313	24.06	6011.4	625.3
353	18.12	5804.2	697.7
395	14.31	4804.6	699.9

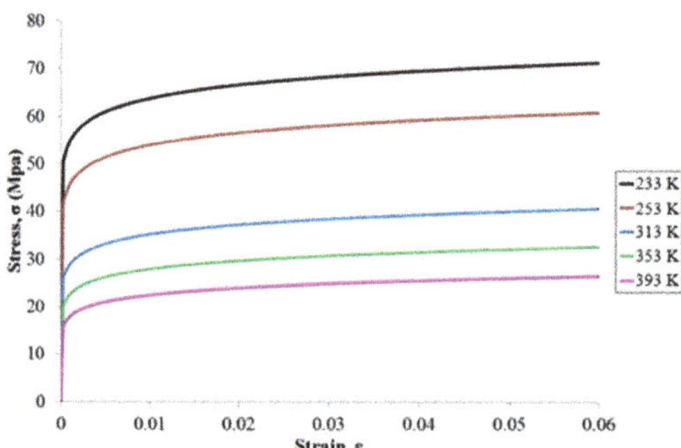

Figure 5. The stress–strain curve for SAC305 [19,21].

Finally, the thermal cycling load [22–26] was applied to our FEA model according to the JEDEC JESD22-A104D Condition G, with a temperature range of −40 °C to 125 °C. We fixed the ramp rate. It is 16.5 °C/min, and the dwell time is 10 min. The total time for a complete temperature cycle is 40 min. The thermal cycling temperature profile is shown in Figure 6.

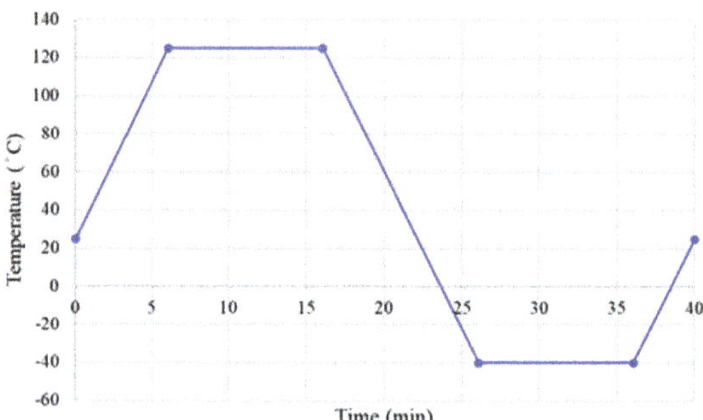

Figure 6. Thermal cycling temperature profile.

After eight cycles, the incremental equivalent plastic strain will be stabilized, and it can be input to the Coffin–Manson equation to calculate the reliability prediction life of WLP. The reliability life between experiment and simulation is shown in Table 5. We can see that the difference between experiment and simulation prediction reliability life falls within an acceptable range for five test vehicles. Therefore, the WLP simulation models can be trusted. This study uses a validated simulation procedure and a controlled mesh size, as well as an automatic model generation technique we developed to create a large database of different design parameters and use it for machine learning.

Table 5. Reliability comparison for six test vehicles.

Test Vehicles	Experiment Reliability Life (Cycles)	Simulation Reliability Life (Cycles)	Difference (Cycles)	Difference (%)
WLP-1	318	319	1	0.3
WLP-2	1013	982	31	3.1
WLP-3	587	571	16	2.7
WLP-4	876	804	72	8.2
WLP-5	904	880	24	2.6
FO-WLP	249	248	1	0.4

4. Machine Learning Prediction Results

4.1. Supervised Learning—KRR

In this study, we select four design parameters that greatly influence the reliability of WLP in building the database. They are chip thickness (CT), SBL thickness (SBLT), upper pad diameter (UPD), and lower pad diameter (LPD). Other structure parameters are fixed as WLP-2. The diagram of structure design parameters is shown in Figure 7.

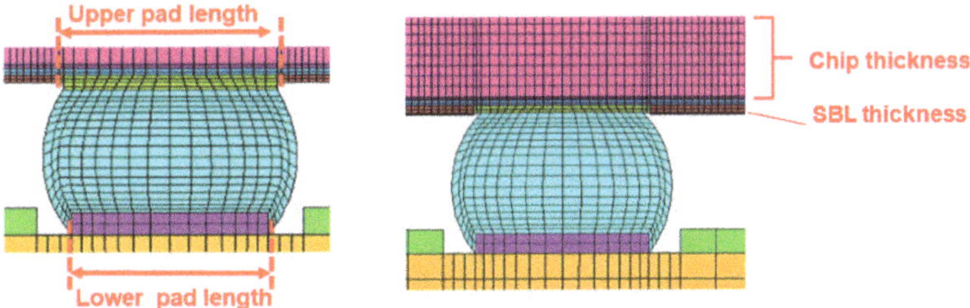

Figure 7. The diagram of WLP design parameters.

The design parameters for different training sets are shown in Tables 6–11.

Table 6. Design parameters for 256 training data.

Design Parameters	Parameter Values
UPD	0.18, 0.2, 0.22, 0.24 (mm)
LPD	0.18, 0.2, 0.22, 0.24 (mm)
CT	0.15, 0.25, 0.35, 0.45 (mm)
SBLT	5, 14.17, 23.33, 32.5 (μm)
Total Number of Training Data	256

Table 7. Design parameters for 625 training data.

Design Parameter	Parameter Values
UPD	0.18, 0.195, 0.21, 0.225, 0.24 (mm)
LPD	0.18, 0.195, 0.21, 0.225, 0.24 (mm)
CT	0.15, 0.225, 0.300, 0.375, 0.45 (mm)
SBLT	5, 11.88, 18.75, 25.63, 32.5 (μm)
Total Number of Training Data	625

Table 8. Design parameters for 1296 training data.

Design Parameter	Parameter Values
UPD	0.18, 0.192, 0.204, 0.216, 0.228, 0.24 (mm)
LPD	0.18, 0.192, 0.204, 0.216, 0.228, 0.24 (mm)
CT	0.15, 0.21, 0.27, 0.33, 0.39, 0.45 (mm)
SBLT	5, 10.5, 16, 21.5, 27, 32.5 (μm)
Total Number of Training Data	1296

Table 9. Design parameters for 2401 training data.

Design Parameter	Parameter Values
UPD	0.18, 0.19, 0.2, 0.21, 0.22, 0.23, 0.24 (mm)
LPD	0.18, 0.19, 0.2, 0.21, 0.22, 0.23, 0.24 (mm)
CT	0.15, 0.2, 0.25, 0.3, 0.35, 0.4, 0.45 (mm)
SBLT	5, 9.58, 14.17, 18.75, 23.33, 27.92, 32.5 (μm)
Total Number of Training Data	2401

Table 10. Design parameters for 4096 training data.

Design Parameter	Parameter Values
UPD	0.18, 0.189, 0.197, 0.206, 0.214, 0.223, 0.231, 0.24 (mm)
LPD	0.18, 0.189, 0.197, 0.206, 0.214, 0.223, 0.231, 0.24 (mm)
CT	0.15, 0.189, 0.197, 0.206, 0.214, 0.223, 0.231, 0.24 (mm)
SBLT	5, 9.58, 14.17, 18.75, 23.33, 27.92, 32.5 (μm)
Total Number of Training Data	4096

Table 11. Design parameters for 1296 testing data.

Design Parameter	Parameter Values
UPD	0.184, 0.194, 0.205, 0.219, 0.226, 0.234 (mm)
LPD	0.184, 0.194, 0.205, 0.219, 0.226, 0.234 (mm)
CT	0.174, 0.221, 0.289, 0.341, 0.379, 0.426 (mm)
SBLT	7.25, 12.55, 17.95, 22.65, 27.35, 30.35 (μm)
Total Number of Testing Data	1296

In Table 11, we randomly pick 100 data as testing data. Five training datasets (Tables 6–10) are used to train the KRR model separately and use the testing dataset to test the model's generalization. This study applies grid search for parameter optimization, and the data preprocessing is MinMaxScaler. The best model performance for each training set is shown in Table 12.

Table 12. The performance of the KRR model.

Item / Training Model	256 Training Data	625 Training Data	1296 Training Data	2401 Training Data	4096 Training Data
α	0	0.001	0.001	0.001	0.001
γ	1	1688	1969	2	2
Maximum training diff. (%)	0%	3.45%	3.88%	3.31%	3.28%
Average training diff. (cycles)	0	6.23	5.64	5.73	5.58
Maximum testing diff. (%)	3.14%	2.95%	2.26%	2.79%	2.41%
Average testing diff. (cycles)	10.30	7.25	5.70	5.31	5.29
Training CPU time (sec.)	0.2	0.8	7	15	138

We have four criteria for measuring the quality of the model. They are maximum training difference, average training difference, maximum testing difference, and average testing difference. The model's generalizability is the main concern of this research, and the testing performance of the model is shown in Figures 8 and 9.

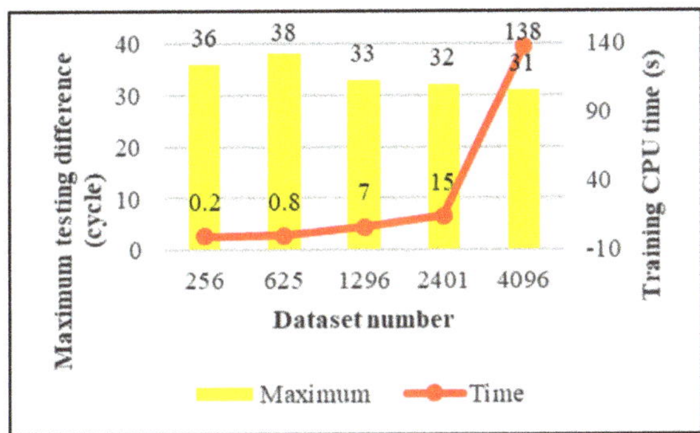

Figure 8. The maximum testing difference of the KRR model.

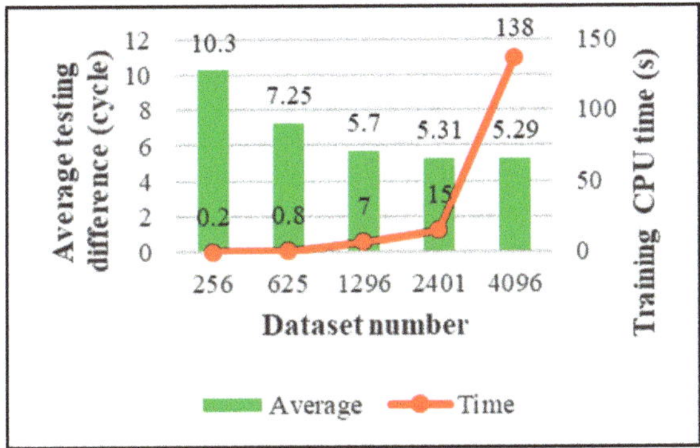

Figure 9. The average testing difference of the KRR model.

From Figure 8, we can see that the maximum testing difference is very stable; it shows that our trained model will not overfit. From Figure 9, we can see that the average testing difference gradually decreases with the increase of training data; this means the accuracy of the model is increasing. Therefore, we can continue to add training data to improve the model performance. On the other hand, judging from the growth curve of the training CPU time, one can conclude that using a larger dataset would cause the model to take a long time to train. In order to reduce the training time of the model, this study introduces the K-means algorithm.

4.2. Supervised/Unsupervised Machine Learning—KRR Mixed with K-Means

First, to demonstrate the effectiveness of K-means, a larger training dataset should be generated. This research mixes five training datasets (Tables 6–10) with 1296 testing data (Table 11), and removes duplicate data; the final total data is equal to 9601. We randomly pick 9000 data as training data, and 601 as testing data. When the pure KRR algorithm is used for training, the total time spent is 1340 s. The specific performance of the model is shown in Table 13.

Table 13. The performance of the KRR model with 9000 data.

Item / Training Model	9000 Training Data with 601 Testing Data
α	0.001
γ	1.0
Maximum training difference (%)	2.49%
Average training difference (cycles)	2.33
Maximum testing difference (%)	2.30%
Average testing difference (cycles)	3.24
Training CPU time (sec.)	1340

In Table 13, we can see that the model's prediction accuracy is further improved. The average training and testing differences are both under five cycles. Our target now is to reduce the model training time. Using K-means as the preprocessing step of KRR, the training data is divided into K clusters. Each cluster corresponds to a sub-model with KRR. When we input the testing data, the test data uses a K-means model to determine which cluster it belongs to, and is trained on that particular sub-model. With K = 4, we show the performance of the K-K model in Tables 14 and 15.

Table 14. The performance of the sub-models.

Item / K Number	1	2	3	4
n	2222	2225	2266	2287
m	145	153	147	156
Maximum training difference (cycles)	25	7	34	26
Average training difference (cycles)	1.88	0.87	2.43	2.03
Maximum testing difference (cycles)	36	4	38	44
Average testing difference (cycles)	4.68	0.98	5.65	4.98

Table 15. The performance of the K-K model in K = 4.

Item / Training Model	9000 Training Data with 601 Testing Data
Maximum training difference (%)	2.10%
Average training difference (cycles)	1.81
Maximum testing difference (%)	3.72%
Average testing difference (cycles)	4.05

In Table 14, "n" represents the number of training data in each cluster; the total data number is 9000. "m" represents the number of testing data in each cluster; the total data number is 601. The final maximum difference is the maximum value among all sub-models. The average training difference is the weighted average. The final training and testing results are shown in Table 15, and the accuracy is similar to the pure KRR model. In this study, the focus of the K-K model is the training CPU time. In Table 16 and Figure 10, it can be seen that as the value of K increases, the training time decreases rapidly.

Table 16. The performance of the K-K model.

Item	K 4	8	16	24	32
Maximum training difference (%)	2.10	2.45	3.73	3.75	45
Average training difference (cycles)	1.81	2.45	2.91	2.06	2.90
Maximum testing difference (%)	3.72	3.44	3.66	4.57	4.57
Average testing difference (cycles)	4.05	3.69	4.13	3.71	4.47
Training CPU time (sec.)	40	15	9	8	7

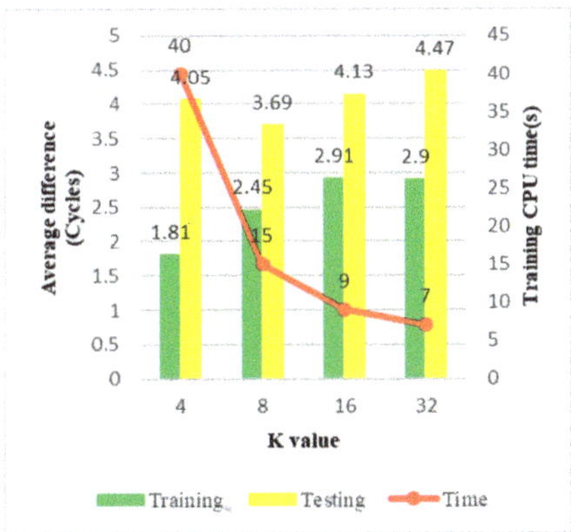

Figure 10. The average difference of the K-K model.

From Figure 10, we can find that the CPU time gradually decreases, and the trend slows down as the value of K increases. The average difference remains stable; both training and testing are under five cycles. In addition, a comparison of different machine learning algorithms using 9000 training data and 601 testing data is presented in Table 17.

Table 17. The comparison of machine learning algorithms.

	KRR	K-K	ANN	RF	SVR
Maximum training difference (%)	2.49%	4.00%	3.99%	2.61%	3.55%
Average training difference (cycles)	2.33	2.90	3.74	3.77	1.16
Maximum testing difference (%)	2.30%	4.57%	2.89%	6.06%	2.43%
Average testing difference (cycles)	3.24	4.47	4.18	10.24	2.51
Training CPU time (sec.)	1340	7	469	8	1799

In Table 17, the performance of the KRR model and the SVR model is very close in terms of training time and training error. The K-K hybrid model achieves similar accuracy as the pure KRR, ANN, and SVR algorithms. However, in terms of training time, it can reach the level of the RF algorithm, which dramatically improves the training efficiency of the algorithm.

5. Conclusions

This study, using validated FEA, created five training datasets and one testing dataset for machine learning. In comparison with FEA, the results indicated that the KRR and the K-K machine learning algorithms are fast and effective in predicting the reliability life of WLP. With the increase of training data, the accuracy of the AI model is gradually improved. However, as the AI dataset grows, training time will increase dramatically, making it necessary to reduce the training time. Using a hybrid model combining K-means and KRR can significantly reduce training time while maintaining similar prediction accuracy. When K is 32, we can obtain a data prediction model with an average error of around four cycles, and the overall CPU training time is 7 s, much less than the pure KRR model's 1340 s training time. As compared with ANN, RF, and SVR, the K-K model is undoubtedly fast and accurate.

Author Contributions: Supervision, Methodology development, K.-N.C.; Simulation, Validation, Q.-H.S. All authors have read and agreed to the published version of the manuscript.

Funding: This research was funded by MOST-Taiwan grant number [Project MOST 108-2221-E-007-007-MY3].

Conflicts of Interest: The authors declare no conflict of interest.

References

1. Chou, C.Y.; Hung, T.Y.; Yang, S.Y.; Yew, M.C.; Yang, W.K.; Chiang, K.N. Solder joint and trace line failure simulation and experimental validation of fan-out type wafer level packaging subjected to drop impact. *Microelectron. Reliab.* **2008**, *48*, 1149–1154. [CrossRef]
2. Liu, C.M.; Lee, C.C.; Chiang, K.N. Enhancing the Reliability of Wafer Level Packaging by Using Solder Joints Layout Design. *IEEE Trans. Compon. Packag. Technol.* **2006**, *29*, 877–885. [CrossRef]
3. Lin, Y.T.; Peng, C.T.; Chiang, K.N. Parametric design and reliability analysis of wire interconnect technology wafer level packaging. *J. Electron. Packag.* **2002**, *124*, 234–239. [CrossRef]
4. Cheng, H.C.; Chiang, K.N.; Lee, M.H. An effective approach for three-dimensional finite element analysis of ball grid array typed packages. *J. Electron. Packag.* **1998**, *120*, 129–134. [CrossRef]
5. Hoerl, A.E.; Kennard, R.W. Ridge Regression: Biased Estimation for Nonorthogonal Problems. *Technometrics* **1970**, *12*, 55–67. [CrossRef]
6. Cristianini, N.; Shawe-Tylor, J. *An Introduction to Support Vector Machines and Other Kernel-Based Learning Methods*; Cambrige University Press: Cambridge, UK, 2000.
7. Macqueen, J. Some Methods for Classification and Analysis of Multi Variate Observations. In Proceedings of the Fifth Berkeley Symposium on Mathematical Statistics and Probability, Berkeley, CA, USA, 18–21 July 1965; pp. 281–297.
8. Chou, P.H.; Liang, S.Y.; Chiang, K.N. Reliability Assessment of Wafer Level Package Using Artificial Neural Network Regression Model. *J. Mech.* **2019**, *35*, 829–837. [CrossRef]
9. Hsiao, H.Y.; Chiang, K.N. AI-Assisted Reliability Life Prediction Model for Wafer-Level Packaging using the Random Forest Method. *J. Mech.* **2021**, *37*, 28–36. [CrossRef]
10. Yuan, C.A.; Han, C.N.; Yew, M.C.; Chou, C.Y.; Chiang, K.N. Design, analysis, and development of novel three-dimensional stacking WLCSP. *IEEE Trans. Adv. Packag.* **2005**, *28*, 3877–3896. [CrossRef]
11. Cawley, G.C.; Talbot, N.L.C.; Foxall, R.J.; Dorling, S.R.; Mandic, D.P. Heteroscedastic kernel ridge regression. *Neurocomputing* **2004**, *57*, 105–124. [CrossRef]
12. Cawley, G.C.; Talbot, N.L.C. Reduced rank kernel ridge regression. *Neural Processing Lett.* **2002**, *16*, 293–302. [CrossRef]
13. Tanaka, A.; Imai, H.; Kudo, M.; Miyakoshi, M. Mathematical interpretations of kernel ridge regression. In Proceedings of the 7th International Conference on Computing Anticipatory Systems (CASYS 05), Liege, Belgium, 8–13 August 2005.
14. Tylavsky, D.J.; Sohie, G.R. Generalization of the matrix inversion lemma. *Proc. IEEE* **1986**, *74*, 1050–1052. [CrossRef]
15. Adnan, R.M.; Parmar, K.S.; Heddam, S.; Shahid, S.; Kisi, O. Suspended Sediment Modeling Using a Heuristic Regression Method Hybridized with Kmeans Clustering. *Sustainability* **2021**, *13*, 4648. [CrossRef]
16. Panigrahy, S.K.; Tseng, Y.C.; Lai, B.R.; Chiang, K.N. An overview of AI-Assisted design-on-Simulation technology for reliability life prediction of advanced packaging. *Materials* **2021**, *14*, 5342. [CrossRef] [PubMed]
17. Hsieh, M.C.; Tzeng, S.L. Solder joint fatigue life prediction in large size and low cost wafer-level chip scale packages. In Proceedings of the IEEE 15th International Conference on Electronic Packaging Technology, Chengdu, China, 12–15 August 2014; pp. 496–501.
18. Hsieh, M.C. Modeling correlation for solder joint fatigue life estimation in wafer-level chip scale packages. In Proceedings of the 10th International Microsystems, Packaging, Assembly and Circuits Technology Conference (IMPACT-IEEE), Taipei, Taiwan,, 21–23 October 2015; pp. 65–68.
19. Wang, P.H.; Lee, Y.C.; Lee, C.K.; Chang, H.H.; Chiang, K.N. Solder Joint Reliability Assessment and Pad Size Studies of FO-WLP with Glass Substrate. *IEEE Trans. Device Mater. Reliab.* **2021**, *21*, 96–101. [CrossRef]
20. Rogers, B.; Scanlan, C. Improving WLP reliability through solder joint geometry optimization. In *Proceedings of the IMAPS—International Symposium on Microelectronics, Orlando, FL, USA, 29 September–3 October 2013*; International Microelectronics Assembly and Packaging Society: Research Triangle Park, NC, USA; pp. 546–550.

21. Motalab, M.; Mustafa, M.; Suhling, J.C.; Zhang, J.W.; Evans, J.; Bozack, M.J.; Lall, P. Thermal cycling reliability predictions for PBGA assemblies that include aging effects. In Proceedings of the ASME International Technical Conference and Exhibition on Packaging and Integration of Electronic and Photonic Microsystems, Burlingame, CA, USA, 16–18 July 2013.
22. Clech, J.P. Acceleration Factors and Thermal Cycling Test Efficiency for Lead-free Sn-Ag-Cu Assemblies. In Proceedings of the SMTA International, Chicago, IL, USA, 25–29 September 2005.
23. Pan, N.; Henshall, G.A.; Billaut, F.; Dai, S.; Strum, M.J.; Benedetto, E.; Rayner, J. An Acceleration Model for Sn-Ag-Cu Solder Joint Reliability Under Various Thermal Cycle Conditions. In Proceedings of the SMTA International Conference, Chicago, IL, USA, 25–29 September 2005.
24. Lau, J.H. Solder joint reliability of flip chip and plastic ball grid array assemblies under thermal, mechanical, and vibrational conditions. *IEEE Trans. Compon. Packag. Manuf. Technol.* **1996**, *19*, 728–735. [CrossRef]
25. Lau, J.H.; Lee SW, R.; Chang, C. Solder joint reliability of wafer level chip scale packages (WLCSP): A time-temperature-dependent creep analysis. *J. Electron. Packag.* **2000**, *122*, 311–316. [CrossRef]
26. Zhai, C.J.; Sidharth; Blish, R. Board level solder reliability versus ramp rate and dwell time during temperature cycling. *IEEE Trans. Device Mater. Reliab.* **2003**, *3*, 207–212. [CrossRef]

Article

Implementation and Performance Evaluation of a Bivariate Cut-HDMR Metamodel for Semiconductor Packaging Design Problems with a Large Number of Input Variables

Yu-Hsiang Yang [1], Hsiu-Ping Wei [1], Bongtae Han [1,*] and Chao Hu [2]

[1] Mechanical Engineering Department, University of Maryland, College Park, MD 20742, USA; yhyangv@umd.edu (Y.-H.Y.); hpwei@umd.edu (H.-P.W.)
[2] Department of Mechanical Engineering, Iowa State University, Ames, IA 50011, USA; chaohu@iastate.edu
* Correspondence: bthan@umd.edu; Tel.: +1-301-405-5255

Abstract: A metamodeling technique based on Bivariate Cut High Dimensional Model Representation (Bivariate Cut HDMR) is implemented for a semiconductor packaging design problem with 10 design variables. Bivariate Cut-HDMR constructs a metamodel by considering only up to second-order interactions. The implementation uses three uniformly distributed sample points ($s = 3$) with quadratic spline interpolation to construct the component functions of Bivariate Cut-HDMR, which can be used to make a direct comparison with a metamodel based on Central Composite Design (CCD). The performance of Bivariate Cut-HDMR is evaluated by two well-known error metrics: R-squared and Relative Average Absolute Error ($RAAE$). The results are compared with the performance of CCD. Bivariate Cut HDMR does not compromise the accuracy compared to CCD, although the former uses only one-fifth of sample points (201 sample points) required by the latter (1045 sample points). The sampling schemes and the predictions of cut-planes and boundary-planes are discussed to explain possible reasons for the outstanding performance of Bivariate Cut HDMR.

Keywords: bivariate cut-HDMR; semiconductor packaging; central composite design; R-squared; relative average absolute error

Citation: Yang, Y.-H.; Wei, H.-P.; Han, B.; Hu, C. Implementation and Performance Evaluation of a Bivariate Cut-HDMR Metamodel for Semiconductor Packaging Design Problems with a Large Number of Input Variables. *Materials* **2021**, *14*, 4619. https://doi.org/10.3390/ma14164619

Academic Editor: Marko Topic

Received: 20 July 2021
Accepted: 13 August 2021
Published: 17 August 2021

Publisher's Note: MDPI stays neutral with regard to jurisdictional claims in published maps and institutional affiliations.

1. Introduction

Numerous metamodeling techniques (also known as response surface methods, surrogate models, or reduced-order models) have been developed and implemented for engineering design optimization [1]. Metamodeling includes two parts: generation of discrete sample points and connection of the discrete sample points. Each metamodeling technique possesses its own characteristics that can be suited for certain applications.

For a typical engineering system, a metamodel considering up to second-order interactions is often sufficient to describe system responses [2,3]. For example, a metamodeling technique called central composite design (CCD) has been implemented widely in the field of semiconductor packaging design community, which uses quadratic polynomial functions for fitting sample points [4–6]. It was implemented for commercial software such as optiSLang [7], Design-Expert [8], etc. The CCD metamodeling technique requires P number of sample points to produce the metamodel for N number of input variables, defined as [3]:

$$P = 1 + 2N + 2^N \tag{1}$$

As the number of input variables increases, the computational cost may become prohibitively high due to an extremely large number of sample points required (this is the well-known "curse of dimensionality"). This situation will be exacerbated when the modeling requires a computationally expensive analysis such as time-dependent and nonlinear analysis that is routinely encountered in complex semiconductor packaging architectures [9–11].

In order to build accurate and efficient metamodels for high dimensional input-output systems, numerous advanced metamodeling techniques such as the high-dimensional model representation (HDMR) technique [12–14], reduced dimensional polynomial chaos expansion [15], and active and rank-adaptive tensor regression [16] have been developed to enhance the efficiency of metamodeling in various engineering fields.

Among these techniques, one family of HDMR, called Cut-HDMR, possesses two unique practical features: (1) it involves function evaluations only at sample points, and, more importantly, (2) it determines the number of sample points from a pre-defined function of the number of input variables regardless of the nature of engineering applications, i.e., selection of sample points is simple and straightforward [14,17,18]. Based on these features, numerous metamodeling techniques based on Cut-HDMR have been developed such as RBF-HDMR [19], Adaptive MLS-HDMR [20], and Kriging-HDMR [21].

HDMR decomposes a multivariate function into multiple lower-order component functions, based on the hierarchical structure of interaction effects of the input variables. The high performance of some of the metamodeling techniques based on Cut-HDMR considering up to second-order component functions (this will be referred to as Bivariate Cut-HDMR) has been confirmed for nonlinear numerical test functions [19,20] and statistical analysis of multiconductor transmission line networks [22].

The objectives of this paper are (1) to introduce the cut-HDMR to the semiconductor packaging design community and to help implement the Bivariate Cut-HDMR for those who are not familiar with the HDMR, and (2) to investigate the performance of Bivariate Cut-HDMR for a complex semiconductor packaging problem (10 design input valuables). The result is compared with the performance of CCD, which has been utilized widely in the semiconductor packaging industry.

2. Background: Bivariate Cut-HDMR

The fundamentals of HDMR are described first. A specific HDMR that uses the Dirac measure located at a cut center, called Cut-HDMR, is presented together with its approximated version, Bivariate Cut-HDMR, which considers up to second-order component functions.

2.1. High-Dimensional Model Representation (HDMR)

The concept of high-dimensional expansion was implemented originally to estimate the sensitivity of a function with respect to arbitrary groups of variables [23]. Later, the term, HDMR, was first introduced by Rabitz and Alis [12]. They detailed and completed the general foundations of HDMR.

The HDMR expansion is performed based on the interaction effects of input variables. The term "interaction" employed here means that more than one variable act together to affect the performance function. This is distinctly different from the term "correlation" employed in statistics, which represents whether and how strongly a pair of random variables are related.

A general form of HDMR is defined as [12]:

$$\begin{aligned} y(\mathbf{x}) \; &= y(x_1, x_2, \ldots, x_N) \\ &\equiv y_0 + \sum_{i=1}^{N} y_i(x_i) + \sum_{1 \leq i < j \leq N} y_{ij}(x_i, x_j) + \cdots + \sum_{1 \leq i_1 < \cdots < i_l \leq N} y_{i_1 i_2 \ldots i_l}(x_{i_1}, x_{i_2}, \ldots, x_{i_l}) + \cdots + y_{12\ldots N}(x_1, x_2, \ldots, x_N) \end{aligned} \tag{2}$$

where $y(\mathbf{x})$ and the bold letter, $\mathbf{x}$, represent the performance function and the vector of input variables, $(x_1, x_2, \ldots, x_N)$, respectively; y_0 is a constant representing the mean of the performance function, which is called zeroth-order effect or mean effect; $y_i(x_i)$ represents the effect when the variable x_i acts independently on $y(\mathbf{x})$, which is called first-order effect or main effect; $y_{ij}(x_i, x_j)$ is the effect on $y(\mathbf{x})$ when the variables x_i and x_j act together, which is called second-order effect or bivariate interaction effect. It should be noted that $y_{ij}(x_i, x_j)$ excludes the main effects of x_i and x_j as well as the mean effect. The subsequent terms indicate the higher order interaction effects of more variables acting together on $y(x)$. The last term $y_{12...N}(x_1, x_2, \ldots, x_N)$ represents the residual influence.

Each effect in the general form of HDMR is called a component function. The general form of component functions can be expressed as [12]:

$$
\begin{aligned}
y_0 &\equiv My(x) \\
y_i(x_i) &\equiv M^i y(x) - y_0 \\
y_{ij}(x_i, x_j) &\equiv M^{ij} y(x) - y_i(x_i) - y_j(x_j) - y_0 \\
&\vdots \\
y_{12...n}(x) &\equiv y(x) - y_0 - \sum_i y_i(x_i) - \sum_{ij} y_{ij}(x_i, x_j) - \ldots - \sum_{12...N} y_{12...N}(x_1, x_2, \ldots, x_N)
\end{aligned}
\tag{3}
$$

The functions in the above equation are defined as:

$$
\begin{aligned}
My(x) &= \int_{K^n} y(x) d\gamma(x) \\
M^{i_1 i_2 \cdots i_l} y(x) &= \int_{K^{n-l}} y(x) \left[\prod_{j \notin \{i_1, \ldots, i_l\}} d\gamma_j(x_j) \right]
\end{aligned}
\tag{4}
$$

where $K^n = \{(x_1, x_2, \ldots, x_n): 0 \leq x_i \leq 1, i = 1, 2, \ldots, n\}$ is an n-dimensional unit cube and γ is a measure [24]. A measure is a function that quantifies the size of sets. A measure assigns a non-negative real number or $+\infty$ to subsets of a certain set. Each distinct measure embodies a different way to assess how big a set is.

There is no unique decomposition of the model output $y(x_1, x_2, \ldots, x_N)$; all HDMR expansions follow the general form in Equation (2). The choice of a particular HDMR expansion depends on the application and the nature of any constraints in sampling input variables. For example, for the uncertainty analysis of a model output (e.g., an analysis of the variance of an output), the component functions in the HDMR should be chosen to represent the independent contributions of input variables to the overall uncertainty of the output. It is known as ANOVA-HDMR [12,25].

The ANOVA-HDMR is typically carried out by multi-dimensional Monte Carlo integration due to its complexity. The Monte Carlo integration needs a large number of sample points to attain good accuracy. It is impractical for the advanced semiconductor packaging applications where computational cost of each sample point is high. However, another approach of HDMR, Cut-HDMR, can tackle the challenge.

2.2. Cut-HDMR and Bivariate Cut-HDMR

Cut-HDMR uses the Dirac measure [26] located at a point $m = (m_1, m_2, \ldots, m_n)$ (also known as cut center):

$$
d\gamma(x) = \prod_{i=1}^{n} \delta(x_i - m_i) dx_i
\tag{5}
$$

By combining it with Equations (3) and (4), the component functions of Cut-HDMR can be expressed as:

$$
\begin{aligned}
y_0 &\equiv Mf(x) = y(m) \\
y_i(x_i) &\equiv M^i y(x) - y_0 = y(m_1, \ldots, m_{i-1}, x_i, m_{i+1}, \ldots, m_n) - y_0 \\
y_{ij}(x_i, x_j) &\equiv M^{ij} y(x) - y_i(x_i) - y_j(x_j) - y_0 = y(m_1, \ldots, x_i, \ldots, x_j, \ldots, m_n) - y_i(x_i) - y_j(x_j) - y_0 \\
&\vdots \\
y_{12...n}(x) &\equiv y(x) - y_0 - \sum_i y_i(x_i) - \sum_{ij} y_{ij}(x_i, x_j) - \ldots - \sum_{12...N} y_{12...N}(x_1, x_2, \ldots, x_N)
\end{aligned}
\tag{6}
$$

where $y(m_1, \ldots, m_{i-1}, x_i, m_{i+1}, \ldots, m_n)$ is a 1D performance function along the x_i direction that passes through m; $y(m_1, \ldots, x_i, \ldots, x_j, \ldots, m_n)$ is a 2D performance function of the (x_i, x_j) plane that passes through m, and so on.

Equation (6) shows that Cut-HDMR is an expression as a superposition of its values on lines, planes, and hyperplanes of higher orders passing through the cut center, m. The expansions of Cut-HDMR do not contain any integral. Cut-HDMR uses only arithmetic computation to determine the component functions, and thus it requires the least amount of computational cost compared to other HDMRs [12,27].

For most well-defined physical systems, the high-order interactions are negligible [14,28], and thus the multivariate performance function of such a physical system can be approximated well by the sum of low-order component functions. Experience shows that an HDMR expansion up to the second-order often provides a satisfactory description of the function for many high-dimensional systems when the input variables are properly chosen [14].

It has been proven that the mean values of input variables, μ, are the optimal cut center m when only the terms up to the second-order are considered [25]. Accordingly, the metamodel based on Bivariate Cut-HDMR can be obtained by substituting Equation (6) into Equation (2) with the cut center being the mean values of input variables. This Bivariate Cut-HDMR metamodel is written as [29]:

$$y(x) \cong \sum_{1 \leq i < j \leq N} y\left(x_i,\ x_j,\ \mu^{\sim ij}\right) - (N-2)\sum_{i=1}^{N} y\left(x_i,\ \mu^{\sim i}\right) + \frac{(N-1)(N-2)}{2}y_0 \quad (7)$$

where $y_0 = \mu = [\mu_1,\ \mu_2,\ \ldots,\ \mu_N]^T$ is the vector of the mean values of N input variables (cut center); $\mu^{\sim i}$ is μ without the element μ_i; $\mu^{\sim ij}$ is μ without the elements μ_i and μ_j; $y\left(x_i,\ \mu^{\sim i}\right)$ is the 1D performance function along the x_i direction that passes through μ (cut-line); and $y\left(x_i,\ x_j,\ \mu^{\sim ij}\right)$ is a 2D performance function on the (x_i, x_j) plane that passes through μ (cut-plane).

Figure 1 illustrates the concept of Bivariate Cut-HDMR using an arbitrary 2D function, which is decomposed into four component functions. Figure 1a shows the 2D function, $x^2 + y^2 + xy - 14x - 16y + 122 = 0$, as the black meshed surface, and the dot in the figure represents the zeroth-order effect (i.e., a constant). In Figure 1b, the blue curve is the 1D performance function along the x_1 direction, in which x_2 is kept as μ_2. The green line is the zeroth-order effect along the x_1 direction. The main effect of x_1 is the red curve, obtained by subtracting the green line from the blue curve.

The same procedure can be applied to obtain the main effect of x_2, as shown in Figure 1c. In Figure 1d, the blue surface is obtained by the superposition of the red curves in Figure 1b,c, which represents the performance function without any interaction effects. The green plane is the zeroth-order effect. By subtracting the blue surface and the green plane from the black surface, the interaction effect of the (x_1, x_2) pair is obtained, which is shown as the red surface.

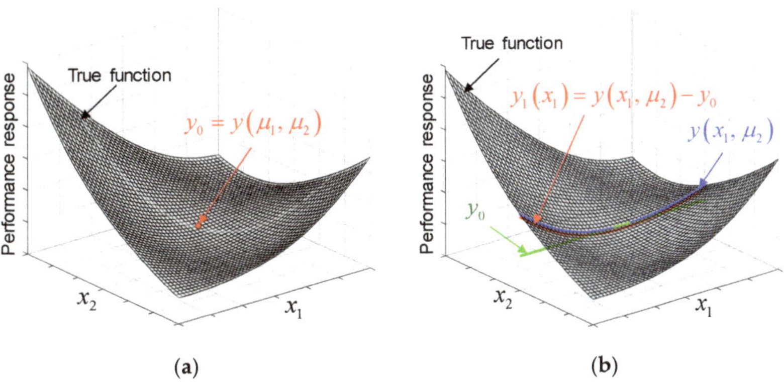

Figure 1. *Cont.*

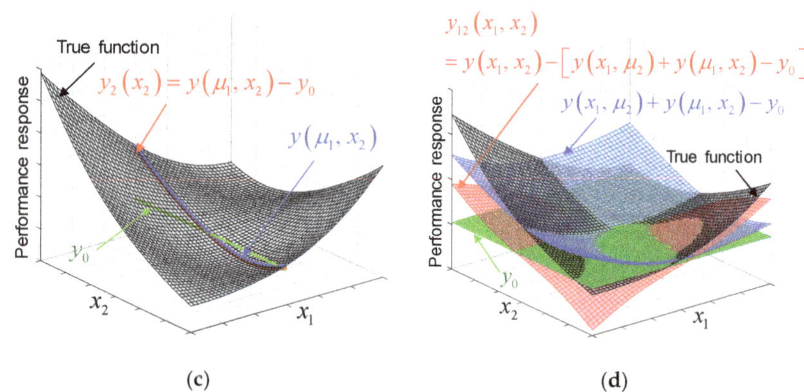

(c)   (d)

Figure 1. Illustration of Bivariate Cut-HDMR using an arbitrary 2D function; (**a**) 2D function and the effect of zeroth-order, (**b**) the main effect of x_1, (**c**) the main effect of x_2, and (**d**) the interaction effect of x_1 and x_2.

3. Implementation for Semiconductor Packaging Application

The Bivariate Cut-HDMR technique is implemented to construct a metamodel for a semiconductor packaging application. The application involves warpage prediction of a thin flat ball grid array (TFBGA) package with 10 design input variables.

3.1. Description of TFBGA Package

Figure 2 shows the schematic diagram of a TFBGA package. The first chip is attached to a substrate by the first die attach film (DAF). The second chip is attached to the first chip by the second DAF. Then, they were encapsulated by epoxy molding compound (EMC). A stacked die TFBGA package is often used as the top package of a Package-on-Package (PoP). Warpage at solder pad areas is one of the most critical factors to high PoP stacking yield [30].

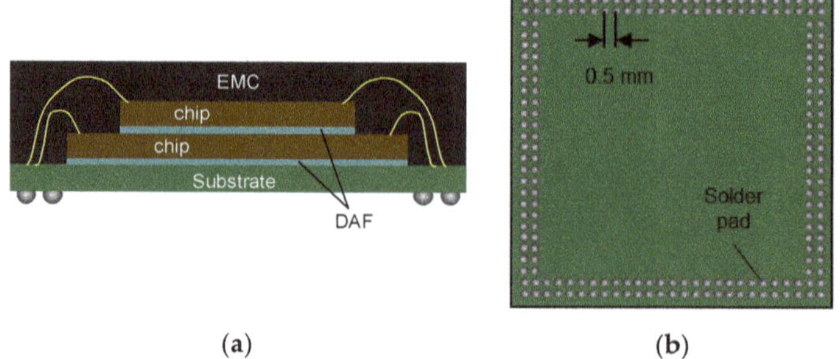

(a)   (b)

Figure 2. TFBGA package: (**a**) cross-sectional view and (**b**) bottom view.

A finite element (FE) model was constructed for warpage prediction. Figure 3 shows details of the FE model built by a commercial FE analysis package (ANSYS®). The quarter symmetry model of boundary conditions and the die stack configuration are shown in (a) and (b); and the enlarged view of cross-section is shown in (c). The material properties and the nominal dimensions used in the model are summarized in Tables 1 and 2. The nominal dimensions of the TFBGA package are adopted from the design in Refs. [10,31].

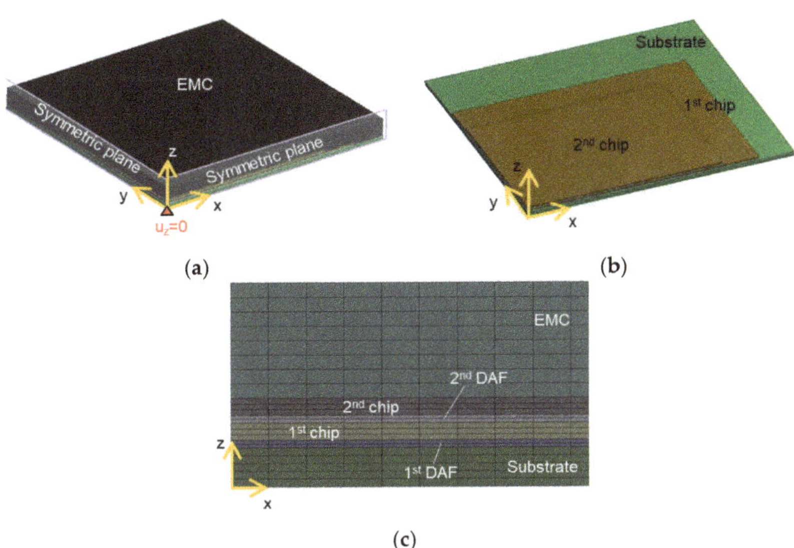

Figure 3. Quarter FE model of TFBGA package: (**a**) boundary conditions; (**b**) die stack configuration; and (**c**) enlarged view of cross-section.

Table 1. Properties of materials used in the TFBGA package.

Material	Young's Modulus (GPa)	Poisson's Ratio	CTE (ppm/°C)		T_g (°C)
			A_1 ($<T_g$)	A_2 ($>T_g$)	
Silicon die	130	0.23	2.8		–
DAF	2.2 @ 25 °C 0.98 @ 100 °C 0.008 @ 200 °C	0.3	65.3	162.9	138
Substrate	17.5	0.3	15 (in-plane) 61.5 (out-of-plane)		–
EMC	29.237 @ 25 °C 14.030 @ 125 °C 1.932 @ 175 °C 1.498 @ 235 °C	0.21	9.12	36	137.5

Table 2. Dimensions of TFBGA package.

Structure	Length × Width × Thickness
1st Die (mm)	13 × 11 × 0.575
1st DAF (mm)	13 × 11 × 0.025
2nd Die (mm)	11 × 9 × 0.575
2nd DAF (mm)	11 × 9 × 0.025
Substrate (mm)	15 × 15 × 0.13
EMC (mm)	15 × 15 × 0.55

The TFBGA package was subjected to the EMC molding process at 175 °C, which was used as a stress-free temperature. The conventional lead-free solder reflow profile with the peak temperature of 260 °C was considered [32].

In this implementation, 10 design variables were considered for the warpage prediction of solder pad areas. The details of design variables are summarized in Table 3. The design spaces of the package dimensions and the material properties were defined by the values found in the literature: package dimensions in [31,33–39] and material properties in [40–45].

Table 3. Design variables of the TFBGA package.

Variable	Physical Meaning	Range of Design Space	Mean
x_1	EMC thickness (mm)	0.25–0.85	0.55
x_2	Substrate thickness (mm)	0.12–0.34	0.23
x_3	1st chip thickness (mm)	0.050–0.075	0.0625
x_4	2nd chip thickness (mm)	0.050–0.075	0.0625
x_5	1st DAF thickness (mm)	0.02–0.025	0.0225
x_6	2nd DAF thickness (mm)	0.01–0.02	0.015
x_7	EMC CTE above T_g (ppm/°C)	25–47	36
x_8	Substrate CTE (ppm/°C)	12–18	15
x_9	Substrate modulus (GPa)	7.5–27.5	17.5
x_{10}	Half of PKG width and length (mm)	7–8	7.5

3.2. Sample Points

The number of sample points to construct a Bivariate Cut-HDMR metamodel can be generally expressed as [12]:

$$R = 1 + N(s-1) + \frac{N(N-1)}{2}(s-1)^2 \tag{8}$$

where N is the number of input variables and s is the number of sample points taken along the direction of each input variable. $N(s-1)$ points are used to construct 1D performance functions, and $N(N-1)(s-1)^2/2$ points are used to construct the 2D performance functions.

For the univariate terms (i.e., s number of sample points distributed along each input variable), the center becomes the reference point, and the remaining $(s-1)$ sample points are evenly distributed on two sides with respect to the reference point. For the bivariate terms, the sample points form a uniform gird on a plane with the center as a reference point.

Cut-HDMR, in its original form [14], states that a set of sample points can be selected to calculate the values of corresponding component functions and to form a look-up table that can be used to interpolate component functions at an arbitrary point in the design domain. There has been no universally accepted sampling strategy and interpolation algorithm. The implementation of this study uses three uniformly distributed sample points ($s = 3$) with quadratic spline interpolation to construct the component functions of Bivariate Cut-HDMR. In this way, the Bivariate Cut-HDMR metamodel can be compared directly with the CCD metamodel.

Figure 4 and Table 4 show the number of sample points required for the CCD and Bivariate Cut-HDMR metamodels. After $N = 7$, the number of sample points for CCD becomes more than double the number of sample points for Bivariate Cut-HDMR. Considering only the number of sample points, Bivariate Cut-HDMR has a significant advantage over CCD when a metamodeling problem has a large number of input variables.

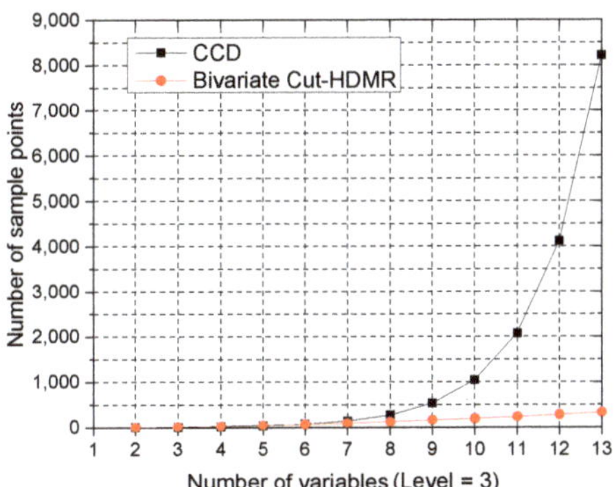

Figure 4. Numbers of sample points required by Bivariate Cut-HDMR and CCD as a function of the number of variables.

Table 4. Number of sample points required for Central composite design (CCD) and Bivariate Cut HDMR with $s = 3$.

		Number of Sample Points		Ratio of Two Numbers of Sample Points R/P
N	s	CCD $P=1+2N+2^N$	Bivariate Cut-HDMR $R=1+2N+\frac{N(N-1)}{2}(2)^2$	
2	3	9	9	100%
3	3	15	19	127%
4	3	25	33	132%
5	3	43	51	119%
6	3	77	73	95%
7	3	143	99	69%
8	3	273	129	47%
9	3	531	163	31%
10	3	1045	201	19%
11	3	2071	243	12%
12	3	4121	289	7%
13	3	8219	339	4%

3.3. Construction of Bivariate Cut-HDMR Metamodel

3.3.1. Obtain Sample Points

For $s = 3$, a total of 201 sample points (Equation (8)) are required to construct the Bivariate Cut-HDMR metamodel. The sample points consist of one mean sample point (cut center), 20 univariate sample points, and 180 bivariate sample points.

The mean sample point is the design point, which is the mean values of each design variables. The 20 univariate sample points are the sample points, where one of the design variables takes either maximum or minimum value in its design space while other design variables keep the mean values. The 180 bivariate sample points are the sample points, where two of the design variables take either maximum or minimum value in their design spaces while other design variables keep the mean values. The warpage values of 201 (=1 + 20 + 180) sample points were obtained by the FE model. Since the dimensions of the FE model varies with sample points, the FE model must regenerate different meshes for each sample point. The Supplementary Materials includes: (1) the warpage values of 201 sample points that were used to construct the Bivariate Cut-HDMR metamodel; (2) the warpage values of 1045 sample points that were used to construct the CCD metamodel; and (3) the Monte Carlo simulation sample points used in Section 4.

3.3.2. Construct Performance Functions

After the warpage values at the 201 sample points are obtained, the metamodel can be constructed by applying Equation (7). The quadratic spline interpolation scheme was adopted with all sample points to form the 1D performance functions (cut-lines), $y\left(x_i,\ \mu^{\sim i}\right)$, and the 2D performance functions (cut-planes), $y\left(x_i,\ x_j,\ \mu^{\sim ij}\right)$ by following the procedures below:

- 1D performance functions:
  1. Select a design variable.
  2. Find the three sample points along the design variable that was obtained earlier, i.e., high, mid, and low values of the design variable and other design variables keep the mean values.
  3. Construct the 1D function of the design variable with the three sample points using quadratic spline interpolation. This can be done by using the built-in function that is available in commercial software (e.g., MATLAB).
  4. Select another design variable and repeat steps 2–3 until all 1D performance functions along each design variable are built.

- 2D performance functions:
  1. Select a pair of design variables.
  2. Find the nine sample points along two design variables that were obtained earlier (other design variables keep the mean values) as shown in the figure.
  3. Construct the 2D function of the design variable with the nine sample points using quadratic spline interpolation. This can be done by using the built-in function that is available in commercial software (e.g., MATLAB).
  4. Select another pair of design variables and repeat steps 2–3 until all 2D performance functions of each pair of design variables are built.

The performance functions are illustrated in Figures 5 and 6. Figure 5 shows the 1D performance functions of EMC thickness and substrate thickness, and Figure 6 shows the 2D performance functions of two pairs of design variables. The pair of substrate thickness and EMC CTE has the strongest second-order interaction effect among other pairs. In contrast, the pair of package width and length and 1st chip thickness has the weakest second-order interaction effect. Red dots indicate the sample points used to construct the cut-lines and the cut-planes.

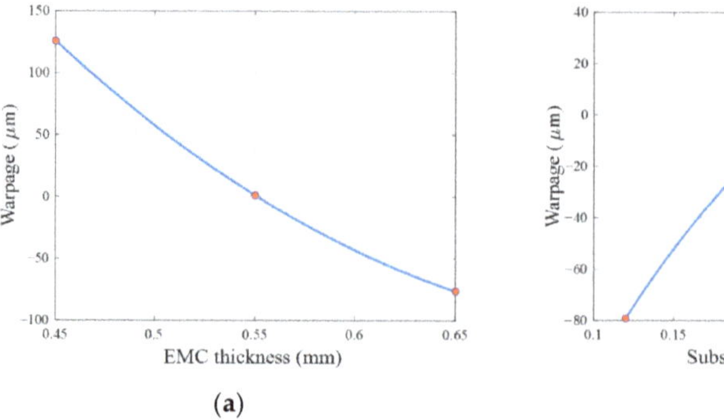

Figure 5. 1D performance functions of two design variables: (**a**) EMC thickness and (**b**) substrate thickness.

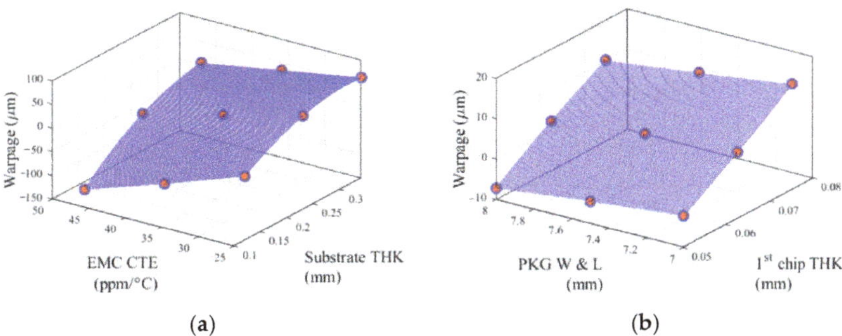

Figure 6. 2D performance functions of two pairs of design variables: (**a**) substrate thickness and EMC CTE, and (**b**) package width and length and 1st chip thickness.

Following is the example of determining the response of a random input by using the constructed metamodel. Assuming that a random input x is (0.81, 0.25, 0.050, 0.064, 0.023, 0.013, 15.3, 10.2, 28.2, 7.69). Equation (7) can be written as:

$$y\left(\begin{array}{c} 0.81,\ 0.25,\ 0.050,\ 0.064,\ 0.023, \\ 0.013,\ 15.3,\ 10.2,\ 28.2,\ 7.69 \end{array}\right) = \sum_{1 \le i < j \le 10} y\left(x_i, x_j, \mu^{\sim ij}\right) - (10-2)\sum_{i=1}^{10} y\left(x_i, \mu^{\sim i}\right) + \frac{(10-1)(10-2)}{2}y_0 \quad (9)$$

where the warpage at the cut center, $y_0 = \mu = [\mu_1, \mu_2, \ldots, \mu_{11}]^{\mathrm{T}}$, is 0.9 μm; $y\left(x_i, \mu^{\sim i}\right)$ and $y\left(x_i, x_j, \mu^{\sim ij}\right)$ are the values of x on all known 1D performance functions and 2D performance functions that were constructed earlier. Thus, the warpage value at the random input x can be calculated; it was −49.2 μm.

The above Bivariate Cut-HDMR procedure was integrated in MATLAB (R2020b) codes, and they are available at https://www.mathworks.com/matlabcentral/fileexchange/9289 0-bivariate-cut-hdmr (accessed on 25 May 2021). Those who are interested in implementing Bivariate Cut-HDMR metamodeling can run the script readily by following the instructions.

4. Performance Evaluation

The performance of Bivariate Cut-HDMR is evaluated using two well-known error metrics. The performance of CCD is also evaluated for comparison.

4.1. Error Metrics

Two error metrics employed to evaluate the performance are: [46]

- Metric 1: R-squared

$$R^2 = 1 - \frac{\sum_{i=1}^{m}[y(x_i) - \hat{y}(x_i)]^2}{\sum_{i=1}^{m}[y(x_i) - \overline{y}(x_i)]^2} \quad (10)$$

where m is the number of total test sample points; $y(x_i)$ is a performance function at the ith new sample point used for validity check; $\hat{y}(x_i)$ is an approximated performance function at the ith new sample point; and $\overline{y}(x_i)$ is the mean of all $y(x_i)$. R-squared indicates the overall accuracy of a metamodel, and its maximum value is 1.

- Metric 2: Relative average absolute error ($RAAE$)

$$RAAE = \frac{\frac{1}{m}\sum_{i=1}^{m}|y(x_i) - \hat{y}(x_i)|}{STD} \quad (11)$$

where STD is the standard deviation of all $y(x_i)$. Similar to R-squared, $RAAE$ quantifies the overall accuracy of a metamodel. The closer a value of $RAAE$ is to zero, the more accurate a metamodel is.

Monte Carlo simulation (MCS) was performed to produce 1000 additional sample points. They were used to evaluate the performance of Bivariate Cut-HDMR using the above metrics. The results of the performance metrics are summarized in Table 5. The values of R-squared and *RAAE* are 0.9855 and 0.0880, respectively.

Table 5. Performance metrics of Bivariate Cut-HDMR and CCD.

N	s	Method	Number of Sample Points	R-Squared	RAAE
10	3	Bivariate Cut-HDMR	201	0.9855	0.0880
		CCD	1045	0.9662	0.1472

A metamodel based on CCD was also constructed for comparison. A total of 1045 sample points were required for the CCD metamodel, which built a 10D quadratic function to define the warpage behavior. The additional sample points obtained from MCS were utilized again to evaluate the performance of CCD metamodel. The results are also shown in Table 5. The values of R-squared and *RAAE* are 0.9662 and 0.1472, respectively.

More direct and quantitative comparisons are shown in Figure 7, where the absolute errors of the MCS sample points are compared. The absolute errors of half the MCS sample points of Bivariate Cut-HDMR are less than 5 µm. The outcome is remarkable. Bivariate Cut HDMR used only one-fifth of sample points (201 sample points) required by CCD (1045 sample points). However, Bivariate Cut-HDMR does not compromise the accuracy when compared to CCD. The following section is intended to provide some insight into this performance of Bivariate Cut HDMR.

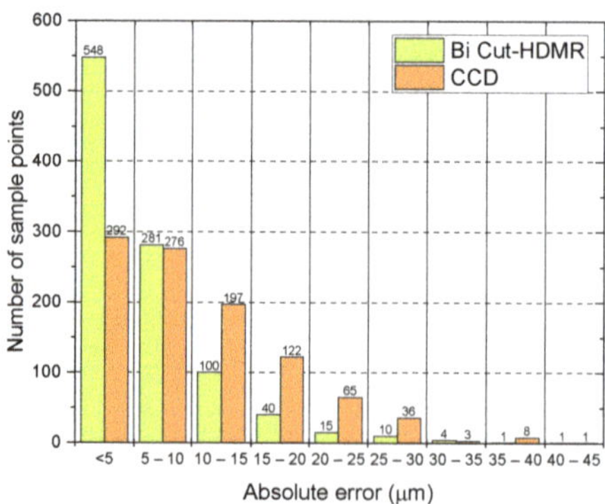

Figure 7. Absolute error of Bivariate Cut-HDMR and CCD of 1000 MCS.

4.2. Discussion: Bivariate Cut-HDMR vs. CCD Metamodel

4.2.1. Sampling Scheme

Figure 8 shows the sampling schemes of Bivariate Cut-HDMR and CCD for a three-variable ($N = 3$, $s = 3$) example. In the figure, the red point is the mean point for both Bivariate Cut-HDMR and CCD; the blue points are used to construct the functions of three lines in the X-, Y- and Z-directions for Bivariate Cut-HDMR and the axial points for CCD; and the yellow points together with the blue points are used to construct the functions of three planes (X-Z plane, Y-Z plane, and X-Y plane) for Bivariate Cut-HDMR and the factorial points for CCD. It also illustrates one of the cut-planes (green planes) and one of the boundary-planes (magenta planes) of both metamodels.

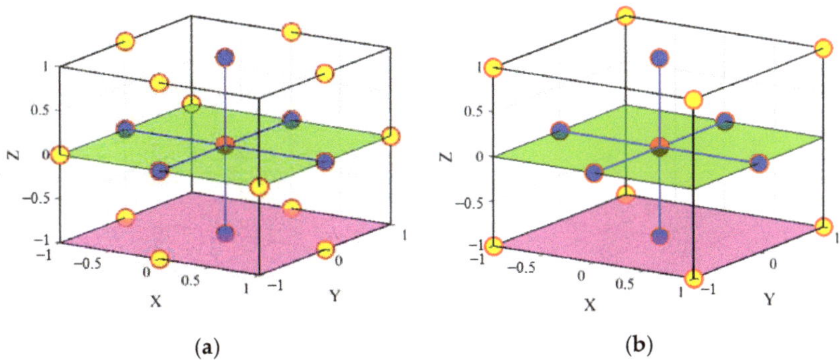

(a)   (b)

Figure 8. Illustration of sample points for $N = 3$: (**a**) Bivariate Cut-HDMR with $s = 3$ and (**b**) CCD, where the red point is the mean point.

The sampling points of Bivariate Cut-HDMR are utilized to construct the first-order and second-order component functions, i.e., every sample point is used to construct the 1D and 2D performance functions (as illustrated in Figures 5 and 6). On the other hand, the sample points of CCD are aimed to cover the boundaries of a design domain.

4.2.2. Prediction of Cut-Planes

As mentioned earlier, the sampling scheme of Bivariate Cut-HDMR is designed to construct the cut-lines and cut-planes. The prediction on the cut-planes performed by both metamodels are compared. As shown in Figure 8, Bivariate Cut-HDMR has more sample points (9) than CCD (5) on the cut-planes (green planes).

Figures 9 and 10 show the two predicted surfaces (cut-planes), which were studied in the TFBGA application. Each figure has the identical nine dots (warpage values obtained from the FE model) in (a) and (b). Red dots are the sample points used to construct for each metamodel. Blank dots are the sample points that were used to construct the Bivariate Cut-HDMR metamodel but not used to construct the CCD metamodel.

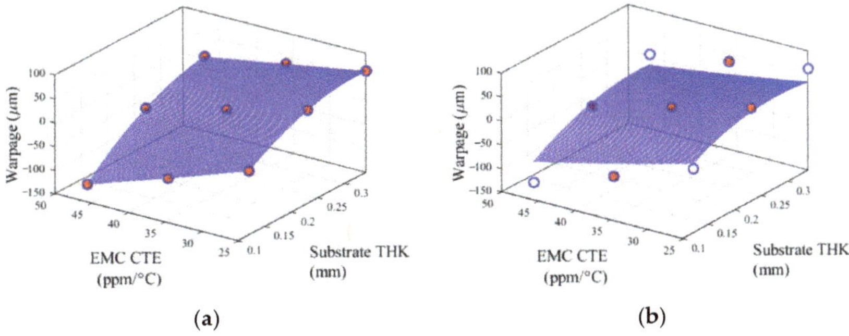

(a)   (b)

Figure 9. 2D performance functions of design variables of (**a**) Bivariate Cut-HDMR and (**b**) CCD, where design variables other than EMC CTE and substrate thickness are kept at their mean values.

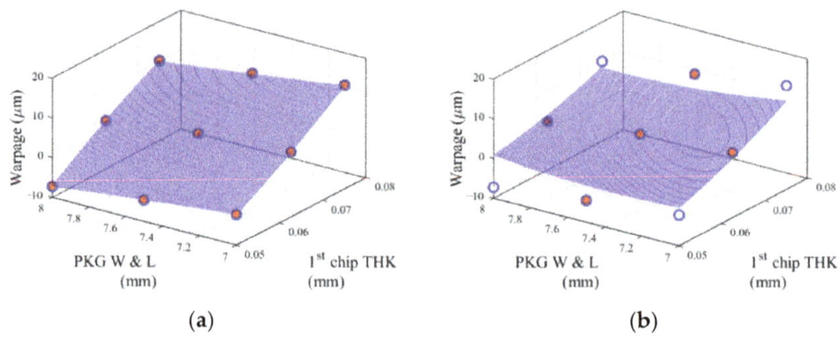

(a) (b)

Figure 10. 2D performance functions of design variables of (**a**) Bivariate Cut HDMR and (**b**) CCD, where design variables other than package width and length and 1st chip thickness are kept at their mean values.

The surfaces (cut-planes) in Figures 9a and 10a were constructed by Bivariate Cut-HDMR (quadratic spline interpolation) with two sets of nine sample points shown in the figures. There is no error between warpage values obtained from FE (dots) modeling and the predicted surfaces.

The surfaces in Figures 9b and 10b were plotted by the CCD metamodel obtained from 1045 sample points. The five sample points shown in Figures 9b and 10b were just a small portion of the total 1045 sample points used to construct the CCD metamodel (a second-order polynomial function). This attempt for CCD to fit all 1045 sample points inevitably produces the discrepancy between true warpage values (dots) and predicted surfaces in the entire design domain, especially in the corners, as shown in Figures 9b and 10b.

4.2.3. Prediction of Boundary-Planes

The example in Figure 8 ($N = 3$) shows five sample points on the boundary-planes of both metamodels. It is important, however, to note that there are lesser or no sample points on the boundary-planes of both metamodels when the number of input variables (N) increases. On the boundary-planes of the TFBGA application ($N = 10$), there were no sample point for Bivariate Cut-HDMR and only four sample points for CCD.

Figure 11 shows two predicted surfaces (boundary-planes) of the TFBGA application. Variables other than the two variables shown in the plots were kept at their maximum values, i.e., it represents one of the boundary-planes in the design domain. Red dots in (b) are the sample points used to construct the CCD metamodel. They also appear in (a), although they are not used for Bivariate Cut-HDMR.

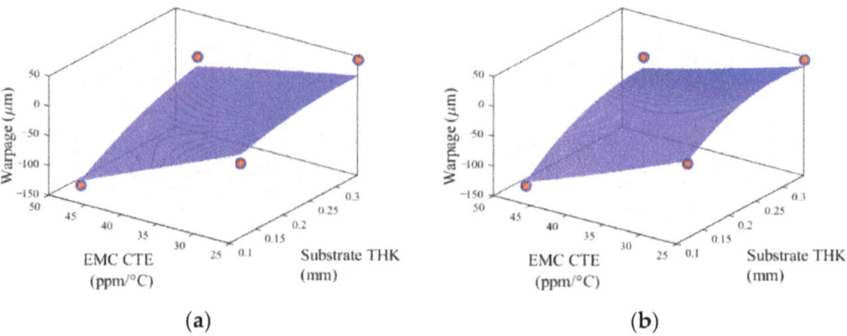

(a) (b)

Figure 11. 2D performance functions of (**a**) Bivariate Cut-HDMR and (**b**) CCD, where design variables other than EMC CTE and substrate thickness are kept at their maximum values.

The Bivariate Cut-HDMR surface (boundary-planes) of Figure 11a are plotted by 201 sample points, while the CCD surface of Figure 11b are plotted by 1045 sample points including the four sample points on the boundary-plane. It is noteworthy that the predicted 2D performance function of Bivariate Cut-HDMR is similar to the CCD surface, despite the fact that CCD utilizes four sample points on the boundary-plane, but Bivariate Cut-HDMR does not.

5. Conclusions

Bivariate Cut-High Dimensional Model Representation (Bivariate Cut-HDMR) was implemented successfully for the warpage problem of a thin flat ball grid array package with 10 design variables. The implementation with three uniformly distributed sample points ($s = 3$) in conjunction with quadratic spline interpolation allowed for comparing its performance with a metamodel based on Central Composite Design (CCD).

The performance of both metamodels were evaluated by two well-known error metrics: R-squared and Relative Average Absolute Error ($RAAE$). The results were compared with the performance of CCD: the R-squared values of CCD and Cut-HDMR were 0.9662 and 0.9855, respectively; the $RAAE$ values of CCD and Cut-HDMR were 0.1472 and 0.0880, respectively.

The outcome was remarkable. Bivariate Cut HDMR used only one-fifth of sample points (201 sample points) required by CCD (1045 sample points); however, Bivariate Cut-HDMR did not compromise the accuracy in both error metrics compared to CCD, which was confirmed by more direct and quantitative comparisons using the absolute errors of the Monte Carlo simulation (MCS) sample points.

Two technical reasons for the outstanding performance of Bivariate Cut-HDMR were discussed:

(1) Sampling scheme: the sample points of Bivariate Cut-HDMR were utilized to construct the first-order and second-order component functions, while the sample points of CCD were aimed to cover the boundaries of a design domain.
(2) Predictions of cut-planes and boundary-planes: Bivariate Cut-HDMR predicted cut-planes more accurately despite the smaller number of sample points, while both techniques produced similar accuracy for boundary-plane predictions.

Supplementary Materials: The following are available online at https://www.mdpi.com/article/10.3390/ma14164619/s1, Table S1: Bivariate Cut-HDMR sample points, Table S2: CCD sample points, Table S3: MCS sample points.

Author Contributions: Conceptualization, Y.-H.Y., H.-P.W. and B.H.; methodology, Y.-H.Y., H.-P.W. and B.H.; software, Y.-H.Y. and H.-P.W.; validation, Y.-H.Y., H.-P.W. and B.H.; formal analysis, Y.-H.Y. and B.H.; investigation, Y.-H.Y. and B.H.; resources, B.H. and C.H.; data curation, Y.-H.Y.; writing—original draft preparation, Y.-H.Y.; writing—review and editing, H.-P.W., B.H. and C.H.; supervision, B.H.; project administration, B.H. All authors have read and agreed to the published version of the manuscript.

Funding: This research received no external funding.

Institutional Review Board Statement: Not applicable.

Informed Consent Statement: Not applicable.

Data Availability Statement: The data presented in this study are available in Supplementary Materials.

Conflicts of Interest: The authors declare no conflict of interest.

References

1. Wang, G.G.; Shan, S. Review of metamodeling techniques in support of engineering design optimization. *J. Mech. Des.* **2007**, *129*, 370–380. [CrossRef]
2. Simpson, T.W.; Poplinski, J.; Koch, P.N.; Allen, J.K. Metamodels for computer-based engineering design: Survey and recommendations. *Eng. Comput.* **2001**, *17*, 129–150. [CrossRef]
3. Montgomery, D.C. *Design and Analysis of Experiments*, 9th ed.; John Wiley & Sons: Hoboken, NJ, USA, 2017.
4. Ahmadi, M.; Vahabzadeh, F.; Bonakdarpour, B.; Mofarrah, E.; Mehranian, M. Application of the central composite design and response surface methodology to the advanced treatment of olive oil processing wastewater using Fenton's peroxidation. *J. Hazard. Mater.* **2005**, *123*, 187–195. [CrossRef]
5. Sasi, A.; Yadur, A.; Gromala, P. Part 1: Selection of the Materials Based on the Virtual DoE. In *Simulation Driven Design of Novel Integrated Circuits. Proceedings of the 2015 16th International Conference on Thermal, Mechanical and Multi-Physics Simulation and Experiments in Microelectronics and Microsystems, Budapest, Hungary, 19–22 April 2015*; IEEE: New York, NY, USA, 2015; pp. 1–7.
6. Ghaedi, M.; Mazaheri, H.; Khodadoust, S.; Hajati, S.; Purkait, M. Application of central composite design for simultaneous removal of methylene blue and Pb2+ ions by walnut wood activated carbon. *Spectrochim. Acta Part A Mol. Biomol. Spectrosc.* **2015**, *135*, 479–490. [CrossRef]
7. OptiSLang Broschuere. Available online: https://www.dynardo.de/fileadmin/Material_Dynardo/dokumente/broschuere/Brosch_optiSLang_eng_P_web.pdf (accessed on 10 December 2020).
8. Design-Expert v12 (CCD). Available online: https://www.statease.com/docs/v12/designs/ccd/ (accessed on 10 December 2020).
9. Liu, S.; Liu, Y. *Modeling and Simulation for Microelectronic Packaging Assembly: Manufacturing, Reliability and Testing*; John Wiley & Sons: Hoboken, NJ, USA, 2011.
10. Wei, H.-P.; Han, B.; Youn, B.D.; Shin, H.; Kim, I.; Moon, H. Assembly yield prediction of plastically encapsulated packages with a large number of manufacturing variables by advanced approximate integration method. *Microelectron. Reliab.* **2017**, *78*, 319–330. [CrossRef]
11. Che, F.; Li, H.Y.; Zhang, X.; Gao, S.; Teo, K.H. Development of wafer-level warpage and stress modeling methodology and its application in process optimization for TSV wafers. *IEEE Trans. Compon. Packag. Manuf. Technol.* **2012**, *2*, 944–955. [CrossRef]
12. Rabitz, H.; Aliş, Ö.F. General foundations of high-dimensional model representations. *J. Math. Chem.* **1999**, *25*, 197–233. [CrossRef]
13. Rabitz, H.; Aliş, Ö.F.; Shorter, J.; Shim, K. Efficient input—Output model representations. *Comput. Phys. Commun.* **1999**, *117*, 11–20. [CrossRef]
14. Li, G.; Rosenthal, C.; Rabitz, H. High dimensional model representations. *J. Phys. Chem. A* **2001**, *105*, 7765–7777. [CrossRef]
15. Prasad, A.K.; Roy, S. Accurate reduced dimensional polynomial chaos for efficient uncertainty quantification of microwave/RF networks. *IEEE Trans. Microw. Theory Tech.* **2017**, *65*, 3697–3708. [CrossRef]
16. He, Z.; Zhang, Z. High-dimensional uncertainty quantification via active and rank-adaptive tensor regression. In Proceedings of the 2020 IEEE 29th Conference on Electrical Performance of Electronic Packaging and Systems (EPEPS), San Jose, CA, USA, 5–7 October 2020; pp. 1–3.
17. Li, G.; Wang, S.-W.; Rosenthal, C.; Rabitz, H. High dimensional model representations generated from low dimensional data samples. I. mp-Cut-HDMR. *J. Math. Chem.* **2001**, *30*, 1–30. [CrossRef]
18. Shim, K.; Rabitz, H. Independent and correlated composition behavior of material properties: Application to energy band gaps for the Ga α In 1 − α P β As 1 − β and Ga α In 1 − α P β Sb γ As 1 – β − γ alloys. *Phys. Rev. B* **1998**, *58*, 1940. [CrossRef]
19. Shan, S.; Wang, G.G. Metamodeling for high dimensional simulation-based design problems. *J. Mech. Des.* **2010**, *132*, 051009. [CrossRef]
20. Wang, H.; Tang, L.; Li, G. Adaptive MLS-HDMR metamodeling techniques for high dimensional problems. *Expert Syst. Appl.* **2011**, *38*, 14117–14126. [CrossRef]
21. Tang, L.; Wang, H.; Li, G. Advanced high strength steel springback optimization by projection-based heuristic global search algorithm. *Mater. Des.* **2013**, *43*, 426–437. [CrossRef]
22. Yücel, A.C.; Bağcı, H.; Michielssen, E. An ME-PC enhanced HDMR method for efficient statistical analysis of multiconductor transmission line networks. *IEEE Trans. Compon. Packag. Manuf. Technol.* **2015**, *5*, 685–696. [CrossRef]
23. Sobol, I.M. Sensitivity estimates for nonlinear mathematical models. *Math. Model. Comput. Exp.* **1993**, *1*, 407–414.
24. Halmos, P.R. *Measure Theory*; Springer: New York, NY, USA, 2013; Volume 18.
25. Sobol, I.M. Theorems and examples on high dimensional model representation. *Reliab. Eng. Syst. Saf.* **2003**, *79*, 187–193. [CrossRef]
26. Banon, G. Distinction between several subsets of fuzzy measures. *Fuzzy Sets Syst.* **1981**, *5*, 291–305. [CrossRef]
27. Mukherjee, D.; Rao, B.; Prasad, A. Cut-HDMR-based fully equivalent operational model for analysis of unreinforced masonry structures. *Sadhana* **2012**, *37*, 609–628. [CrossRef]
28. Aliş, Ö.F.; Rabitz, H. Efficient implementation of high dimensional model representations. *J. Math. Chem.* **2001**, *29*, 127–142. [CrossRef]
29. Xu, H.; Rahman, S. Decomposition methods for structural reliability analysis. *Probabilistic Eng. Mech.* **2005**, *20*, 239–250. [CrossRef]
30. Wei, H.-P.; Yang, Y.-H.; Han, B. Stacking Yield prediction of package-on-package assembly using advanced uncertainty propagation analysis: Part I stochastic model development. *J. Electron. Packag.* **2020**, *142*, 011001. [CrossRef]

31. Tzeng, Y.L.; Kao, N.; Chen, E.; Lai, J.Y.; Wang, Y.P.; Hsiao, C. Warpage and stress characteristic analyses on package-on-package (pop) structure. In Proceedings of the 2007 9th Electronics Packaging Technology Conference, Singapore, 10–12 December 2007; pp. 482–487.

32. Standard, J.I. *Moisture/Reflow Sensitivity Classification for Nonhermetic Solid State Surface Mount Devices*; IPC/JEDEC J-STD-020D.1; JEDEC Solid State Technology Association: Arlington, VA, USA, 2008.

33. Carson, F.; Lee, S.M.; Vijayaragavan, N. Controlling top package warpage for POP applications. In Proceedings of the 2007 57th Electronic Components and Technology Conference, Sparks, NV, USA, 29 May–1 June 2007; pp. 737–742.

34. Lin, W.; Lee, M.W. PoP/CSP warpage evaluation and viscoelastic modeling. In Proceedings of the 2008 58th Electronic Components and Technology Conference, Lake Buena Vista, FL, USA, 27–30 May 2008; pp. 1576–1581.

35. Sun, P.; Leung, V.; Yang, D.; Shi, D. Development of a novel cost-effective Package-on-Package (PoP) solution. In Proceedings of the 2009 International Conference on Electronic Packaging Technology & High Density Packaging, Beijing, China, 10–13 August 2009; pp. 46–51.

36. Bin, L.H.; Eu, O.K.; Azid, I.A. Solder paramater sensitivity for Package-on-Package (POP) on fatigue life prediction. In Proceedings of the 2012 35th IEEE/CPMT International Electronics Manufacturing Technology Conference (IEMT), Ipoh, Malaysia, 6–8 November 2012; pp. 1–6.

37. Eslampour, H.; Joshi, M.; Park, S.; Shin, H.; Chung, J. Advancements in Package-on-Package (PoP) technology, delivering performance, form factor & cost benefits in next generation Smartphone processors. In Proceedings of the 2013 IEEE 63rd Electronic Components and Technology Conference, Las Vegas, NV, USA, 28–31 May 2013; pp. 1823–1828.

38. Hu, D.-C.; Lin, C.-T.; Chan, Y.-C. A PoP structure to support I/O over 1000. In Proceedings of the 2013 IEEE 63rd Electronic Components and Technology Conference, Las Vegas, NV, USA, 28–31 May 2013; pp. 412–416.

39. Kim, J.; Ahn, Y.; Han, G.; Cho, B.; Park, D.; Yoon, J.; Lee, C.; Nicholls, L.; Wen, S. Strip grinding introduction for thin PoP. In Proceedings of the 2014 IEEE 64th Electronic Components and Technology Conference (ECTC), Lake Buena Vista, FL, USA, 27–30 May 2014; pp. 1361–1365.

40. Tzeng, Y.L.; Chen, E.; Lai, J.Y.; Wang, Y.P.; Hsiao, C. Stress and thermal characteristic analyses for advanced FCBGA packages. In Proceedings of the 2006 International Microsystems, Package, Assembly Conference, Taipei, Taiwan, 18–20 October 2006; pp. 1–4.

41. Sun, P.; Leung, V.C.-K.; Xie, B.; Ma, V.W.; Shi, D.X.-Q. Warpage reduction of package-on-package (PoP) module by material selection & process optimization. In Proceedings of the 2008 International Conference on Electronic Packaging Technology & High Density Packaging, Shanghai, China, 28–31 July 2008; pp. 1–6.

42. Che, F.; Ho, D.; Ding, M.Z.; MinWoo, D.R. Study on process induced wafer level warpage of fan-out wafer level packaging. In Proceedings of the 2016 IEEE 66th Electronic Components and Technology Conference (ECTC), Las Vegas, NV, USA, 31 May–3 June 2016; pp. 1879–1885.

43. Zhang, Q.; Lo, J.C.; Lee, S.R.; Xu, W.; Yang, W. Characterization of orthotropic CTE of BT substrate for PBGA warpage evaluation. In Proceedings of the 2016 15th IEEE Intersociety Conference on Thermal and Thermomechanical Phenomena in Electronic Systems (ITherm), Las Vegas, NV, USA, 31 May–3 June 2016; pp. 1312–1319.

44. Hou, F.; Lin, T.; Cao, L.; Liu, F.; Li, J.; Fan, X.; Zhang, G. Experimental verification and optimization analysis of warpage for panel-level fan-out package. *IEEE Trans. Compon. Packag. Manuf. Technol.* **2017**, *7*, 1721–1728. [CrossRef]

45. Yip, L.; Hariharan, G.; Chaware, R.; Singh, I.; Lee, T. Board Level Reliability Optimization for 3D IC Packages with Extra Large Interposer. In Proceedings of the 2017 IEEE 67th Electronic Components and Technology Conference (ECTC), Orlando, FL, USA, 30 May–2 June 2017; pp. 1269–1275.

46. Hajikolaei, K.H.; Gary Wang, G. High dimensional model representation with principal component analysis. *J. Mech. Des.* **2014**, *136*, 011003. [CrossRef]

 *materials*

Article

Coefficient Extraction of SAC305 Solder Constitutive Equations Using Equation-Informed Neural Networks

Cadmus Yuan [1], Qinghua Su [2] and Kuo-Ning Chiang [2,3,*]

[1] Department of Mechanical and Computer-Aided Engineering, Feng Chia University, Taichung 40724, Taiwan; cayuan@fcu.edu.tw
[2] Department of Power Mechanical Engineering, National Tsing Hua University, Hsinchu City 30013, Taiwan; 0967356474shq@gmail.com
[3] College of Semiconductor Research, National Tsing Hua University, Hsinchu City 30013, Taiwan
* Correspondence: knchiang@pme.nthu.edu.tw; Tel.: +886-03-574-2925

Abstract: Equation-Informed Neural Networks (EINNs) are developed as an efficient method for extracting the coefficients of constitutive equations. Subsequently, numerical Bayesian Inference (BI) iterations were applied to estimate the distribution of these coefficients, thereby further refining them. We could generate coefficients optimally aligned with the targeted application scenario by carefully adjusting pre-processing mapping parameters and identifying dataset preferences. Leveraging graphical representation techniques, the EINNs formulation is implemented in temperature- and strain-rate-dependent hyperbolic Garofalo, Anand, and Chaboche constitutive models to extract the corresponding coefficients for lead-free SAC305 solder material. The performance of the EINNs-based extracted coefficients, obtained from experimental results of SAC305 solder material, is comparable to existing studies. The methodology offers the dual advantage of providing the coefficients' value and distribution against the training dataset.

Keywords: Equation-Informed Neural Networks; advanced electronic packaging; numerical Bayesian Inference; constitutive equations; Pb-free SAC305 solders

 check for **updates**

Citation: Yuan, C.; Su, Q.; Chiang, K.-N. Coefficient Extraction of SAC305 Solder Constitutive Equations Using Equation-Informed Neural Networks. *Materials* **2023**, *16*, 4922. https://doi.org/10.3390/ma16144922

Academic Editor: Balázs Illés

Received: 28 May 2023
Revised: 30 June 2023
Accepted: 5 July 2023
Published: 10 July 2023

1. Introduction

Proper material constitutive models and related coefficients are fundamental for reliable finite element predictions, encompassing the performance prediction model [1], the manufacturing process [2,3], and the reliability prediction models. Non-linear material properties, based on the temperature- and strain-rate-dependent material models, are often necessary for modeling critical sections of electronic packaging [4,5] and further influence the accuracy and predictability of the surrogate AI models [6–8].

Solder, a key component in electronic packaging, is often associated with potential fatigue failures. Wilde et al. conducted a study on the rate-dependent constitutive relationship of Pb-rich material [9], resulting in extracting Anand-based coefficients and identifying kinematic hardening, also known as the Bauschinger effect. To gain a better understanding of the creep characteristics of Pb-free solders, Xiao and Armstrong [10] performed tensile tests on both eutectic PbSn and Sn3.9Ag0.6Cu solder. Their findings revealed substantial microstructural alterations in the Sn3.9Ag0.6Cu with significantly lower absolute creep rates than the PbSn eutectic. The creep measurement data were successfully fitted into the Garofalo model [11], and the corresponding Garofalo coefficient was extracted.

Furthermore, Motalab et al. [12,13] conducted creep tests under meticulous control of the microstructure of the SAC305 solder without an oxidized surface, yielding a set of nine parameters for the Anand model. Basit et al. [14] utilized the Anand constitutive model with the extracted coefficients for solder joint lifetime prediction. The Chaboche material model [15], which considers the Bauschinger effect, was applied by Xie and

Chen [16], Deshpande et al. [17], Wang et al. [5], and Yan et al. [18] for life prediction using Manson–Coffin type equations.

Ma and Suhling reviewed the constitutive equation and the corresponding coefficients of lead-free solder joint [19], and significant coefficient discrepancies have been reported. On the other hand, finite element engineers frequently face difficulties in selecting an appropriate material model and its parameters, as the measurement conditions may differ from those in practical applications. Kuczynska et al. [20] performed mechanical/dynamic tests against the solder joint to verify the ability of these material models and their coefficients to map the lifetime differences depending on the temperature rate under field and testing conditions, as well as on the mean operating temperature.

Considering the many application scenarios, which may range from low to high temperatures and strain rates, an emerging trend encourages users to obtain their own material coefficients [5]. This approach emphasizes the importance of tailoring the coefficients to the specific conditions encountered in each unique application. The least squares method and its derivatives are frequently employed in extracting coefficients. Although this approach is well established, matrix multiplication and inversion may diminish computational efficiency when handling extensive datasets. Moreover, the obtained coefficients based on the least square method are sensitive to the outliers, and this method does not apply to the censored data.

Historically, machine learning has harbored a certain resistance to rule-based inference. However, the efficacy of neural networks in symbolic computation is gaining recognition by integrating symbolic reasoning with continuous representations. Pioneers such as Zaremba et al. [21] and Allamanis [22] have explored the application of neural networks in handling mathematical objects. A significant advancement came from Lample and Charton [23] when they proposed a representation for mathematical expressions. Sharma et al., Chhabra et al., and Yadav et al. have applied neural network method for material optimization studies [24–26]. As a result, the theoretical basis for representing mathematical and symbolic equations using neural networks is well-established.

In this research, we have developed the Equation-Informed Neural Networks (EINNs) method, synergistically incorporating the Bayesian Inference (BI) iteration technique to extract the coefficients of constitutive equations from measurement data. As visualized in Figure 1, the foundational concept of EINNs begins with constructing an artificial neural network to embody the constitutive equation f, where coefficient α_k is designated as their respective weights. Subsequently, EINNs deploy a pre-processing mapping technique on the input/output data pairs which are obtained from the experiments, a strategy rooted in neural network learning theory, and enables exploring coefficient fitting across various domains in-depth.

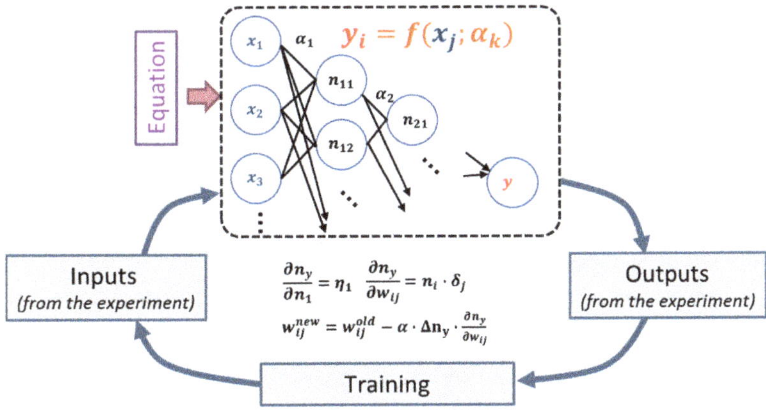

Figure 1. The concept of Equation-Informed Neural Networks (EINNs).

This neural network can be incrementally trained using input and output data pairs, facilitating the simultaneous approximation of coefficient α_k. Theoretically, the steepest descent algorithm of the neural network backpropagation bolsters the computation efficiency and fosters the selective learning of data pairs. The final coefficients are obtained by the post-processing conversion. Utilizing the coefficients obtained by EINNs as initial values, Bayesian Inference (BI) is applied to obtain the distribution of the coefficients against the training datasets and further enhance the accuracy of coefficient extraction.

This paper is organized as follows: the "Theory" section provides an introduction to the framework of Equation-Informed Neural Networks (EINNs) and the numerical Bayesian Inference (BI) method. The subsequent section, "EINN Formulation", presents the conversion process of constitutive equations from their conventional mathematical forms to their EINN equivalents, complete with pre-processing mapping and post-processing functions. In the "Applications" section, we apply the EINN formulation to the coefficient extraction of the material constitutive equation pertinent to Pb-free SAC305 solder joints. Detailed discussions and numerical results pertaining to the EINN formulations of the Chaboche, hyperbolic Garofalo, and Anand material models are also included. The paper concludes with a concise summary of our findings.

2. Theory

2.1. The Framework of Equation-Informed Neural Networks (EINNs)

Assume a constitutive equation is given by the function:

$$y_i = f(x_j; \alpha_k), \tag{1}$$

where y_i, x_j, and α_k are vectors in real space with dimensions i, j, and k, respectively. The x_j and y_i represent the input and output of the functions, while α_k refers to the coefficients. Design pre-processing mapping functions:

$$\mathcal{M}_x(x_j) = X_j \text{ and } \mathcal{M}_y(y_i) = Y_i, \tag{2}$$

which serve to effectively modify the domains of x_j and y_i to optimize the precision of coefficient extraction. Consequently, a new function can be formulated as $Y_i = F(X_j, A_k)$. Meanwhile, the corresponding neural network representations of Y_i are formulated, and the coefficient A_k is assigned as the weighting.

The learning process of the neural network involves continuous adjustment of these weights or coefficients. These adjustments can be computed for each known data pair using steepest-descent-based backpropagation as $A_k^{new} = A_k^{new} - \eta \Delta_y \frac{\partial y_i}{\partial A_k}$. Since these updates are independent of each data pair, the computationally expensive matrix multiplication and inversion inherent in the least squares-based approaches can be avoided. Furthermore, incorporating ratios into the adjustments allows for user emphasis on specific data pairs. This can be implemented as $A_k^{new} = A_k^{new} - \eta \sum_l r_l \cdot \left(\Delta_y \frac{\partial y_i}{\partial A_k} \right)^{(l)}$, where l is the coefficient adjustment from each data pair and $\sum_l r_l = 1$.

Following several learning iterations with satisfactory accuracy, the coefficient A_k of the constitutive equation can be obtained. However, due to the pre-processing mapping function (2) being applied, counteractions are required to reverse its effect. Therefore, we define the post-processing conversion functions as follows.

$$a_k = g_k(x_j, y_i, A_k). \tag{3}$$

Through the combined application of pre-processing mapping functions and post-processing conversion of coefficients, the EINN framework gains an additional degree of freedom, bolstering the accuracy of coefficient extraction. Additionally, the steepest descent method offers a unique opportunity to prioritize specific data pairs while maintaining high computational efficiency.

2.2. The Numerical Bayesian Inference (BI) Iteration

We define the mean square error (MSE) function of Equation (1) with respect to the coefficients α_k, as

$$\epsilon(\alpha_k) = \sum_l \left\| y_{t,i}^{(l)} - y_i\left(x_j^{(l)}; \alpha_k\right) \right\|^2,\tag{4}$$

where x_j^l and $y_{t,i}^l$ denote the input and ground truth of the l-th datapair, respectively.

Assume that the distribution of the data pairs $y_{t,i}^{(l)}$ and $x_j^{(l)}$ are normal, and so is the error function $\epsilon(\alpha_k)$, denoted as $\epsilon(\alpha_k) \sim N(\mu, \tau)$. Because the parameter τ cannot be negative, we assume it follows the gamma distribution, so that $\tau \sim G(a_0, b_0)$, where a_0 and b_0 are the gamma distribution parameters of τ. Moreover, assume that all the coefficients follow the normal distribution, say $\alpha_k \sim N(\mu_k, \tau_k)$, and μ_k and τ_k are the average and precision, respectively. The posterior distribution after the BI remains normal distribution. In practice, we set μ_k equal to α_k.

Consequently, the probabilities of the coefficient τ and α_k can be derived as $\mathcal{P}(\tau) = \frac{b_0^{a_0} \tau^{a_0-1} e^{-b_0\tau}}{\Gamma(a_0)}$ and $\mathcal{P}(\alpha_k) = (2\pi)^{-\frac{1}{2}} \tau_k^{\frac{1}{2}} e^{-\frac{1}{2}\tau_k(\alpha_k-\mu_k)^2}$, respectively. The likelihood with respect to coefficient τ and α_k is $\mathcal{L}(\alpha_k, \tau) \equiv P(data|\alpha_k, \tau) = \prod_{l=1}^{L} \frac{1}{\sqrt{2\pi}} \tau^{\frac{n}{2}} e^{-\frac{\left(y_{t,i}^{(l)} - y_i^{(l)}\right)^2}{2\sigma^2}} = (2\pi)^{-\frac{n}{2}} \tau^{\frac{n}{2}} e^{-\frac{\tau}{2}\epsilon(\alpha_k)}$ [27].

The posterior of the τ distribution can be updated by the gamma–gamma conjugate:

$$a_0^{new} = a_0^{old} + \frac{n}{2} \text{ and } b_0^{new} = b_0^{old} + \frac{1}{2}\epsilon,\tag{5}$$

As Equation (1) is not always a linear function, the posterior of coefficient α_k cannot always be computed by conjugate. Therefore, under the assumption that the value of $\Delta\alpha_k$ is relatively small, a numerical integration approach is applied:

$$\int_0^\infty \mathcal{L}(\alpha_k) \cdot P(\alpha_k) d\alpha_k \sim \sum_{n=1}^{n=N} \mathcal{L}\left(\alpha_k^{(0)} + n\cdot\Delta\alpha_k\right) \cdot P\left(\alpha_k^{(0)} + n\cdot\Delta\alpha_k\right) \cdot \Delta\alpha_k,\tag{6}$$

where $\alpha_k^{(0)}$ is the minimal value of α_k and n is the number of the equal split between the assigned maximum and minimum α_k with a total of N splits. The posterior can then be obtained using normal distribution approximation.

We employ the Markov Chain Monte Carlo (MCMC) method to compute large hierarchical models requiring integration over many parameters. By applying the Gibbs sampling, the τ distribution parameters a_0 and b_0 are first updated through the conjugate (Equation (5)), and a new τ value will be sampled from the gamma distribution. Each α_k will be updated sequenently, and the new value will be accepted. Following thousands of iterations, every α_k exhibits a normal distribution. The mean value of this distribution is computed and assigned as the updated value for α_k.

3. EINN Formulation

This section outlines the development of Equation-Informed Neural Network (EINN) formulations for the hyperbolic Garofalo, nine-parameter Anand, and Chaboche models, including pre-processing mapping and post-processing coefficient functions.

3.1. Hyperbolic Garofalo Model

The conventional hyperbolic Garofalo constitutive equation can be written as:

$$\dot{\epsilon}_p = C_1 \cdot [\sinh(C_2\sigma)]^{C_3} \cdot e^{-\frac{Q}{RT}},\tag{7}$$

where $\dot{\varepsilon}_p$, σ, Q, R, and T represent the plastic strain rate, stress, activation energy, gas constant, and temperature, respectively. C_1, C_2, and C_3 are the coefficients that need to be extracted from the experimental data.

We introduce $e = \dot{\varepsilon}_p \cdot e^{\frac{Q}{RT}}$ and accumulate the data pairs of $\{e\}$ and $\{\sigma\}$ from the experimental results. In order to proportional convert the original data to the $[a, b+a]$ domain, the pre-processing matching functions are defined as follows:

$$\begin{aligned} \mathcal{M}_x(\sigma) &= \frac{\sigma - \sigma_m}{\Delta\sigma} b + a = x \text{ and} \\ \mathcal{M}_y(e) &= \frac{e - e_m}{\Delta e} b + a = y, \end{aligned} \tag{8}$$

where σ_m and $\Delta\sigma$ represent the minimal and maximum different values of set $\{\sigma\}$, and e and Δe correspond to set $\{e\}$. Parameters a and b are parts of pre-processing mapping, and $a = 0.001$ and $b = 1$ are assigned for this case. The values after the pre-processing are defined as x and y, respectively. Subsequently, a new function can be derived as:

$$y = C[\sinh Ax]^n, \tag{9}$$

The corresponding neural network can be defined in Figure 2. The definition of the neurons is given in Table 1.

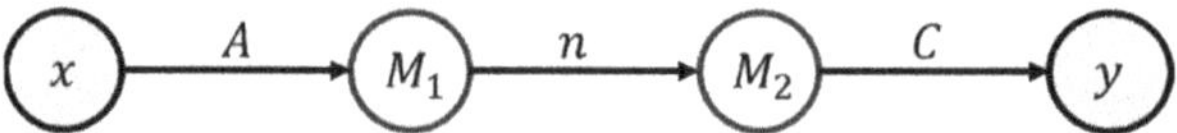

Figure 2. The EINNs for hyperbolic Garofalo model.

Table 1. The neuron definition of the EINN representation of the hyperbolic Garofalo equation.

Neuron	Net Value	Activation
M_1	$M_{1,net} = A \cdot x$	$M_1 = \ln(\sinh M_{1,net})$
M_2	$M_{2,net} = n \cdot M_1$	$M_2 = e^{M_{2,net}}$
y	$y_{net} = C \cdot M_2$	$y = y_{net}$

Accordingly, the post-processing conversion of the coefficients can be approximated as $C_1 = C\Delta e \cdot r_2^n$, $C_2 = A \cdot b / \Delta\sigma$, and $C_3 = n$, where $r_2 = \left(\frac{Ab}{\Delta\sigma} - Aa\right)$.

3.2. Anand Model

Anand et al. [28] proposed a set of viscoplastic constitutive equations for the rate-dependent deformation of metals. Recently, the Anand model has been extensively applied to microelectronic solders exhibiting large viscoplastic deformations. In addition to the activation energy, there are eight coefficients in the Anand model. A two-step approach is commonly employed to extract these eight coefficients [9,12,13].

The governing equation for the first step of the Anand model, including the ultimate tensile stress (σ^*), plastic strain rate ($\dot{\varepsilon}_p$), activation energy (Q), and temperature (T), is expressed in Equation (10). $\hat{s}$, ξ, A, n, and m are the coefficients that need to be extracted.

$$\sigma^* = \frac{\hat{s}}{\xi}\left(\frac{\dot{\varepsilon}_p}{A} \cdot e^{\frac{Q}{RT}}\right)^n \sinh^{-1}\left[\left(\frac{\dot{\varepsilon}_p}{A} e^{\frac{Q}{RT}}\right)^m\right], \tag{10}$$

Utilizing the same method as in the previous section, we assume $e_0 = \dot{\varepsilon}_p \cdot e^{\frac{Q}{RT}}$. Since the value of the strain rate is relatively small compared to other input parameters, a scaling factor R is applied, such that $e = \frac{e_0}{R}$. For consistency within this paper, the same activation function as in the previous section is assumed. The data pair of $\{e\}$ and $\{\sigma^*\}$ is collected from the Motalab et al. [12,13]. An additional y and x are introduced to represent the output and input parameters, and the pre-processing mapping functions are defined as

$$x = \frac{e - e_m}{\Delta e} b_e + a_e \text{ and } y = \frac{\sigma^* - \sigma_m^*}{\Delta \sigma^*} b_\sigma + a_\sigma, \tag{11}$$

where e_m and Δe are the minimal and maximum difference among set $\{e\}$, and so are σ_m^* and $\Delta \sigma^*$ in $\{\sigma^*\}$. a_e, b_e, a_σ, and b_σ are the mapping coefficients. By defining $\beta = \frac{\$}{\zeta}$, the new function can be written as

$$y = \beta^* \left(\frac{x}{A^*}\right)^{n^*} \sinh^{-1}\left[\left(\frac{x}{A^*}\right)^{m^*}\right], \tag{12}$$

Based on Equation (12), the EINN representation can be formulated as Figure 3. This network's definitions are listed in Table 2.

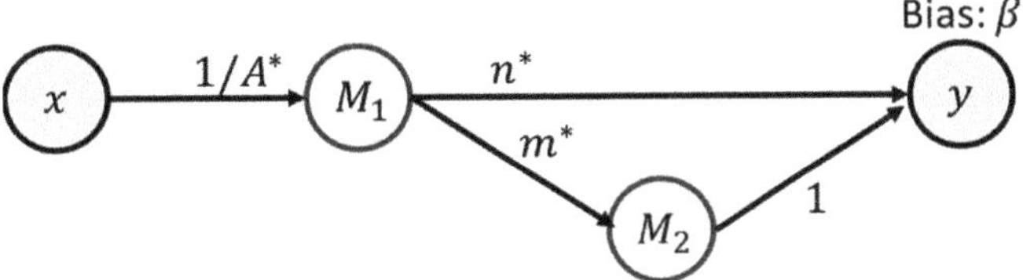

Figure 3. The EINNs for step 1 Anand equation.

Table 2. The neuron definition of the EINN representation of the step 1 Anand equation.

Neuron	Net Value	Activation
M_1	$M_{1,net} = 1/A^* \cdot x$	$M_1 = \ln(M_{1,net})$
M_2	$M_{2,net} = m^* \cdot M_1$	$M_2 = \ln\left(\sinh^{-1} e^{M_{2,net}}\right)$
y	$y_{net} = n^* \cdot M_1 + M_2 + \beta$	$y = e^{y_{net}}$

By defining $r = \sigma_m^* - a_\sigma \frac{\Delta \sigma^*}{b_\sigma}$, the post-processing of the coefficients can be written as follows:

$$\frac{1}{A} = \frac{1}{A^*}\left(\frac{bb_e}{\Delta e} + \frac{a_e}{avg(y)}\right) \cdot \frac{1}{R}, \; n = n^*, \; m = m^* \text{ and } \beta = \left[\frac{\Delta \sigma}{b_\sigma} \cdot \beta^* + \frac{r}{\beta^* \cdot avg(x)}\right] = \frac{\$}{\zeta}, \tag{13}$$

where $avg(x)$ and $avg(y)$ are the averges of $\{e\}$ and $\{\sigma^*\}$.

The governing equation of the second step of the Anand model is listed in (14), and s_0, a, and h_0 are the three remaining coefficients. The parameter c is defined in (15), and ζ is defined as the smallest positive real number to keep $c < 1$.

$$\sigma = \sigma^* - \left[(\sigma^* - cs_0)^{1-a} + (a - 1)\left\{(ch_0)(\sigma^*)^{-a}\right\}\varepsilon_p\right]^{1/(1-a)}, \tag{14}$$

$$c = \frac{1}{\zeta}\sinh^{-1}\left(\frac{\dot{\varepsilon}_p}{A} e^{\frac{Q}{RT}}\right)^m, \tag{15}$$

We assume that $x = \sinh^{-1}\left[\left(\frac{\dot{\varepsilon}}{A} e^{\frac{Q}{RT}}\right)^m\right]$, $y = (\sigma^* - \sigma)$, $l = \sigma^*$, and $z = \varepsilon_p$, and the pre-processing mapping functions are defined as

$$y = \frac{y - y_m}{\Delta y} b_y + a_y, \; l = \frac{l - l_m}{\Delta l} b_l + a_l, \; x = \frac{x - x_m}{\Delta x} b_x + a_x \text{ and } z = \frac{z - z_m}{\Delta z} b_z + a_z, \tag{16}$$

By assuming $1 - a = a\prime$, the new function can be written as

$$y = \left[\left(\bar{1} + \left(-\frac{s_0}{\varsigma}\right)^* \cdot x\right)^{a\prime*} - a\prime* \left\{\left(\left(\frac{h_0}{\varsigma}\right)^* \cdot \bar{y}\right)(1)^{a\prime* - 1}\right\} \cdot \bar{z}\right]^{1/a\prime*}, \tag{17}$$

Based on Equation (17), the EINN representation can be formulated as Figure 4. This network's definitions are listed in Table 3. Moreover, the post-processing of coefficients can be derived as

$$a\prime* = a\prime, \ s_0' = \frac{r_y}{r_l} \cdot (s_0), \ h_0' = \frac{r_y}{r_l} \cdot r_z(h_0), \tag{18}$$

where $r_l = \frac{b_l}{\Delta l}, r_x = \frac{b_x}{\Delta x}, r_y = \frac{b_y}{\Delta y}$, and $r_z = \frac{b_z}{\Delta z}$.

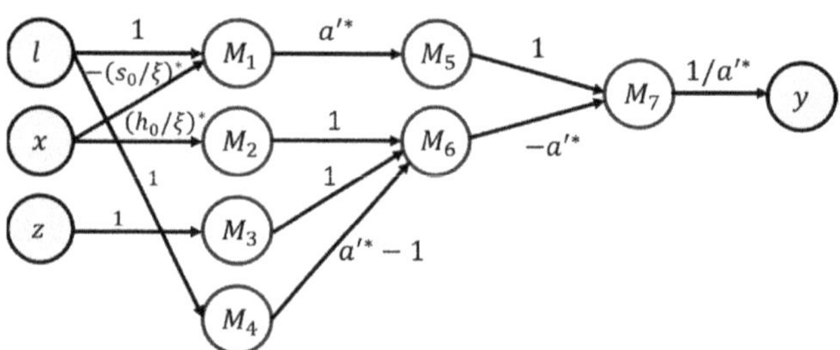

Figure 4. The EINNs for the step 2 Anand equation.

Table 3. The neuron definition of the EINN representation of the step 2 Anand equation.

Neuron	Net Value	Activation
M_1	$M_{1,net} = l + \left(-\frac{s_0}{\varsigma}\right)^* x$	$M_1 = \ln(M_{1,net})$
M_2	$M_{2,net} = \left(\frac{h_0}{\varsigma}\right)^* \cdot M_1$	$M_2 = \ln(M_{2,net})$
M_3	$M_{3,net} = z$	$M_3 = \ln(M_{3,net})$
M_4	$M_{4,net} = l$	$M_4 = \ln(M_{4,net})$
M_5	$M_{5,net} = a\prime* \cdot M_1$	$M_5 = e^{M_{5,net}}$
M_6	$M_{6,net} = M_2 + M_3 + (a\prime* - 1) \cdot M_4$	$M_6 = e^{M_{6,net}}$
M_7	$M_{7,net} = M_5 - a\prime* \cdot M_6$	$M_7 = \ln(M_{7,net})$
y	$y_{net} = \frac{1}{a\prime*} \cdot M_2$	$y = e^{y_{net}}$

3.3. Chaboche Model

The Chaboche model [15,29] is often applied for presenting the metallic material with the Bauschinger effect under cyclic loading. The original function can be written as

$$\alpha = \frac{C}{\gamma}\left(1 - e^{-\gamma \cdot \varepsilon_p}\right) + \sigma_0 \tag{19}$$

where α and ε_p are the back tensile stress and the plastic strain. σ_0 is the initial yielding stress, and C and γ are the fitting coefficients. To simplify the equation, we substitute and C/γ as β. Let $x = \varepsilon_p, y = \alpha$, as the parameters, with the pre-processing mapping functions:

$$x = \frac{\varepsilon_p - \varepsilon_{p,m}}{\Delta \varepsilon_p} \text{ and } y = \frac{\alpha - \alpha_m}{\Delta \alpha}, \tag{20}$$

where $\varepsilon_{p,m}$ and $\Delta \varepsilon_p$ are the minimal and maximum differences among set $\{\varepsilon_p\}$, and so are α_m and $\Delta \alpha$ in $\{\alpha\}$, and $s = \sigma_0$. Hence, the new function can be re-written as

$$y = \beta^* \left(1 - e^{-\gamma^* \cdot x}\right) + s, \tag{21}$$

with the EINN formulation shown in Figure 5 and the neuron definition listed in Table 4.

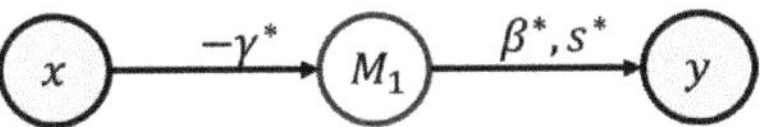

Figure 5. The EINN formulation for the Chaboche model.

Table 4. The neuron definition of the EINN representation of the Chaboche equation.

Neuron	Net Value	Activation
M_1	$M_{1,net} = (-\gamma^*) \cdot x$	$M_1 = 1 - e^{M_{1,net}}$
y	$y_{net} = \beta^* \cdot M_1 + s^*$	$y = y_{net}$

Furthermore, the post-processing of coefficients can be derived as

$$\sigma_0 = x_m + s^* \cdot \Delta\alpha, \quad \gamma = \frac{\gamma^*}{\Delta\varepsilon_p}, \text{ and } C = \frac{\Delta\alpha}{\Delta\varepsilon_p} \cdot (\beta^* \cdot \gamma^*) \tag{22}$$

4. Applications

Building on the EINN formulation and Bayesian Inference (BI) iteration described in the preceding section, this chapter discusses the extraction of coefficients from the hyperbolic Garofalo, nine-parameter Anand, and Chaboche models for the SAC305 solder material.

4.1. Hyperbolic Garofalo Model

The experimental dataset is drawn from Xiao and Armstrong [10]. To determine the coefficient C in Equation (9), we employ a grid search combined with a bisection optimization technique, whereas the EINN structure for coefficients A and n is addressed using standard backpropagation. To emphasize coefficient extraction for low temperatures (both 318 and 353 K) and low strain rates, ratios are assigned to the datapairs, as shown in Table 5. Table 5 also lists the input (plastic strain) and output (stress) of the EINN learning. Utilizing the post-processing conversion formula, the EINN coefficients, C_1, C_2, and C_3, are obtained and presented in the middle column of Table 6. The hyperbolic model, when compared to the experimental data, is depicted in Figure 6. The data at 388 K exhibits a more significant difference than the others, primarily due to the ratio setting outlined in Table 5.

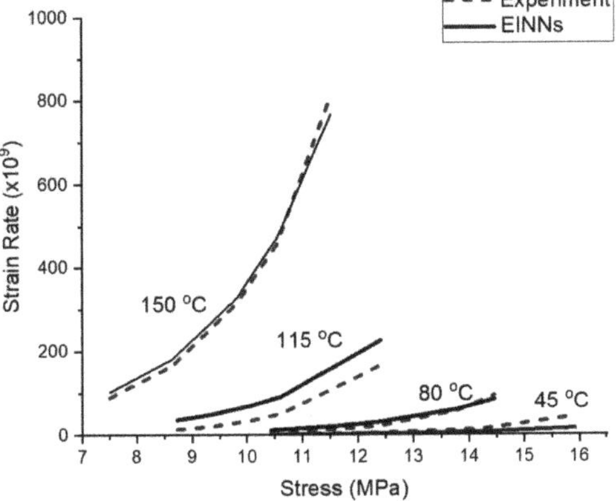

Figure 6. The obtained hyperbolic Garofalo curves for different temperatures. The experimental data are from Xiao and Armstrong [10].

Table 5. The ratios applied to the data pair to emphasize the preference.

Data ID	Temperature (°C)	Plastic Strain Rate ($\dot{\varepsilon}_p$, 10^{-9} 1/s)	Stress (MPa)	Ratio
1	45	1.4	10.54	18.0
2	45	4.0	12.30	6.0
3	45	13.6	14.25	5.0
4	45	43.9	15.92	0.5
5	80	6.6	10.43	18.0
6	80	13.1	11.59	6.0
7	80	21.2	12.40	5.0
8	80	57.8	13.80	3.0
9	80	95.0	14.46	1.0
10	115	13.4	8.73	15.0
11	115	19.8	9.35	12.0
12	115	34.2	10.07	6.0
13	115	50.5	10.61	6.0
14	115	166.0	12.41	1.0
15	150	90.1	7.51	1.0
16	150	165.0	8.63	1.0
17	150	315.0	9.82	1.0
18	150	454	10.54	0.5
19	150	810	11.52	0.5

Table 6. The comparison of the extracted coefficients of Hyperbolic Garofalo Model.

	Xiao and Armstrong [10]	EINNs	EINNs + BI
Q (kJ/mol)	62,000	65,000	65,000
C_1	0.184	0.539	0.443
C_2	0.221	0.473	0.482
C_3	2.89	1.055	1.073
MSE *	37,188.5	11,794.9	10,967.4

*: defined by Equation (4).

A total of 1000 Bayesian Inference interactions were performed to obtain the distribution of the extracted coefficients. The distributions are displayed in Figure 7, represented as the ratio of each coefficient value to the average, and are expanded by the precision τ of the error function. As denoted by the dashed lines in Figure 7, which signify a 5% difference, stable distributions of coefficients C_2 and C_3 are observed, while the large variation in C_1 is attributed to the ratio setting, which induces a higher discrepancy among the 388 K data.

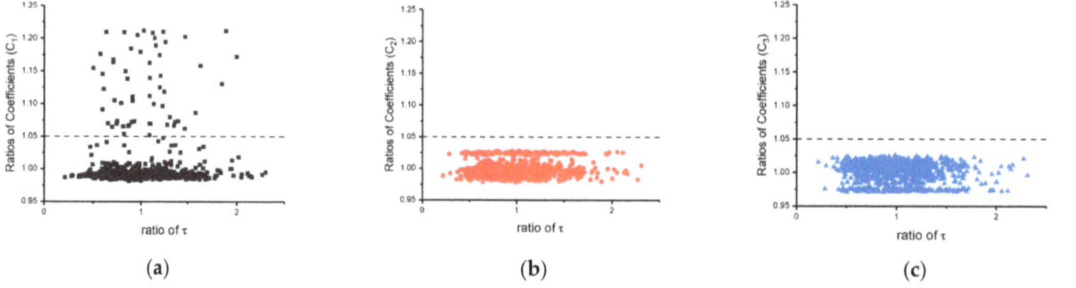

Figure 7. The distribution of the coefficients from the BI iteration. (**a**–**c**) are coefficients C_1, C_2, and C_3, respectively.

The coefficient extraction of the hyperbolic Garofalo constitutive equation highlights the flexibility of the EINN framework, as it allows for assigning ratios to data pairs to prioritize specific data. The fitting accuracy of the EINN results demonstrates a significant improvement compared to the original reports [10] as indicated by the mean square error (MSE) of Table 6, followed by Equation (4). Although the distribution of the C_1 coefficients demonstrate a small fraction of the outliers from BI integration as Figure 7, both C_2 and C_3 show statistical difference within $\pm 5\%$ difference. Over 1000 iterations, only 58 instances of C_1 shows more than $\pm 5\%$ difference of the average value. Consequently, a robust set of coefficients for the hyperbolic Garofaolo constitutive model is achieved.

4.2. Anand Model

In this section, the Anand constitutive model coefficients extraction is implemented for the lead-free SAC305 solder. The same activation energy as in the previous section is applied for the sake of research consistency. To extract the remaining eight coefficients of the Anand constitutive model, the first step involves utilizing temperature and strain rate-dependent ultimate tensile stresses to determine the initial four coefficients. Subsequently, the second step defines the remaining parameters based on temperature and strain rate-dependent stress and plastic strain.

The experimental data are sourced from Motalab et al. [12]. The EINN formulation, following Equation (12), is applied with the pre-processing mapping coefficients a_e, b_e, a_σ, and b_σ (Equation (13)) which are 0.8, 0.15, 0.9, and 0.1. It is vital to note that the selection of these mapping coefficients depends on the numerical characteristics of the dataset, and it is essential for preventing numerical errors during the backpropagation-based machine learning of the EINN formulation.

During the learning phase of the EINN formulation, the coefficients n^*, m^*, A^*, and β of Equation (12) and Figure 3 are constrained to be positive. A grid search technique is employed to identify optimal initial values concerning the experimental data.

Furthermore, learning ratios are implemented to emphasize the learning preference for low strain rates and temperatures close to the working temperature of electronic components. After hundreds of iterations, the EINN coefficients are reported in Table 7. The MSE values indicate that the coefficients obtained from the EINN formulation exhibit similar accuracy to those obtained using conventional methods. The obtained step 1 Anand model is plotted in Figure 8.

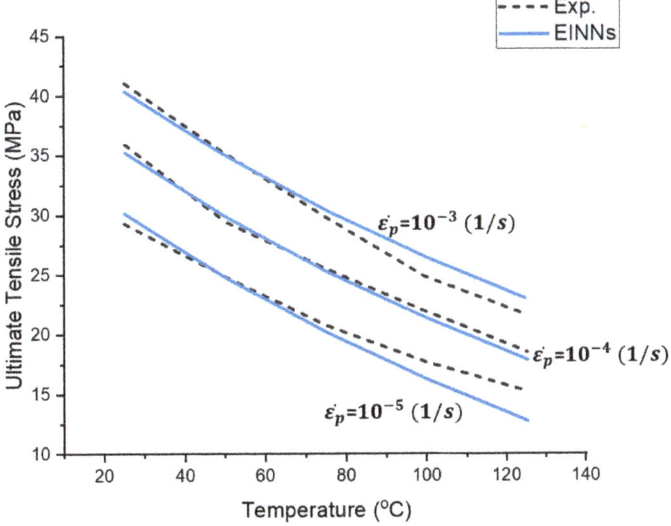

Figure 8. Step 1 Anand model at different temperatures. The experimental data are based on Motalab et al. [12].

Table 7. The comparison of the extracted coefficients of step 1 Anand model.

	Motalab et al. [12]	EINNs	EINNs + BI
A	3501	1650	1649
n	1.00×10^{-2}	1.54×10^{-4}	1.64×10^{-4}
m	0.25	0.54	0.53
$\beta = \hat{s}/\zeta$	7.55	4.11	4.16
MSE *	17.03	15.96	15.78

*: defined by Equation (4).

The EINN formulation coefficients serve as initial inputs for Bayesian Inference (BI) to analyze the statistical distribution of the coefficients. Figure 9 illustrates the distribution of the coefficients, with dashed lines indicating differences within ±5%. Due to its low value, the coefficient n was not examined. Both coefficients A and β exhibit distribution within ±5% difference. Out of 1500 values, only 61 cases of coefficient m exceed ±5% difference, which can be attributed to the preference settings during the EINN learning process. The average coefficients obtained from BI are presented in the last column of Table 7 and are utilized for the subsequent coefficient extraction step in the Anand model.

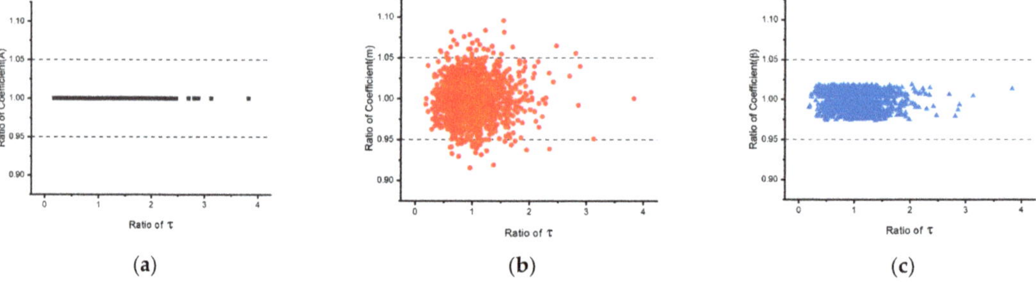

(a)   (b)   (c)

Figure 9. The distribution of the coefficients from the BI iteration. (**a**–**c**) are coefficients A, m, and β, respectively.

The temperature and strain rate dependent stress–strain curves are obtained from Motalab [12]. The EINN formulation of the step 2 Anand model, as indicated in Equation (17) and Figure 4, is applied with the pre-processing mapping parameters shown in Table 8, based on Equation (16), while in the EINN learning procedure, the values of s_0 and h_0 are forced to be positive. A grid search technique is applied to define the optimal initial coefficients. The learning ratios are implemented to emphasize the learning preference for low strain rates and temperatures close to the working temperature of electronic components, following the coefficient extraction strategy of Motalab et al. [12]. With Equation (18), the optimized coefficients can be obtained, as listed in Table 9, and the stress–strain curves at different strain rates from the Anand model are plotted against the experiment [12], as shown in Figure 10.

Table 8. The pre-processing mapping parameters.

	y	l	x	z
a	1	0.8	0.1	0.8
b	0	0.4	0.05	0.1

Table 9. The comparison of the extracted coefficients of step 2 Anand model.

	Motalab et al. [12]	EINNs	EINNs + BI
ζ	4	17.66	17.66
s_0	21	55.96	57.45
a	1.78	2.30	2.26
h_0	18,000	828,822	828,822

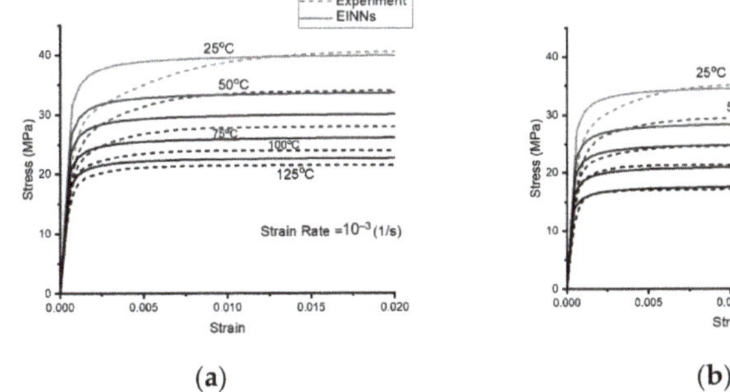
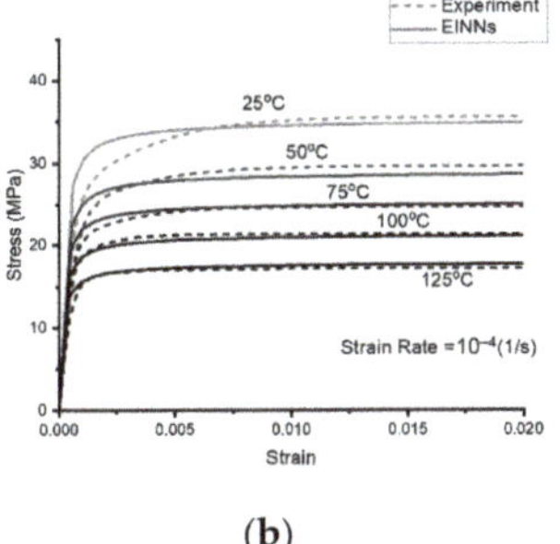
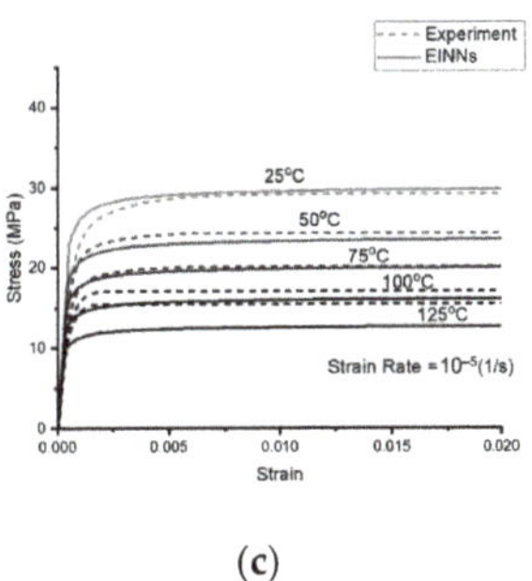

(a) (b) (c)

Figure 10. The obtained step 2 Anand model curves for different temperatures. (**a**–**c**) are the obtained Anand model with strain rates of 10^{-3}, 10^{-4}, and 10^{-5} (1/s). The experimental data are based on Motalab et al. [12].

The dataset with high preference is applied to the BI iteration to mitigate the large coefficient shifting. Figure 11 plots the MSEs of EINNs and EINNs with BI against the Anand coefficient obtained by Motalab et al. [12], under different temperatures and strain rates. By adjusting the ratio of EINN network learning, the coefficient extraction can be fine-tuned to perform better in the room to the working temperature at a low strain rate, as indicated in Figure 11.

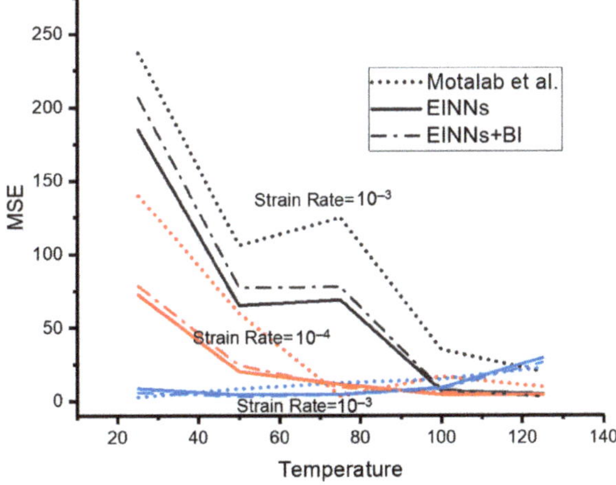

Figure 11. The MSE plot of step 2 Anand model coefficient extraction results Motalab et al. [12].

4.3. Chaboche Model

To study the lifetime of the ball-grid-array-type of advanced electronic packaging, the Chaboche material model is often applied [5,8]. The Chaboche model and its coefficients can be extracted from the temperature-dependent stress–strain curves by a given strain rate.

Unlike the previous sectors, this section investigates the extraction of Chaboche coefficients from the Anand model.

The Anand coefficients from Tables 6 and 7, adjusted via Bayesian Inference (BI), are utilized to generate inputs for the Chaboche model. A strain rate of 10^{-5} (1/s) is maintained, given that the Anand coefficients have been optimized for lower strain rates, as demonstrated in the previous section. Stress–strain curves can be generated by the Anand model (as Equations (14) and (15)) for each temperature point, including $-40\,°C$, $-20\,°C$, $40\,°C$, $80\,°C$, and $122\,°C$.

The temperature-dependent stress–strain data serve as the training datasets. With the pre-processing mapping established by Equation (20), we apply the EINN formulation for the Chaboche model as Equation (21). Following this, the steepest-descent coefficient optimization is applied to the EINN formulation (as illustrated in Figure 5) with the neural definitions outlined in Table 4. The post-processing of the coefficients Equation (22) allows for the acquisition of Chaboche coefficients at various temperatures. The resultant data are documented in Table 10, with the mean square errors (MSE) compared to the input dataset.

Table 10. Temperature-dependent Chaboche coefficients of EINNs.

Temperature	σ_0	C	γ	MSE *
$-40\,°C$	39.32	9174.1	1004.7	1.11
$-20\,°C$	33.80	7535.7	964.0	0.84
$40\,°C$	21.35	4216.0	840.0	0.50
$80\,°C$	15.43	2988.0	824.3	0.28
$122\,°C$	10.50	1886.5	759.1	0.18

*: defined by Equation (4).

The coefficients derived from the EINN formulation are subsequently incorporated into Bayesian Inference (BI) iterations for the temperature-dependent Chaboche model. Figure 12 delineates the distribution of coefficients σ_0, C, and γ across different temperatures, magnified by the precision τ of the error function. The vertical axes in this figure represent the ratio of the coefficient value obtained at each BI iteration to the averaged value. Table 11 contains the averaged coefficient post-BI.

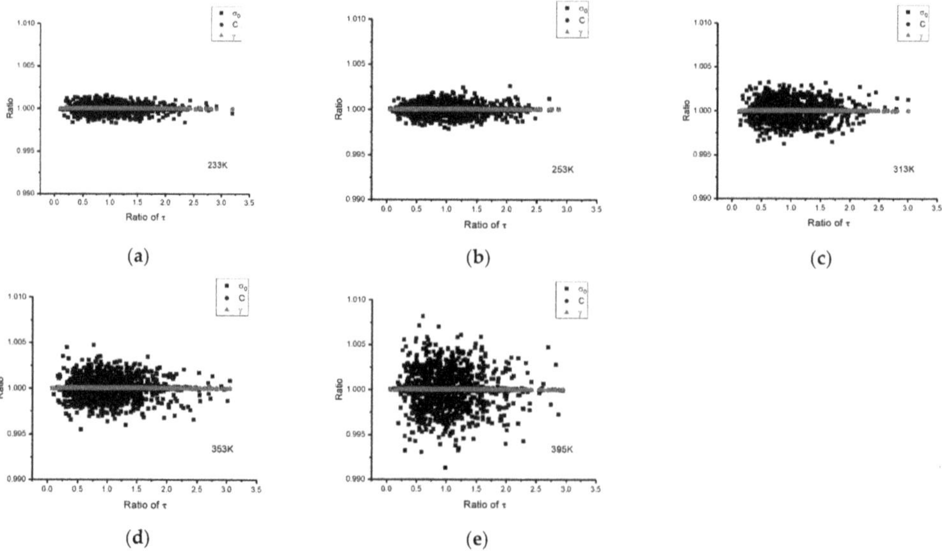

Figure 12. The distributions of the temperature-dependent Chaboche coefficients obtained by BI. (**a–e**) represent the distributions of -40, -20, 40, 80, and 122 °C, respectively.

Table 11. Temperature-dependent Chaboche coefficients of EINNs and BI.

Temperature	σ_0	C	γ	MSE *
−40 °C	39.30	9174.1	1004.7	1.11
−20 °C	33.78	7535.7	964.0	0.84
40 °C	21.39	4216.0	840.0	0.49
80 °C	15.45	2988.0	824.3	0.27
122 °C	10.53	1886.5	759.1	0.17

*: defined by Equation (4).

While variations in all coefficients lie within a ±5% difference, a larger variety, coupled with a lower MSE, as listed in Tables 10 and 11, is evident at higher temperatures. This suggests a reduced coefficient sensitivity at these elevated temperatures. By introducing Young's modulus obtained by linear extrapolation from the experiment [12], the temperature-dependent stress–strain curves are plotted in Figure 13.

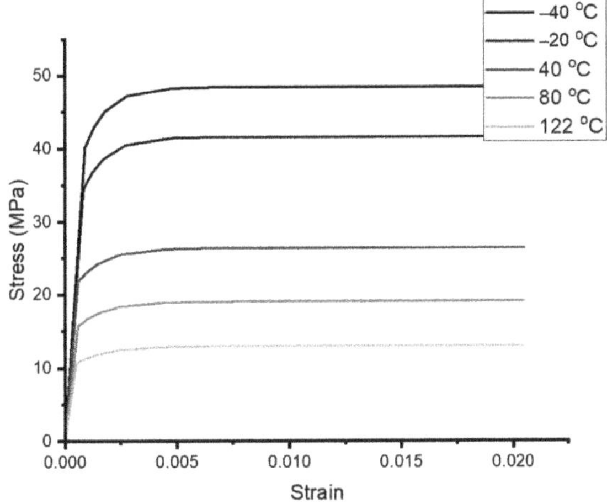

Figure 13. The temperature-dependent stress–stress curves from Chaboche model using the coefficients in Table 11.

5. Conclusions

In this study, we developed the concept of Equation-Informed Neural Networks (EINNs) as an efficient method for extracting the coefficients of constitutive equations. Subsequently, the MCMC with numerical Bayesian Inference (BI) iterations was applied to estimate the distribution of these coefficients, thereby further refining them.

The EINN formulation was derived by leveraging graphical representation techniques to convert the mathematical form of constitutive equations into an equivalent EINN format. By carefully adjusting pre-processing mapping parameters and identifying dataset preferences, we could generate coefficients optimally aligned with the targeted application scenario.

The EINN formulation has been successfully applied to the hyperbolic Garofalo, Anand, and Chaboche constitutive models. This paper details the EINN formulation with its neural network format, the definition of each neuron, the appropriate pre-processing techniques, and the post-processing of the coefficients.

The extraction of coefficients for the hyperbolic Garofalo and Anand models was conducted using experimental results from lead-free SAC305 solder material studies by Xiao and Armstrong [10] and Motalab et al. [12,13]. Our report includes the employed pre-processing mapping techniques and parameters. With the dataset preference, the constitutive equations with extracted coefficients performed better in the interested zone.

Comparisons with coefficients of the constitutive equations from the aforementioned studies demonstrated that those extracted from the EINN formulation were alike. Importantly, the mean square error (MSE) of the EINN formulation learning was comparable to those from the literature [10,12,13]. The performance of the MSE depends on many factors, such as the prescription capability of the material model and experimental measurement accuracy. In this research, the MES is applied as a comparison of how the coefficients extracted by the EINNs perform to the ones obtained by the original methods.

Moreover, the MCMC with numerical Bayesian Inference (BI) iteration technique was employed to analyze the robustness of the extracted coefficients against the experiment data, as shown in Figures 7, 9 and 12. A slightly higher variation was observed when the dataset preference was applied to the EINN learning. Nevertheless, the coefficients derived from EINNs remained within a $\pm 5\%$ confidence interval.

In conclusion, the combined use of EINNs with BI provides a powerful tool for extracting coefficients from temperature- and strain-rate-dependent constitutive equations with dataset preference. This is under the assumption that the SAC305 solder material characteristics can be described by the material model and that the experimental measurement is accurate enough. This approach provides the coefficients' value and the distribution of coefficients against the training dataset.

This study's potential limitations may include the dataset preference assumption, which may not universally apply across all scenarios. Additionally, the applicability of the EINN formulation to all forms of constitutive equations remains to be fully determined, necessitating further exploration of potential limitations. Moreover, advanced neural network backpropagation methods, such as Levenberg–Marquardt (LM) algorithm, will be applied to EINN frameworks.

Author Contributions: Conceptualization, methodology, software, formal analysis, investigation, data curation, and writing—original draft preparation, C.Y.; Formula checking, writing—review and editing, and discussion during research, Q.S.; Problem identification, concept initialization, writing—review and editing, discussion during research, and supervision, K.-N.C. All authors have read and agreed to the published version of the manuscript.

Funding: This research is partially supported by the project of National Tsing Hua University, sponsored by the Ministry of Science and Technology, under the grand no. MOST109-2221-E-007-001-MY3, the "Research on the system reliability modeling of the high power LED packaging based by the online machine learning methods" project of Feng Chia University, sponsored by the National Science and Technology Council under the grand no. MOST 111-2221-E-035-043, and the "AI-Assisted Design-on-Simulation Technology for Advanced Packaging" project of National Tsing Hua University, sponsored by the Semiconductor Research Cooperation (SRC) under contract no. 2023-PK-3180.

Institutional Review Board Statement: Not applicable.

Informed Consent Statement: Not applicable.

Data Availability Statement: The data presented in this research study are available in this article.

Conflicts of Interest: The authors declare no conflict of interest.

References

1. Liu, D.S.; Chao, Y.C.; Wang, C.H. Study of wire bonding looping formation in the electronic packaging process using the three-dimensional finite element method. *Finite Elem. Anal. Des.* **2004**, *40*, 263–286. [CrossRef]
2. Yuan, C.C.A.; Chang, H.M.; Chiang, K.N. Investigation of the mechanical characteristics of the Cu/low-k BEOL under wire bonding process loading. *J. Mech.* **2022**, *38*, 539–551. [CrossRef]
3. Cheng, H.-C.; Wu, Z.-D.; Liu, Y.-C. Viscoelastic Warpage Modeling of Fan-out Wafer-Level Packaging during Wafer-Level Mold Cure Process. *IEEE Trans. Compon. Packag. Manuf. Technol.* **2020**, *10*, 1240–1250. [CrossRef]
4. Qiu, B.; Xiong, J.; Wang, H.; Zhou, S.; Yang, X.; Lin, Z.; Liu, M.; Cai, N. Survey on Fatigue Life Prediction of BGA Solder Joints. *Electronics* **2022**, *11*, 542. [CrossRef]
5. Wang, P.H.; Lee, Y.C.; Lee, C.K.; Chang, H.H.; Chiang, K.N. Solder Joint Reliability Assessment and Pad Size Studies of FO-WLP with Glass Substrate. *IEEE Trans. Device Mater. Reliab.* **2021**, *21*, 96–101. [CrossRef]

6.   Tauscher, M.; Merk, T.; Adsule, A.; Linnemann, A.; Wilde, J. Surrogate Modeling for Creep Strain-Based Fatigue Prediction of a Ball Grid Array Component. *J. Electron. Packag.* **2024**, *146*, 011003. [CrossRef]

7.   Yuan, C.C.A.; Fan, J.; Fan, X. Deep machine learning of the spectral power distribution of the LED system with multiple degradation mechanisms. *J. Mech.* **2021**, *37*, 172–183. [CrossRef]

8.   Hsiao, H.Y.; Chiang, K.N. AI-assisted reliability life prediction model for wafer-level packaging using the random forest method. *J. Mech.* **2020**, *37*, 28–36. [CrossRef]

9.   Wilde, J.; Becker, K.; Thoben, M.; Blum, W.; Jupitz, T.; Wang, G.; Cheng, Z.N. Rate dependent constitutive relations based on Anand model for 92.5 Pb5Sn2. 5Ag solder. *IEEE Trans. Adv. Packag.* **2000**, *23*, 404–414. [CrossRef]

10.   Xiao, Q.; Armstrong, W.D. Tensile creep and microstructural characterization of bulk $Sn_{3.9}Ag_{0.6}Cu$ lead-free solder. *J. Electron. Mater.* **2005**, *34*, 196–211. [CrossRef]

11.   Garofalo, F.A. *Fundamentals of Creep and Creep-Rupture in Metals*; MacMillian: New York, NY, USA, 1965.

12.   Motalab, M.; Cai, Z.; Suhling, J.C.; Lall, P. Determination of Anand constants for SAC solders using stress-strain or creep data. In Proceedings of the 13th InterSociety Conference on Thermal and Thermomechanical Phenomena in Electronic Systems (ITHERM), San Diego, CA, USA, 30 May–1 June 2012.

13.   Motalab, M.; Mustafa, M.; Suhling, J.C.; Zhang, J.; Evans, J.; Bozack, M.J.; Lall, P. Thermal Cycling Reliability Predictions for PBGA Assemblies that Include Aging Effects. In Proceedings of the ASME 2013 International Technical Conference and Exhibition on Packaging and Integration of Electronic and Photonic Microsystems (InterPACK2013), Burlingame, CA, USA, 16–18 July 2013.

14.   Basit, M.M.; Motalab, M.; Suhling, J.C.; Hai, Z.; Evans, J.; Bozack, M.J.; Lall, P. Thermal cycling reliability of aged PBGA assemblies—Comparison of Weibull failure data and finite element model predictions. In Proceedings of the 2015 IEEE 65th Electronic Components and Technology Conference (ECTC), San Diego, CA, USA, 26–29 May 2015; pp. 106–117.

15.   Chaboche, J.L. Constitutive equations for cyclic plasticity and cyclic viscoplasticity. *Int. J. Plast.* **1989**, *5*, 247–302. [CrossRef]

16.   Xie, M.; Chen, G. A developed crystal plasticity model for viscoplastic mechanical behavior of SAC305 solder under thermomechanical coupled cyclic loading. *Int. J. Plast.* **2022**, *159*, 103465. [CrossRef]

17.   Deshpande, A.; Jiang, Q.; Dasgupta, A.; Becker, U. Role of Nominal Stress State on Cyclic Fatigue Durability of SAC305 Grain-Scale Solder Joints. *J. Electron. Packag.* **2022**, *144*, 031006. [CrossRef]

18.   Yan, L.; Yao, J.; Dai, Y.; Zhang, S.; Bai, W.; Gao, K.; Yang, H.; Wang, Y. Study of Thermal Stress Fluctuations at the Die-Attach Solder Interface Using the Finite Element Method. *Electronics* **2021**, *11*, 62. [CrossRef]

19.   Ma, H.; Suhling, J.C. A review of mechanical properties of lead-free solders for electronic packaging. *J. Mater. Sci.* **2009**, *44*, 1141–1158. [CrossRef]

20.   Kuczynska, M.; Schafet, N.; Becker, U.; Metasch, R.; Roellig, M.; Kabakchiev, A.; Weihe, S. Validation of different SAC305 material models calibrated on isothermal tests using in-situ TMF measurement of thermally induced shear load. *Microelectron. Reliab.* **2018**, *91*, 67–85. [CrossRef]

21.   Zaremba, W.; Kurach, K.; Fergus, R. Learning to Discover Efficient Mathematical Identities. In Proceedings of the the the 27th International Conference on Neural Information Processing Systems, Montréal, QC, Canada, 8–13 December 2014; MIT Press: Cambridge, MA, USA, 2014; pp. 1278–1286.

22.   Allamanis, M.; Chanthirasegaran, P.; Kohli, P.; Sutton, C. Learning Continuous Semantic Representations of Symbolic Expressions. In Proceedings of the 34th International Conference on Machine Learning, Proceedings of Machine Learning Research, Sydney, Australia, 6–11 August 2017; pp. 80–88.

23.   Lample, G.; Charton, F. Deep Learning for Symbolic Mathematics. *arXiv* **2019**, arXiv:1912.01412.

24.   Sharma, A.; Chhabra, D.; Sahdev, R.; Kaushik, A.; Punia, U. Investigation of wear rate of FDM printed TPU, ASA and multi-material parts using heuristic GANN tool. *Mater. Today Proc.* **2022**, *63*, 559–565. [CrossRef]

25.   Chhabra, D.; Deswal, S.; Kaushik, A.; Garg, R.K.; Kovács, A.; Khargotra, R.; Singh, T. Analysis of fused filament fabrication parameters for sliding wear performance of carbon reinforced polyamide composite material fabricated parts using a hybrid heuristic tool. *Polym. Test.* **2023**, *118*, 107910. [CrossRef]

26.   Yadav, M.; Kaushik, A.; Garg, R.K.; Yadav, M.; Chhabra, D.; Rohilla, S.; Sharma, H. Enhancing dimensional accuracy of small parts through modelling and parametric optimization of the FDM 3D printing process using GA-ANN. In Proceedings of the 2022 International Conference on Computational Modelling, Simulation and Optimization (ICCMSO), Bangkok, Thailand, 23–25 December 2022; pp. 89–94.

27.   Watanabe, S. *Mathematical Theory of Bayesian Statistics*; Chapman & Hall: London, UK, 2020.

28.   Anand, L. Constitutive equations for hot-working of metals. *Int. J. Plast.* **1985**, *1*, 213–231. [CrossRef]

29.   Chaboche, J.L. On some modifications of kinematic hardening to improve the description of ratchetting effects. *Int. J. Plast.* **1991**, *7*, 661–678. [CrossRef]

Article

Effect of Kaolin Geopolymer Ceramics Addition on the Microstructure and Shear Strength of Sn-3.0Ag-0.5Cu Solder Joints during Multiple Reflow

Nur Syahirah Mohamad Zaimi [1,2], Mohd Arif Anuar Mohd Salleh [1,2,*], Mohd Mustafa Al-Bakri Abdullah [1,2,*], Nur Izzati Muhammad Nadzri [1,2], Andrei Victor Sandu [3,4], Petrica Vizureanu [3,5], Mohd Izrul Izwan Ramli [1,2], Kazuhiro Nogita [6], Hideyuki Yasuda [7] and Ioan Gabriel Sandu [3,*]

1. Faculty of Chemical Engineering Technology, Universiti Malaysia Perlis (UniMAP), Arau 02600, Perlis, Malaysia; syahirahzaimi25@gmail.com (N.S.M.Z.); izzatinadzri@unimap.edu.my (N.I.M.N.); izrulizwan@unimap.edu.my (M.I.I.R.)
2. Center of Excellence Geopolymer & Green Technology (CeGeoGTech), Universiti Malaysia Perlis (UniMAP), Taman Muhibbah, Jejawi, Arau 02600, Perlis, Malaysia
3. Faculty of Materials Science and Engineering, "Gheorghe Asachi" Technical University of Iasi, 41 "D. Mangeron" Street, 700050 Iasi, Romania; sav@tuiasi.ro (A.V.S.); peviz@tuiasi.ro (P.V.)
4. Romanian Inventors Forum, Str. Sf. P. Movila 3, 700089 Iasi, Romania
5. Technical Sciences Academy of Romania, Dacia Blvd 26, 030167 Bucharest, Romania
6. Nihon Superior Centre for the Manufacture of Electronic Materials, School of Mechanical and Mining Engineering, The University of Queensland, Brisbane, QLD 4072, Australia; k.nogita@uq.edu.au
7. Department of Materials Science and Engineering, Kyoto University, Sakyo-Ku, Kyoto 606-8501, Japan; yasuda.hideyuki.6s@kyoto-u.ac.jp
* Correspondence: arifanuar@unimap.edu.my (M.A.A.M.S.); mustafa_albakri@unimap.edu.my (M.M.A.-B.A.); ioan-gabriel.sandu@academic.tuiasi.ro (I.G.S.)

check for updates

Citation: Zaimi, N.S.M.; Salleh, M.A.A.M.; Abdullah, M.M.A.-B.; Nadzri, N.I.M.; Sandu, A.V.; Vizureanu, P.; Ramli, M.I.I.; Nogita, K.; Yasuda, H.; Sandu, I.G. Effect of Kaolin Geopolymer Ceramics Addition on the Microstructure and Shear Strength of Sn-3.0Ag-0.5Cu Solder Joints during Multiple Reflow. *Materials* **2022**, *15*, 2758. https://doi.org/10.3390/ma15082758

Academic Editor: Kuo-Ning Chiang

Received: 2 March 2022
Accepted: 6 April 2022
Published: 8 April 2022

Publisher's Note: MDPI stays neutral with regard to jurisdictional claims in published maps and institutional affiliations.

Abstract: Solder interconnection in three-dimensional (3D) electronic packaging is required to undergo multiple reflow cycles of the soldering process. This paper elucidates the effects of multiple reflow cycles on the solder joints of Sn-3.0Ag-0.5Cu (SAC305) lead (Pb)-free solder with the addition of 1.0 wt.% kaolin geopolymer ceramics (KGC). The samples were fabricated using powder metallurgy with the hybrid microwave sintering method. Apart from using conventional cross-sectioned microstructure imaging, advanced synchrotron real-time in situ imaging was used to observe primary IMC formation in SAC305-KGC solder joints subjected to multiple reflow soldering. The addition of KGC particles in SAC305 suppressed the Cu_6Sn_5 IMC's growth as primary and interfacial layers, improving the shear strength after multiple reflow soldering. The growth rate constant for the interfacial Cu_6Sn_5 IMC was also calculated in this study. The average growth rate of the primary Cu_6Sn_5 IMCs decreased from 49 µm/s in SAC305 to 38 µm/s with the addition of KGC particles. As a result, the average solidified length in the SAC305-KGC is shorter than SAC305 for multiple reflow soldering. It was also observed that with KGC additions, the growth direction of the primary Cu_6Sn_5 IMC in SAC305 changed from one growth to two growth directions. The observed results can be attributed to the presence of KGC particles both at grains of interfacial Cu_6Sn_5 IMCs and at the surface of primary Cu_6Sn_5 IMC.

Keywords: multiple reflows; synchrotron; composite solder

1. Introduction

Solders play a crucial role in electronic packaging via their provision of mechanical support and continuous electrical connection between the substrates and electronic components. Reflow soldering has been commonly used to form solder interconnection at the component and board-level assemblies. The emergence of complex electronic packagings such as System-in-Package (SiP) and Package-on-Package (POP) require multiple reflow soldering to form all the solder interconnections. In advanced electronic packaging, additional solder rework is required [1], and this process will result in interconnections undergoing more than one reflow cycle. Intermetallic compounds (IMC) will be formed during the reactions between Cu substrates and Sn solder alloys during the soldering process [2,3]. The formation of IMC is crucial as it influences the reliability of the solder joints. It is also inevitable that the thickness and morphology of the IMC layer will grow and evolve with increasing time and temperature [4,5]. Therefore, in the case of multiple reflow cycle processes, the solder joints in the first reflow cycle will undergo further changes in the thickness and morphology of the IMC layer during the subsequent reflow cycles. Researchers reported that the thickness of the IMC layer would increase during the multiple reflow soldering process, affecting the solder joint's reliability [6–8]. Additionally, a thicker formation of the IMC layer could result in a brittle fracture, which would degrade the strength of the solder joints [9–12]. The IMCs such as Cu_6Sn_5 and Ag_3Sn are inherently brittle in nature. It consists of one or more covalent compounds and will be deformed in a brittle manner under mechanical loads [13]. Thus, with a thicker layer of IMC, it can increase the likelihood of failure in the solder joints [14]. Sn–Ag–Cu (SAC) solder alloy is commonly used in the electronics industry [15,16] and is touted as a viable substitute for Sn–Pb solder alloy due to its low melting point. Moreover, the Solder Value Product Council (SPVC) has approved SAC solder alloy as one of the Pb-free solder alloys that can replace Sn–Pb solder alloy [12]. However, a significant concern in using the SAC solder alloy is that the IMC layer's growth is faster than Sn–Pb solders due to the higher service temperature in the SAC solder [9]. Therefore, controlling growth on the formation of the IMC layer in SAC solder alloy, especially during the multiple reflow soldering process, is vitally important to preserve the reliability of the solder joints.

As the quality and reliability of solder joints are dependent on the formation of IMC layers, many researchers took the initiative to enhance the performance of existing Pb-free solder alloys. One of the feasible and viable approaches was using ceramics materials to form composite solders [17–21]. To date, there are various ceramic materials that had been successfully added into the solder matrix, such as silicon carbide (SiC) [12], titanium oxide (TiO_2) [17,18,22], titanium carbide (TiC) [22], samarium oxide (Sm_2O_3) [23], alumina (Al_2O_3) [24], and cerium oxide (CeO_2) [25,26]. These ceramic particles did not react with the phase of the solder matrix, thus forming no new compounds within the solder during the melting process [27]. The added ceramic particles also functioned as a second-strengthening phase in the solder matrix, as their properties remain intact within the solder matrix, which strengthens the solder alloys [27]. The dispersion of ceramic particles in the solder matrix increases nucleation rates that result in grain refinement [28]. Tang et al. [29] reported that the grain size of Cu_6Sn_5 IMC increased with increasing reflow time, but the addition of TiO_2 suppressed the growth of the IMC due to TiO_2 particles preventing the diffusion between Cu and Sn atoms. It was also reported that the solder properties improved due to the enhancement in the growth of the interfacial IMC layer [20,24,26,30]. Therefore, it can be surmised that the growth of the IMC layers during multiple reflow soldering needs to be controlled or limited as it improves solder properties and ensures the reliability of the solder joints.

Geopolymers are inorganic polymers that are formed through the geopolymerization process [31]. The process of geopolymerization occurs as the aluminosilicate sources, which consist of SiO_2 and Al_2O_3 are dissolute in a highly alkaline activated solution. The geopolymerization process results in the formation of a semi-crystalline structure with Si–O–Al and Si–O–Si bonds. The geopolymers are advantageous as they can transform to a crystalline structure using slightly low sintering temperature during the sintering process with excellent mechanical properties compared to typical ceramics. The fabrication of the geopolymer ceramics seems to be advantageous as it is energy efficient. Moreover, geopolymer ceramics consists of several elements such as Si and Al which may also contribute to the properties of the solder alloy. In our previous research [30,32] the effect of the addition of kaolin geopolymer ceramic (KGC) onto the properties of Sn-3.0Ag-0.5Cu

(SAC305) under as-reflowed and isothermal aging conditions was investigated. The results confirmed that the addition of KGC as reinforcement particles in SAC305 enhanced its properties in as-reflowed and isothermal aging conditions. Furthermore, the segregation of KGC in the SAC305 matrix refined the microstructure. It suppressed excessive growth of the interfacial IMC even in high isothermal aging temperatures for more extended periods, which improved solder properties such as solder joint strength and the solderability of SAC305. There are many works in the literature focussing on the interfacial reactions between Sn–Cu-based alloys and substrates during multiple reflow soldering [8,33,34]. As reported by S.Tikale et al. [24], the addition of Al_2O_3 effectively suppressed the growth of Cu_6Sn_5 IMC [24]. Owing to the ability of Al_2O_3 particles to hinder the diffusion of Cu to the liquid solder, results in suppression of the IMC layer under multiple reflow soldering [24]. M.A.A Mohd Salleh et al. [1] discovered that the addition of TiO_2 suppressed Cu_6Sn_5 IMC both as primary and interfacial during multiple reflow soldering and thus can improve the shear strength of solder. Nevertheless, limited studies have been reported on the behavior of primary IMC in the solder alloys with the addition of reinforcement particles during multiple cycles of reflow soldering [1]. Primary IMC such as Cu_6Sn_5 and Ag_3Sn are relatively brittle in the solder bulk [35,36]. Therefore, their behavior and distributions in the bulk solder, specifically during multiple reflow soldering, significantly influence the reliability of the joints. This paper also analyses the formation of IMC Cu_6Sn_5 as primary crystals and interfacial layers in the solder during multiple reflow soldering using advanced techniques such as in situ synchrotron X-ray imaging.

2. Materials and Methods

2.1. Materials

Figure 1 shows the flowchart of the kaolin geopolymer ceramic and composite solder fabrication process. Sn-3.0Ag-0.5Cu (SAC305) powders were used as the solder matrix material. It has a spherical morphology and an average size of ~25–45 μm, purchased from Nihon Superior Co. Ltd. (Osaka, Japan). The kaolin geopolymer ceramic (KGC) powders with an average particle size of ~18 μm were used as the reinforcement material.

2.2. Fabrication of Kaolin Geopolymer Ceramic

The kaolin geopolymer ceramic (KGC) fabrication began with the formation of kaolin geopolymer via the geopolymerization process. Kaolin was purchased from Associated Kaolin Industries Sdn. Bhd. and used as the raw material to produce KGC. The kaolin was geopolymerized using an alkaline activator solution, then cured in an oven at 80 °C for 24 h to produce kaolin geopolymer. Next, the product was crushed using a mechanical crusher and compacted at a load of 4.5 tons. The compacted pellets were sintered at 1200 °C at 3 h of soaking time to produce KGC. Then, the KGC pellets were ball-milled for 10 h in a planetary mill at a speed of 450 rpm with a ball to powder ratio of 10:1 to produce KGC particles with an average size of ~18 μm.

2.3. Fabrication of Composite Solder

A composite solder was developed by reinforcing 1 wt.% of kaolin geopolymer ceramic (KGC) with SAC305 solder powder. The composite solder was fabricated using powder metallurgy with a hybrid microwave sintering method. The SAC305 solder powder and 1 wt.% KGC were weighed, then the mixture was mixed in an airtight container using a planetary mill machine at 200 rpm. The product was uniaxially compacted at a load of 4.5 tons. The spherical compacted pellets were then sintered using the hybrid microwave sintering method at ~185 °C under ambient conditions for ~3 min in a 50 Hz microwave oven. A microwave susceptor material of SiC was used for sintering. A sample of SAC305 without the addition of KGC particles was also fabricated using the same approach.

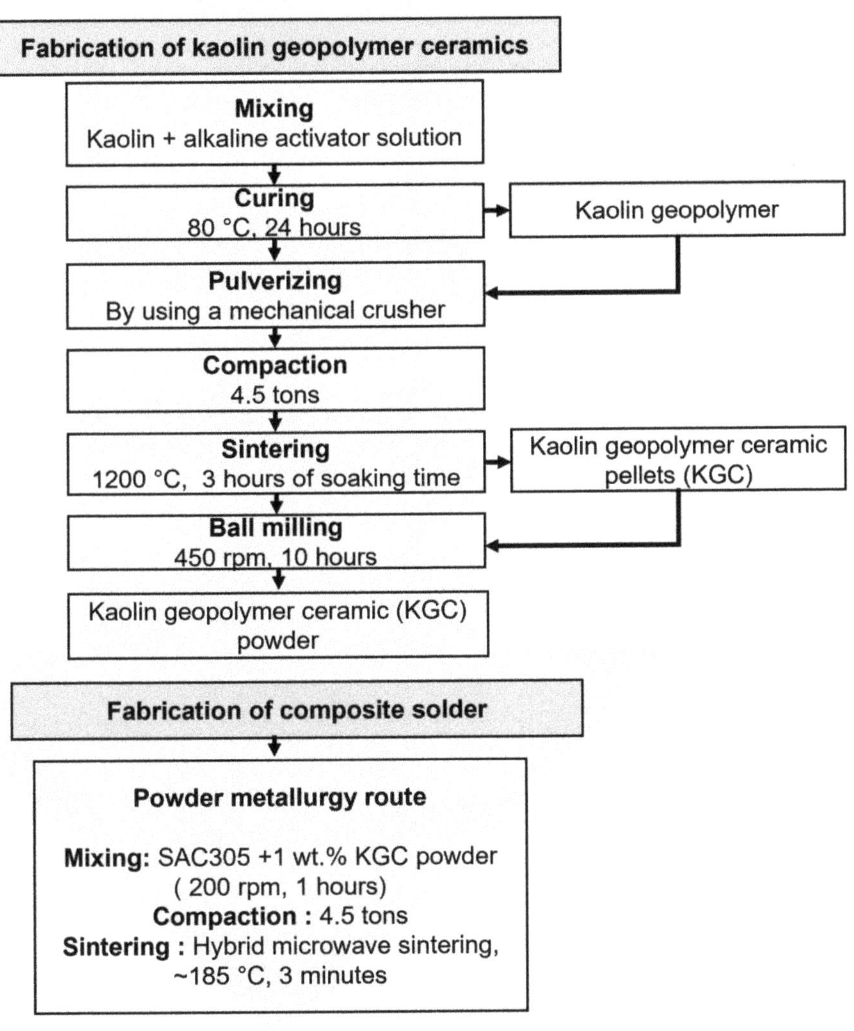

Figure 1. Fabrication process of kaolin geopolymer ceramics and composite solder.

2.4. Microstructure Analysis

The microstructure of the SAC305 and SAC305-KGC was analyzed using a scanning electron microscope (SEM). A sintered pellet was cold-rolled using a rolling machine until the thickness of the sheets was ~50 μm to produce a solder ball. Then, the sheets were punched using a 3.0 mm metal puncher, dipped with rosin mildly activated flux, and reflowed on a Pyrex glass to produce a solder ball with a diameter of ~900 μm. The solder balls were sieved to standardize their average size, then reflowed on a Cu substrate printed circuit board (PCB) with an organic solderability preservative (OSP) using an F4N desktop reflow oven. A small amount of rosin mildly activated flux was applied onto the sample's surface. The flux helped eliminate any contaminations and oxidation before and during the melting process. After that, the reflowed samples SAC305 and SAC305-KGC were cross-sectioned, cold mounted, and grounded with a different grit size of SiC papers. The samples were polished using alumina and colloidal silica suspension to obtain a clearer image of the microstructure under SEM. The average thickness of the IMC layer was measured on the cross-sectioned samples using ImageJ. The IMC thickness (x) was calculated according to Equation (1). The 3D primary intermetallic in the solder joint was microstructurally analyzed as well, where it was etched using an etchant solution from a mixture of 2% 2-nitrophenol, 5% sodium hydroxide, and 93% distilled water to prepare it for analyses.

$$x = A/L \tag{1}$$

where x is IMC thickness, A is the area of the IMC layer and L is the length of the IMC layer.

2.5. In Situ Synchrotron X-ray Radiography Imaging

In situ synchrotron X-ray radiography imaging was conducted using beamline BL20XU at Spring-8 synchrotron in Hyogo, Japan. The experiment was conducted according to the solidification observation setup developed and reported in [1,37,38]. In the experiment, thin sheets of SAC305 and SAC305-KGC were aligned vertically on the 100 µm thick copper (Cu) printed circuit board (PCB), and a small amount of flux was applied. Then, the samples were sandwiched between two glass plates of silica, SiO_2, and polytetrafluoroethylene (PTFE) spacer sheets with an observation window area of 10×10 mm^2, as depicted in Figure 2a. The PTFE sheets were also cut to form vents for flux outgassing purposes during the soldering process. To mimic the soldering process, a furnace equipped with graphite heating elements was used, and the reflow profile JEDEC standard (JESD22-A113D) was used. During the soldering process, the samples were heated from room temperature to 250 °C at a rate of 0.33 °C/s, held for 30 s at the peak temperature, before being cooled at 0.33 °C/s for 6 cycles as in Figure 2c–f. The X-ray energy used was 21 keV. A planar undulator was used, acting as a light source, and the radiations produced were then monochromatized using Si double crystal monochromators. The image detector located at ~2.5–3.0 m away from the samples collected the image signals, which were converted into a digital format of 2000×2000 pixels, resulting in a resolution of 0.47 µm/pixel and a viewing field of 1 mm $\times$ 1 mm. The parameters used in this experiment were selected to provide a high degree of coherence, absorption, and phase contrast, allowing the boundaries of the samples to be observed in the transmitted images.

2.6. Single Lap Shear Testing

Solder joint strength after multiple reflow cycles was evaluated using a single lap shear test, performed using an Instron Machine. The specifications of the copper substrate (PCB-FR4 type) followed the ASTM D1002 standard, as shown in Figure 2b. The fractography of the solder joint after the test was imaged using a scanning electron microscope (SEM) equipped with energy-dispersive X-ray spectroscopy (EDS) under secondary imaging mode to investigate the possible fracture surface mechanism after shearing loads.

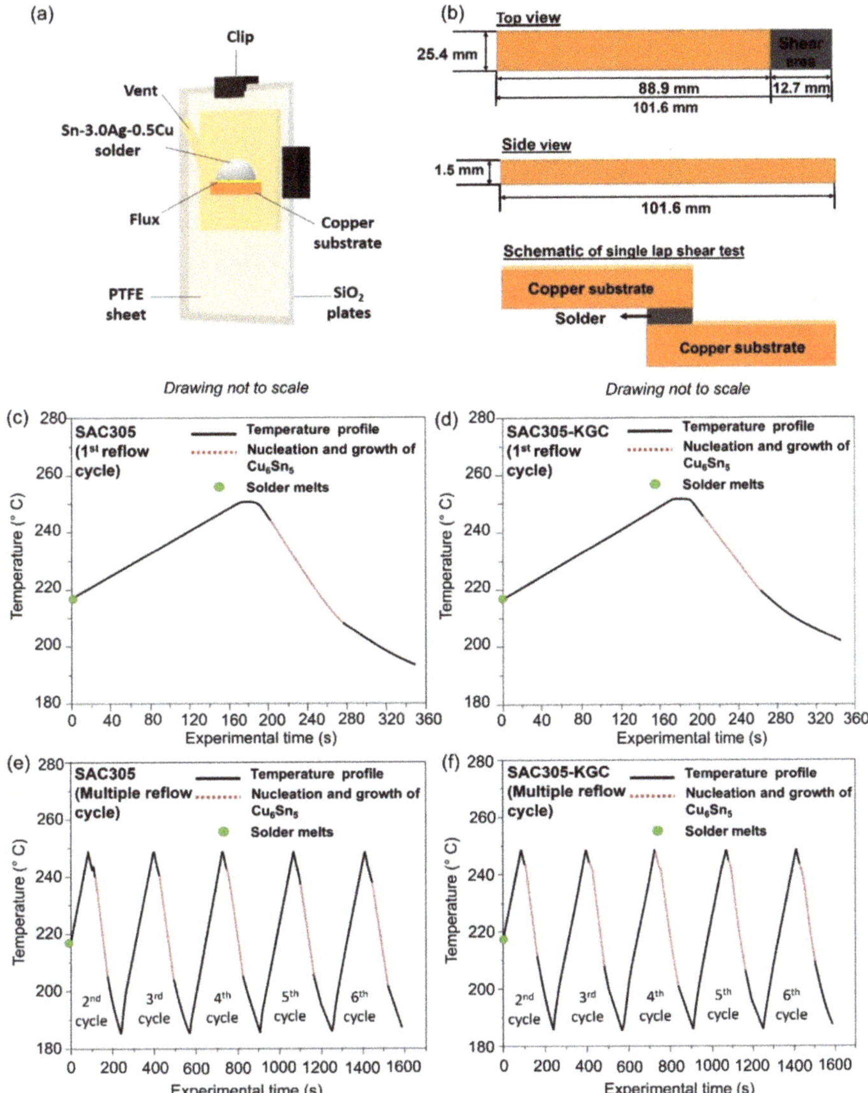

Figure 2. (**a**) Schematic diagram of sample cell for the in situ soldering synchrotron observation, (**b**) Schematic diagram illustrating the configuration for single lap shear test, temperature profile, and growth behavior of primary Cu_6Sn_5 for (**c**) SAC305 1st reflow cycle, (**d**) SAC305-KGC 1st reflow cycle, (**e**) SAC305 multiple reflow cycle (2nd to 6th cycle), and (**f**) SAC305-KGC multiple reflow cycle (2nd to 6th cycle).

3. Results and Discussions

3.1. Microstructure Analysis

3.1.1. Ex Situ Microstructure Analysis of Solder Joints after Multiple Reflows

The microstructure of the solidified SAC305 and SAC305-KGC solder joints after the first, third, and sixth cycles of reflow soldering is shown in Figure 3. The microstructure of SAC305 solder alloys consists of fractions of β-Sn phase and eutectic phases. Based on Figure 3, the β-Sn and eutectic areas were observed in both materials of SAC305 and SAC305-KGC solder joints for multiple cycles of reflow soldering. Fine dots Cu_6Sn_5 and needle-like Ag_3Sn IMC formed in the eutectic area as observed in Figure 3. In this study, hypoeutectic SAC305 solder was soldered on a copper substrate and as a result of Cu diffusion and dissolution during soldering, primary Cu_6Sn_5 will form in the solder joint [26,33]. Per Figure 3a,c,e, IMC particles in the eutectic areas of the SAC305 solder joints are coarse compared to the SAC305-KGC solder, suggesting that the addition of kaolin geopolymer ceramic (KGC) in SAC305 solder alloy suppresses further coarsening of IMCs in the eutectic area after multiple cycles of reflow soldering.

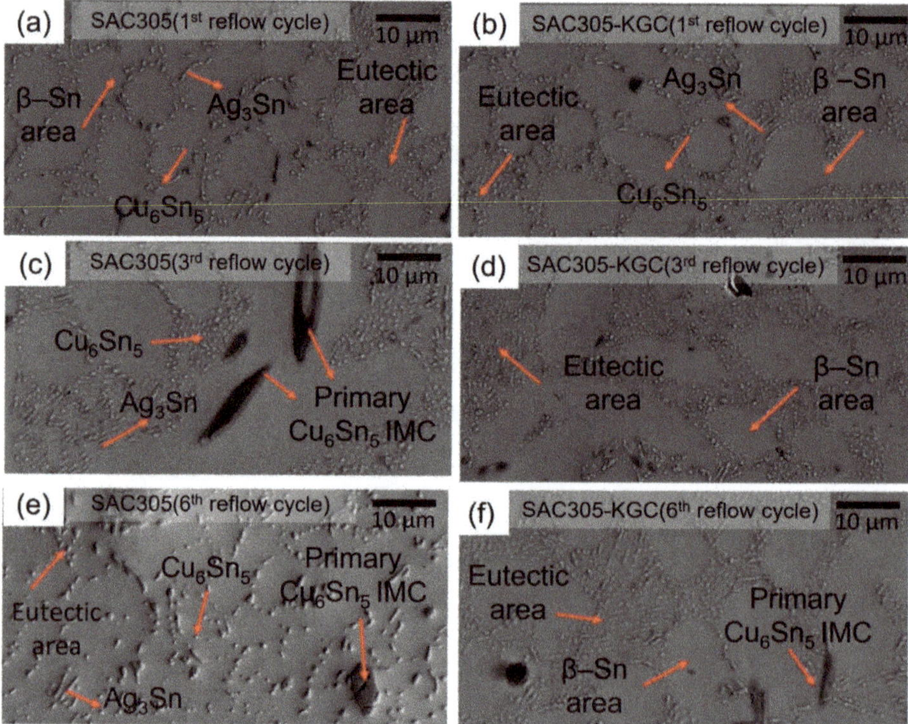

Figure 3. Cross-sectioned microstructure at the bulk solder joints (**a**) SAC305 at 1st reflow cycle, (**b**) SAC305-KGC at 1st reflow cycle, (**c**) SAC305 at 3rd reflow cycle, (**d**) SAC305-KGC at 3rd reflow cycle, (**e**) SAC305 at 6th reflow cycle, and (**f**) SAC305-KGC at 6th reflow cycle.

During the soldering process, the interfacial reaction between molten solder alloy and copper substrate will form an interfacial intermetallic compound (IMC) layer. The cross-sectional images of solder joints were analyzed, and the thickness of the IMC layer was measured per Figures 4 and 5 to elucidate the effects of KGC addition on the interfacial IMC layer for multiple cycles of reflow soldering. The elongated scallop of Cu_6Sn_5 in the SAC305 solder joints was observed to form after reflow soldering, as shown in Figure 4a, suggesting increased concentrations of copper atoms from the substrates to the Sn matrix [26,33]. In the SAC305-KGC solder joints, the small and scalloped shape was formed after reflow soldering; however, with an increasing reflow cycle, the elongated scallop in the SAC305 solder joints became coarser and grew into the solder matrix, as observed in Figure 4c,e. The formation of the elongated scallop IMC layer in SAC305 solder joints is unfavorable, as it could compromise the reliability of the joints by inducing crack formation [32]. However, this trend was not observed in the solder joints of SAC305-KGC since the small and scalloped IMC layer was shorter and became more faceted after multiple cycles of reflow soldering, as can be seen in Figure 4d,f.

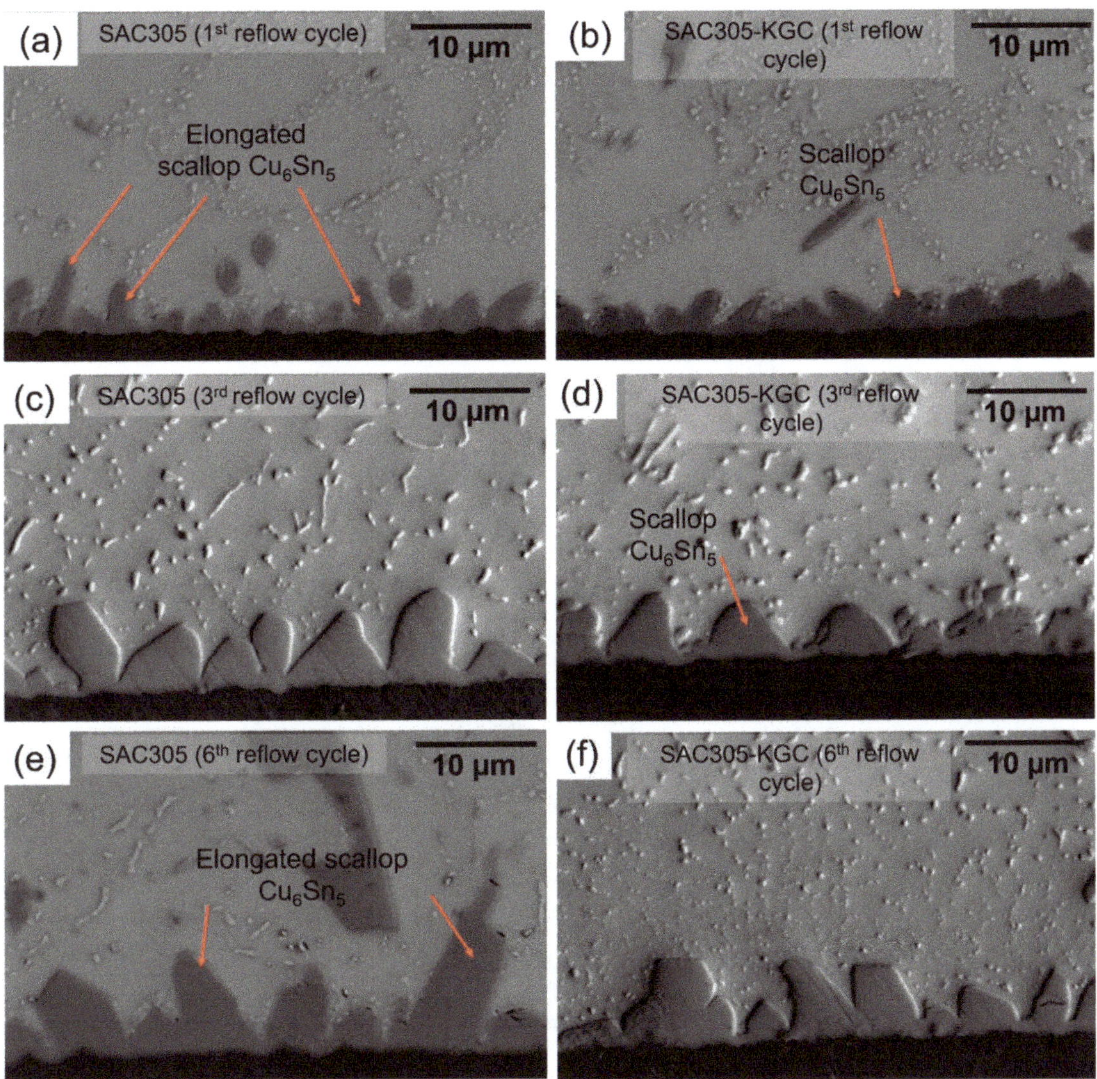

Figure 4. Cross-sectioned microstructure at the interfacial of solder joints (**a**) SAC305 at 1st reflow cycle, (**b**) SAC305-KGC at 1st reflow cycle, (**c**) SAC305 at 3rd reflow cycle, (**d**) SAC305-KGC at 3rd reflow cycle, (**e**) SAC305 at 6th reflow cycle, and (**f**) SAC305-KGC at 6th reflow cycle.

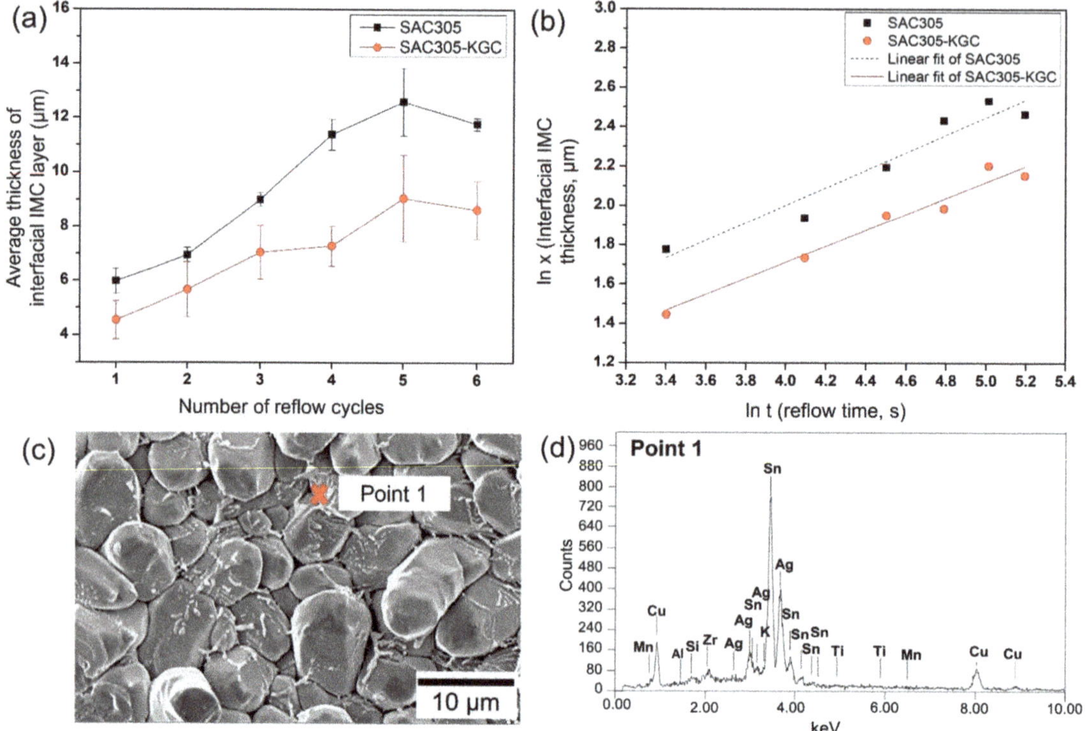

Figure 5. (**a**) Average thickness of interfacial intermetallic compound (IMC) layer at different reflow cycles, (**b**) ln plot growth of interfacial IMC layer respected to different reflow cycles, (**c**) Top view of interfacial IMC layer in SAC305-KGC after 6th cycle reflow, and (**d**) EDX point analysis at Point 1.

The measured average thickness of the interfacial IMC layer for different reflow cycles was plotted and shown in Figure 5a. Initially, the interfacial IMC layer in the SAC305 solder joints grows to an average thickness of ~5.9 μm. After multiple cycles of reflow soldering, the interfacial IMC layer grows to a maximum of ~12.6 μm in SAC305 solder joints. Meanwhile, in the SAC305-KGC solder joints, the interfacial IMC layer grows to an initial average thickness of ~4.5 μm and a maximum of ~9.4 μm after multiple cycles of reflow soldering. The average thickness of the SAC305-KGC solder joints was thinner than SAC305 solder joints, inferring that the addition of KGC might play a role in suppressing the increasing thickness of the IMC layer during multiple cycles of reflow soldering. The thickness of the interfacial IMC layer after multiple cycles of reflow soldering can be generally described per the empirical power law equation [1,6,36,39]:

$$x = kt^n \tag{2}$$

where x is the thickness of the IMC layer at reaction time t, k is the growth rate constant, and n is the time exponent. In this study, the reaction time was based on the time above 250 °C, which is 30 s at each reflow cycle.

According to Liu et al. [40], the interfacial IMC layer's growth could either be controlled by the grain boundary diffusion at the interface, bulk diffusion, or chemical reaction with the values of time exponent, n, of 0.33, 0.50, or 1.0, respectively. In this study, the values of k and n for SAC305 and SAC305-KGC for multiple cycles of reflow soldering can be obtained by the linear fitting method of the ln-ln graph. Figure 5b shows the graph of linear fitting obtained from the experimental data. The results revealed that the time exponent, n, for the growth of interfacial IMC layer in SAC305 and SAC305-KGC solder were 0.45 and 0.41, respectively. The values obtained were near 0.5, which explains the growth of the interfacial IMC layer during multiple cycles of reflow soldering as controlled via bulk diffusion. Regarding the growth rate constant (k), the interfacial IMC in SAC305-KGC has a k value of 0.37 $\mu m^2/s$, compared to SAC305, which is 0.81 $\mu m^2/s$ for multiple reflow cycles. This proved that the growth of the interfacial IMC layer in SAC305 is faster than SAC305-KGC, thus leading to a thicker formation of the interfacial IMC layer after multiple cycles of reflow soldering. Salleh et al. [1] also reported that the addition of TiO_2 in Sn-0.7Cu solder resulted in the growth exponent of 0.5 with $t^{1/2}$ dependence. A top view with high magnification images of the interfacial IMC layer in the SAC305-KGC after the sixth cycle of reflow soldering is shown in Figure 5c. EDX point analysis was performed on the grains of Cu_6Sn_5 IMC. The results from the EDX point analysis at "Point 1", per Figure 5d, confirmed the presence of KGC particles on the surface of Cu_6Sn_5 IMC grains. This observation suggests that KGC particles remained in contact with Cu_6Sn_5 IMC grains after the sixth cycle of reflow soldering, thus suppressing the growth of interfacial Cu_6Sn_5 IMC layer during multiple reflow soldering processes. As mentioned in [34], the channels between the Cu_6Sn_5 scallops provide a path for the rapid diffusion and dissolution of copper atoms from the substrates to the molten solder, resulting in the formation of the Cu_6Sn_6 interfacial layer. As the growth rate constant calculated in SAC305 was faster than SAC305-KGC, this explains that the rapid diffusion of copper atoms during the solid–liquid process could result in a thicker interfacial Cu_6Sn_5 IMC layer with an elongated scalloped shape. Meanwhile, the growth rate constant of the SAC305-KGC solder joints was lower since the presence of the KGC particles on the surface of Cu_6Sn_5 grains might block channels between the Cu_6Sn_5 scallop for rapid diffusion of copper atoms from the substrate and tin atoms from the molten solder. This explains the thinner interfacial IMC layer in the SAC305-KGC solder joints during the stipulated multiple reflow soldering. Moreover, Tang et al. [9] suggested that the theory of adsorption of surface-active materials can be used to determine the role of the reinforcement particles on the interfacial IMC layer and the IMCs at the bulk solder.

3.1.2. In Situ Observation on Primary Cu_6Sn_5 IMC during Multiple Reflows

The growth behavior of primary Cu_6Sn_5 during multiple reflow soldering in SAC305 and SAC305-KGC solder joints was in situ visualized using synchrotron X-ray imaging at Spring8, Japan. Figures 6 and 7 show the synchrotron radiation images for both SAC305 and SAC305-KGC during multiple reflow soldering. The darker rods in the images are primary Cu_6Sn_5, and a slightly brighter is the Sn liquid. Both figures showed the distribution of primary Cu_6Sn_5 formed in the SAC305 and SAC305-KGC solder joints for multiple cycles of reflow soldering. In this experiment, SAC305 and SAC305-KGC solder started to melt at ~217 °C (t = 0 when the solder melts). After a peak temperature of 250 °C for 30 s, the solders began cooling down. During the cooling process for the first reflow cycle of SAC305, the primary Cu_6Sn_5 IMC nucleated at experimental times of 202 s to 272 s and temperatures of ~244 °C to 210 °C as in Figure 2c. Meanwhile, in SAC305-KGC (Figure 2d), the primary Cu_6Sn_5 IMC starts to nucleate at experimental times of 203 s to 262 s and temperatures of ~245 °C to 220 °C. This implies that during the first cycle of reflow soldering in SAC305-KGC, the primary Cu_6Sn_5 IMC took a shorter time to grow from the first IMC nucleation until the completion of the solidification process. The shorter time is taken for the primary Cu_6Sn_5 IMC in the SAC305-KGC to grow suggested the occurrence of rapid solidification, which affects the nucleation growth time with inhibits the tip growth of primary IMC [41]. Besides that, there are also a number of interfacial voids as in Figures 6–8. The formation of interfacial voids was caused by flux outgassing during soldering [37,42].

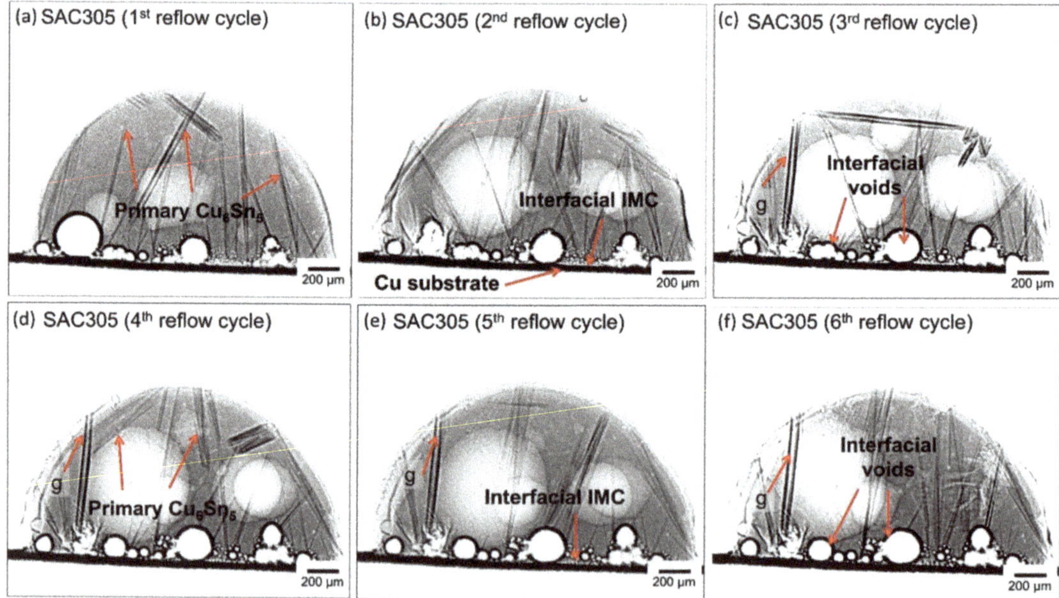

Figure 6. Synchrotron radiation images of SAC305 showing the formation of primary Cu_6Sn_5 IMC for multiple cycles of reflow soldering.

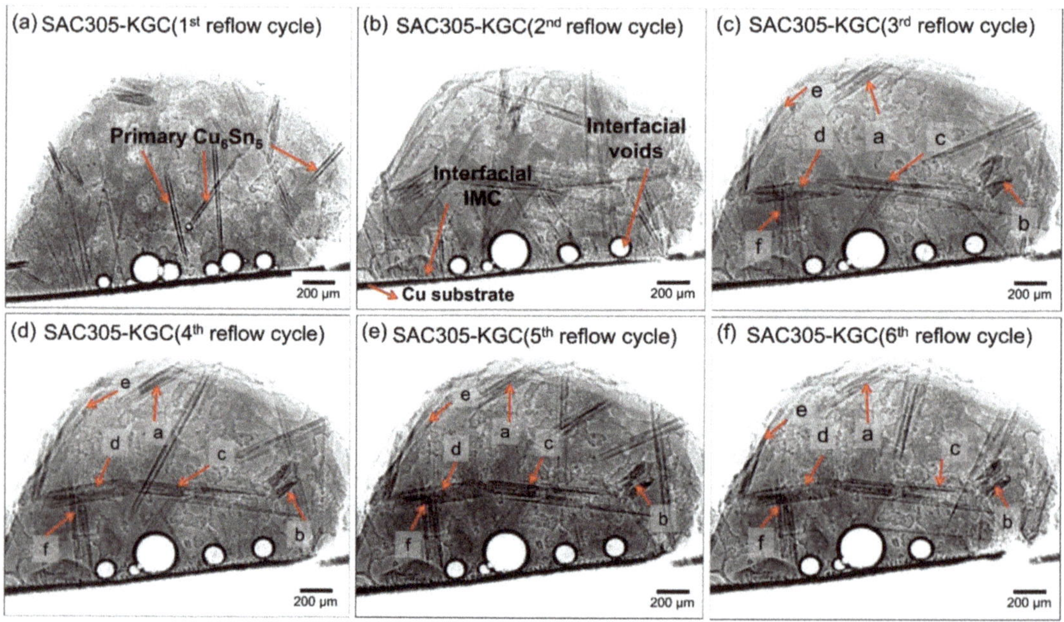

Figure 7. Synchrotron radiation images of SAC305-KGC showing the formation of primary Cu_6Sn_5 IMC with respect to multiple cycles of reflow soldering.

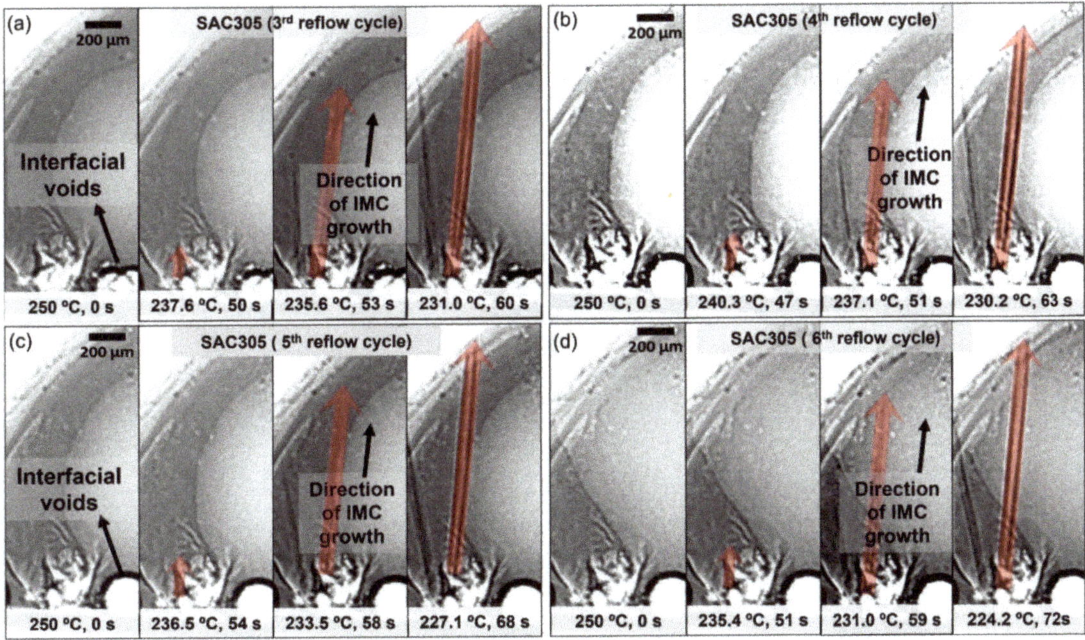

Figure 8. Sequences of the image showing the growth of IMC (denoted by 'g') in SAC305 solder joints during (**a**) 3rd reflow cycle, (**b**) 4th reflow cycle, (**c**) 5th reflow cycle, and (**d**) 6th reflow cycle (t = 0 s when the sample begins to cool).

It can be seen in Figure 6c–f that there is one primary Cu_6Sn_5 IMC (denoted with 'g') nucleated at the exact locations during the third to the sixth cycle of reflow soldering. Meanwhile, in the SAC305-KGC, most primary IMCs marked as 'a', 'b', 'c', 'd', 'e', and 'f' were observed to form at the same locations during the third to sixth cycle reflow soldering (per Figure 7c–f). To further elucidate the finding, snapshot images for primary Cu_6Sn_5 IMC (denoted by 'g') in SAC305 and one of the primaries in SAC305-KGC are shown in Figures 8 and 9, respectively. Primary Cu_6Sn_5 IMC (denoted by 'g') was first nucleated at 227.6 °C, 50 s during the cooling from peak temperature in the third cycle reflow soldering (Figure 8a). Then, this primary IMC was fully melted during the heating process of the fourth reflow cycle and nucleated again at a similar location during the cooling at the following fifth and sixth reflow cycles. Additionally, this primary IMC has one growth direction, per Figure 8, during the multiple reflow cycle. This observation can be caused by the orientation indexed of the crystal structure. Cu_6Sn_5 IMC existed as a close-packed hexagonal crystal structure at a temperature above 186 °C with the orientations index of <0001> [11], and it can be inferred that the primary Cu_6Sn_5 IMCs in SAC305 will preferably grow in one growth direction. Meanwhile, in the SAC305-KGC, the primary Cu_6Sn_5 IMC was first nucleated after 19 s of cooling from a peak temperature of 250 °C, per Figure 9a. However, it should be pointed out that during the subsequent heating process in the fourth, fifth, and sixth cycles, the primary Cu_6Sn_5 IMC was not fully melted. However, during cooling, the primary Cu_6Sn_5 IMC will instantly grow in two growth directions until it solidifies, as depicted in Figure 9b–d. Generally, the fewer the crystal orientations of the IMC, the easier for the IMC to grow and increase in size due to the lower energy consumption during melting [11,36]. A key finding in this work is that (i) primary Cu_6Sn_5 IMC in SAC305-KGC nucleated earlier compared to SAC305, (ii) primary Cu_6Sn_5 IMC in SAC305-KGC do not fully melt during the heating stages of fourth, fifth, and sixth reflow cycle, and yet the primary IMCs will instantly grow during the cooling stages, and (iii) the additions of KGC in SAC305 causes the primary Cu_6Sn_5 IMCs growth in two growth directions. Despite that, the heating and cooling conditions (time and temperature) used during the soldering of both samples, SAC305 and SAC305-KGC, were similar. Therefore, this observation must be explained during the multiple reflow soldering of SAC305 and SAC305-KGC. Deep etching metallographic technique on the solder joints of SAC305 and SAC305-KGC was conducted to evaluate the possible effects of KGC addition on the growth of the primary Cu_6Sn_5 IMC.

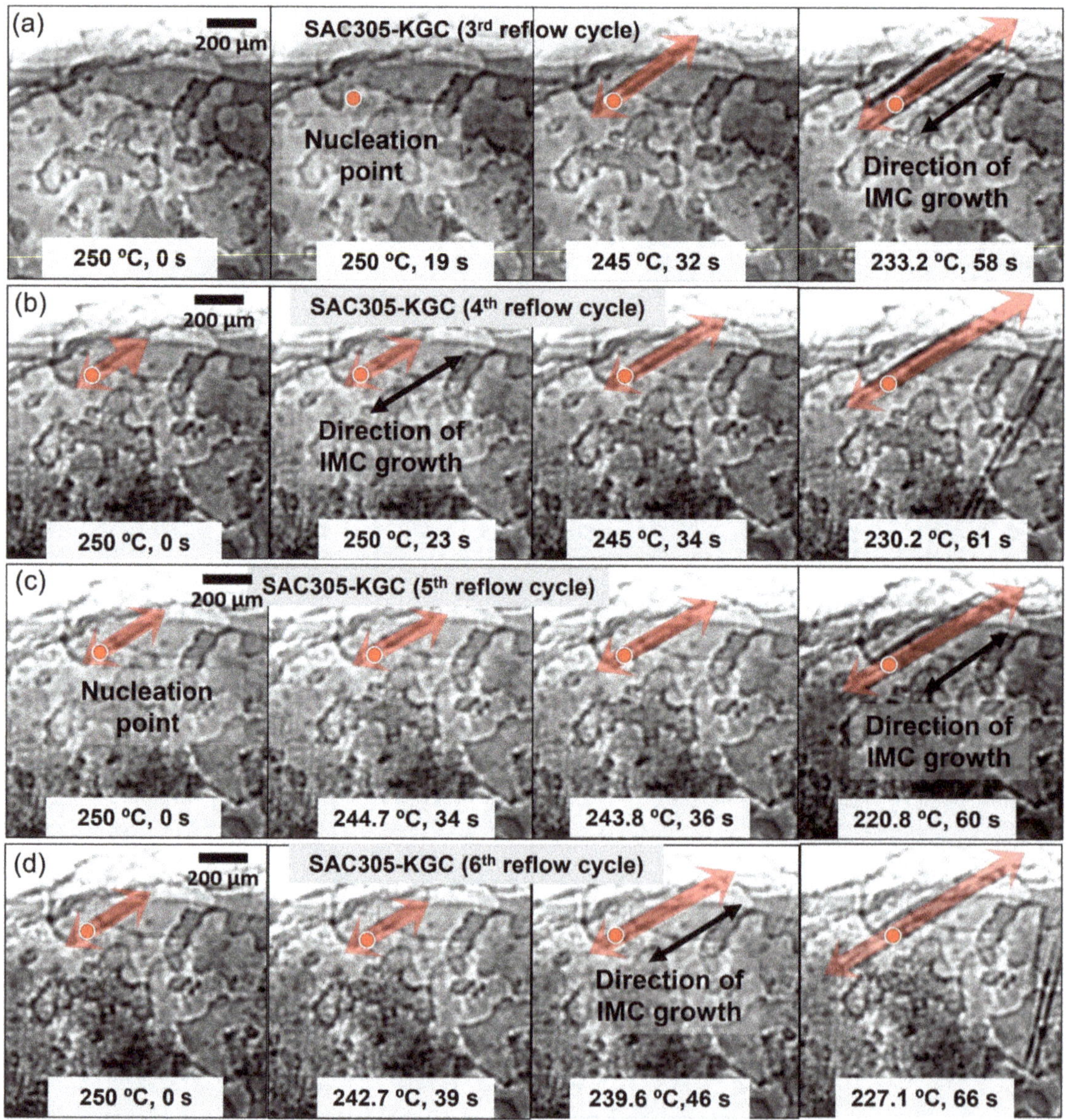

Figure 9. Sequences of the image showing the growth of IMC in SAC305-KGC solder joints during (**a**) 3rd reflow cycle, (**b**) 4th reflow cycle, (**c**) 5th reflow cycle, and (**d**) 6th reflow cycle (t = 0 s when the sample begins to cool).

The deep etching technique removed tin and partially exposed primary Cu_6Sn_5 IMC. Figure 10 shows the top-down images for both the SAC305 and SAC305-KGC solder joints after the sixth cycle of reflow soldering. It can be seen that both solder joints consist of hexagonal rod and "in-plane" branched type of primary IMC, which aligned with the synchrotron radiation images, per Figures 6 and 7. EDX analysis was performed on the primary Cu_6Sn_5 IMC in SAC305 and SAC305-KGC solder joints, and it was determined that the small agglomerations at the edge of primary Cu_6Sn_5 in SAC305-KGC are kaolin geopolymer ceramic (KGC) particles, as confirmed by EDX analysis results shown in Figure 10d. Al, Si, K, Mn, Fe, K, and Zr originated from the KGC systems. Based on this observation, it is possible that the KGC particles can be in contact with primary Cu_6Sn_5 IMC during multiple reflow soldering. It is hypothesized that KGC particles at the primary Cu_6Sn_5 IMC can explain the earlier nucleation in the SAC305-KGC solder joints. It can also be hypothesized that the presence of KGC particles can disturb the fully dissolved primary Cu_6Sn_5 IMC during the subsequent cycles of reflow soldering.

The growth behavior of primary Cu_6Sn_5 IMC at SAC305 and SAC305-KGC was quantified and shown in Figure 11. The final solidified primary Cu_6Sn_5 IMCs' length in SAC305 and SAC305-KGC is shown in Figure 11. The length of the primary Cu_6Sn_5 was measured from its first nucleation until it was completely solidified. Figure 11b shows the growth rate of primary Cu_6Sn_5 for SAC305 and SAC305-KGC solder joints. Based on the graph in Figure 11a, the final solidified length of primary Cu_6Sn_5 in SAC305-KGC is relatively smaller than SAC305. The primary Cu_6Sn_5 in SAC305-KGC grew to a maximum average length of ~602 µm. In the case of SAC305, the primary Cu_6Sn_5 could grow to a maximum average length of ~ 654 µm. The differences in the maximum value of the final average solidified length of primary Cu_6Sn_5 in SAC305-KGC and SAC305 were ~8%. The long primary Cu_6Sn_5 in the solder joints compromised its reliability, as discussed in [36,37]. The differences in the size of primary IMCs can also be linked to the indexed orientations of the crystal structure [36]. As mentioned previously, the fewer growth orientations for IMCs, the easier for the IMCs to grow and increase in size due to their lower energy consumption. Similar to the case of the SAC305-KGC, the addition of KGC can slightly change the growth orientations of the primary Cu_6Sn_5, resulting in much smaller-sized primary IMCs. Additionally, in the case of the SAC305-KGC, the primary Cu_6Sn_5 IMC grew at the maximum average growth rate of ~38 µm/s, while in the case of SAC305, the maximum average growth rate of the primary Cu_6Sn_5 was ~49 µm/s. This led to the conclusion that the primary Cu_6Sn_5 IMC in SAC305-KGC grew to a shorter length at a slower rate. A key finding in this work is that the solidified length of primary Cu_6Sn_5 IMC in SAC305-KGC was relatively smaller, forming at a slower growth rate whilst experiencing earlier nucleation during multiple reflow soldering compared to SAC305. This can be attributed to the addition of KGC particles suppressing the growth of primary intermetallic. Salleh et al. [1] reported that the reinforcement particles in Sn-0.7Cu solder decreased the number density and total length per unit area of the primary Cu_6Sn_5 during multiple reflow soldering.

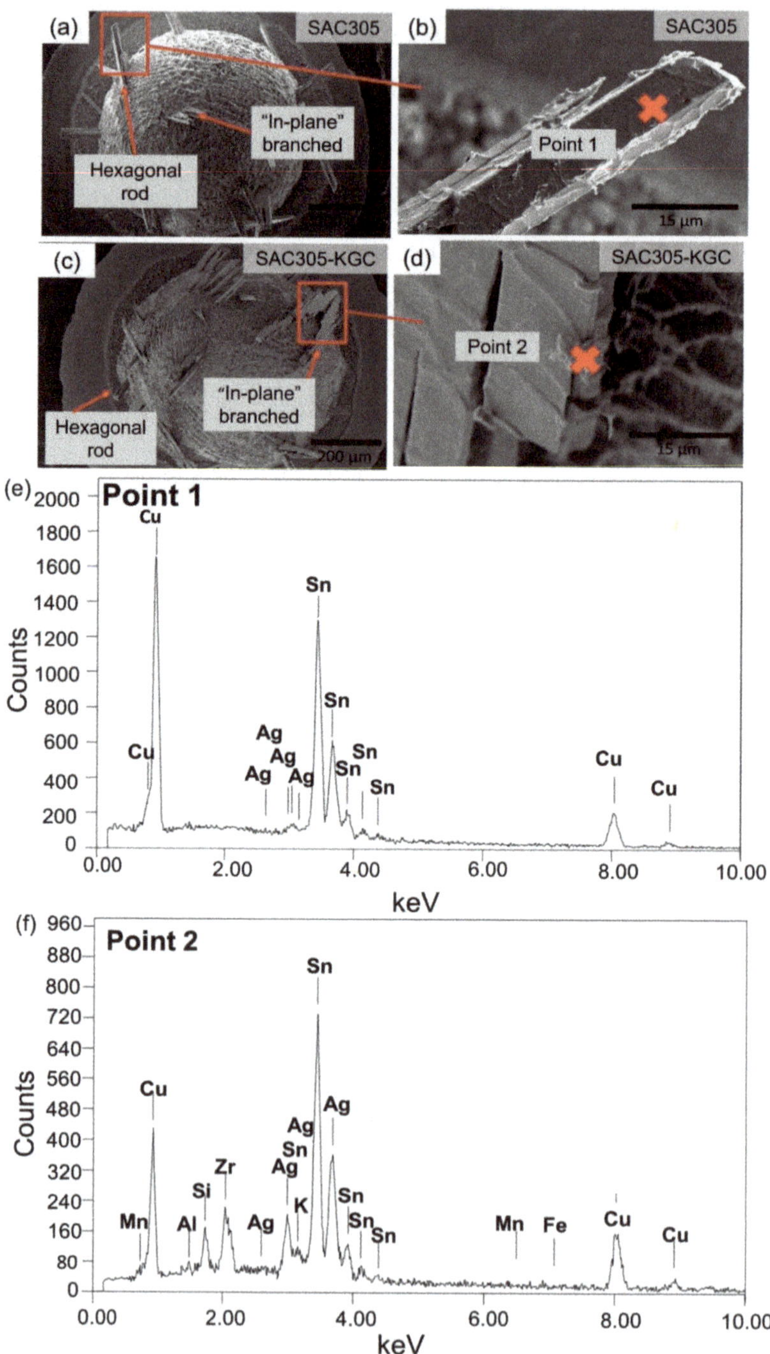

Figure 10. Top-down view of primary IMC in the bulk solder (**a**) Low magnification of SAC305, (**b**) Low magnification of SAC305-KGC, (**c**) High magnification of SAC305, (**d**) High magnification of SAC305-KGC. EDX point analysis at primary IMC in (**e**) SAC305, and (**f**) SAC305-KGC.

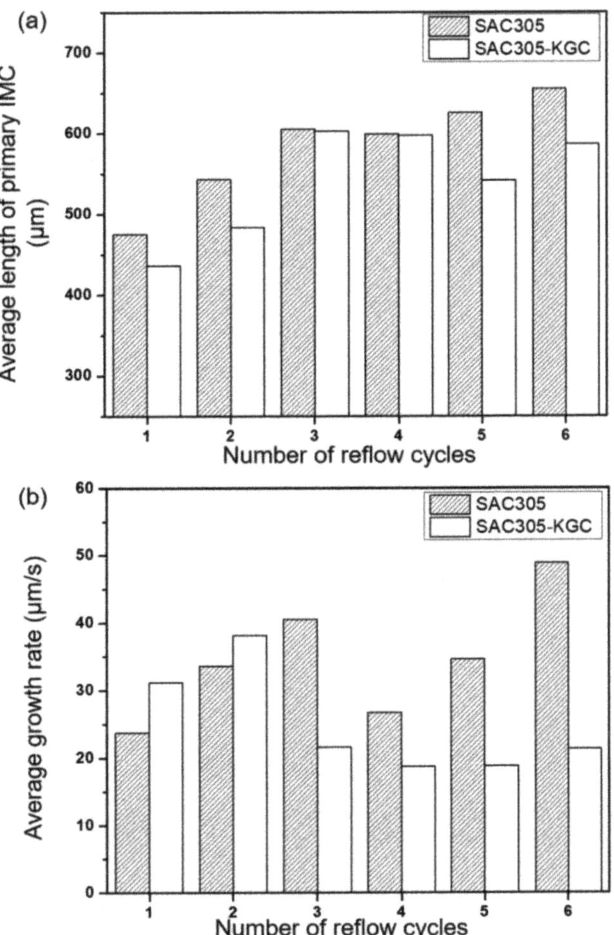

Figure 11. (**a**) Average length of the primary IMC in SAC305 and SAC305-KGC, and (**b**) Average growth rate of primary IMC in SAC305 and SAC305-KGC.

The results indicated that the suppression of intermetallic both as primary crystals and interfacial intermetallic layer is evident in materials of SAC305-KGC. As suggested by Gu et al. [24,38], the suppression of IMC was attributed to the ability of reinforcement particles to act as surface-active materials adsorbed onto solid surfaces. The reinforcement particles in the composite solder can be surface-active materials due to their high surface tension [38]. It is known that the smaller the size of the particles, the larger the surface tension and the specific surface area is. According to adsorption theory, the surface energy of Cu_6Sn_5 can be expressed as follows:

$$\sum_N \gamma_C^N S_N = \sum_N \gamma_O^N S_N - RT \sum_N S_N \int_O^C \frac{\Pi^N}{c}\, dc \rightarrow \min \tag{3}$$

where c is the concentration of KGC particles, γ^N is the surface tension of Cu_6Sn_5 particle N, γ_O^N is the surface tension of Cu_6Sn_5 particle without adsorption of KGC, γ_C^N is the surface tension of Cu_6Sn_5 particle with adsorption of KGC, S_N is the area of Cu_6Sn_5 particle N, Π^N is the number of KGC particles adsorbed by Cu_6Sn_5 particle N, R is gas constant, and T is the absolute temperature. From Equation (3), $\sum_N \gamma_O^N S_N$ is constant since it is not dependent on the concentration of KGC. Thus, the surface energy of Cu_6Sn_5 can be expressed as:

$$\sum_N \gamma_C^N S_N = RT \sum_N S_N \int_O^C \frac{\Pi^N}{c} \, dc \to \max \tag{4}$$

Based on the relationship in Equation (4), it can be inferred that with an increasing amount of adsorbed KGC, the surface energy of Cu_6Sn_5 decreases. As indicated by the Gibs free energy, the decrease in the surface energy in Cu_6Sn_5 decreases the growth velocities of Cu_6Sn_5 and the growth rate for each of Cu_6Sn_5.

The proposed mechanisms are shown in Figure 12 for the effects of KGC particles, described as follows; during the reflow soldering, the KGC particles were likely to segregate into the molten solder [32]. As a result, some KGC particles were adsorbed on the primary Cu_6Sn_5 IMC and onto the copper substrate. Increasing the reflow cycle resulted in more KGC particles adsorbed onto the primary Cu_6Sn_5 IMC, which causes the growth rate of the primary Cu_6Sn_5 IMC to decrease from the third until the sixth cycle of reflow soldering, as depicted in Figure 11b, compared to SAC305. The growth orientations are likely to change with adsorbed KGC since primary Cu_6Sn_5 IMC exhibited two growth directions compared to SAC305, with one growth direction along with <0001> during multiple reflow soldering. Additionally, the adsorbed KGC is likely to disturb the melting of the primary Cu_6Sn_5 IMC during the subsequent heating cycle from the third to the sixth cycle of reflow soldering.

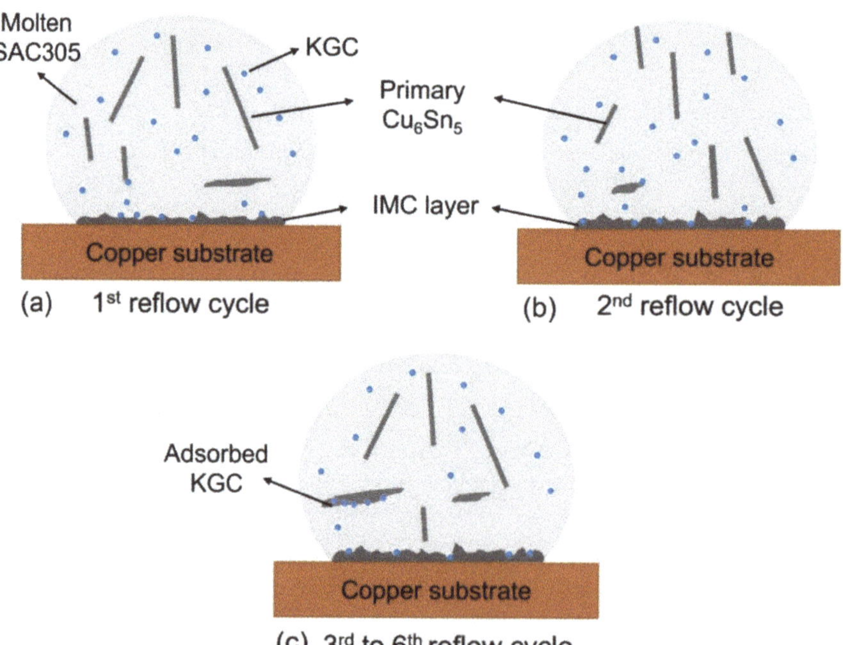

Figure 12. Propose mechanism on adsorption of KGC particles on the surface of Cu_6Sn_5 during multiple reflow soldering.

3.2. Shear Strength of Solder Joints after Multiple Reflows

The mechanical performance of the solder joints was determined using a single lap shear test. Figure 13a shows the plot of the average shear strength for SAC305 and SAC305-KGC subjected to multiple cycles of reflow soldering. Overall, the average shear strength in SAC305 and SAC305-KGC decreased with increasing cycles of reflow soldering. However, the average shear strength in SAC305-KGC solder joints is higher than SAC305 regardless of the reflow soldering cycle. In the SAC305-KGC solder joints, the average shear strength showed ~13% reduction after the sixth cycle of reflow soldering compared with SAC305, which exhibited a 27% reduction after the sixth cycle of reflow soldering. A plausible explanation for the decrease in the average shear strength of the SAC305 is due to the formation of coarser microstructure and thicker interfacial IMC layer during the multiple cycles of reflow soldering. Meanwhile, the SAC305-KGC solder joints exhibited a lower reduction in the average shear strength after the sixth cycle of reflow soldering. This can be explained by the existence of KGC particles on both solder matrix and interfacial, as discussed in previous sections. The abovementioned results showed that the controllable coarsening in the microstructure and thinner interfacial IMC layer in SAC305-KGC benefitted the strength of solder during multiple cycles of reflow soldering. In addition, the relatively finer distribution of the IMCs in the solder bulk strengthened the solder matrix via dispersion strengthening [24]. Additionally, the smaller size of the primary IMCs in SAC305-KGC during multiple reflow soldering contributes to the strength of the solder joints.

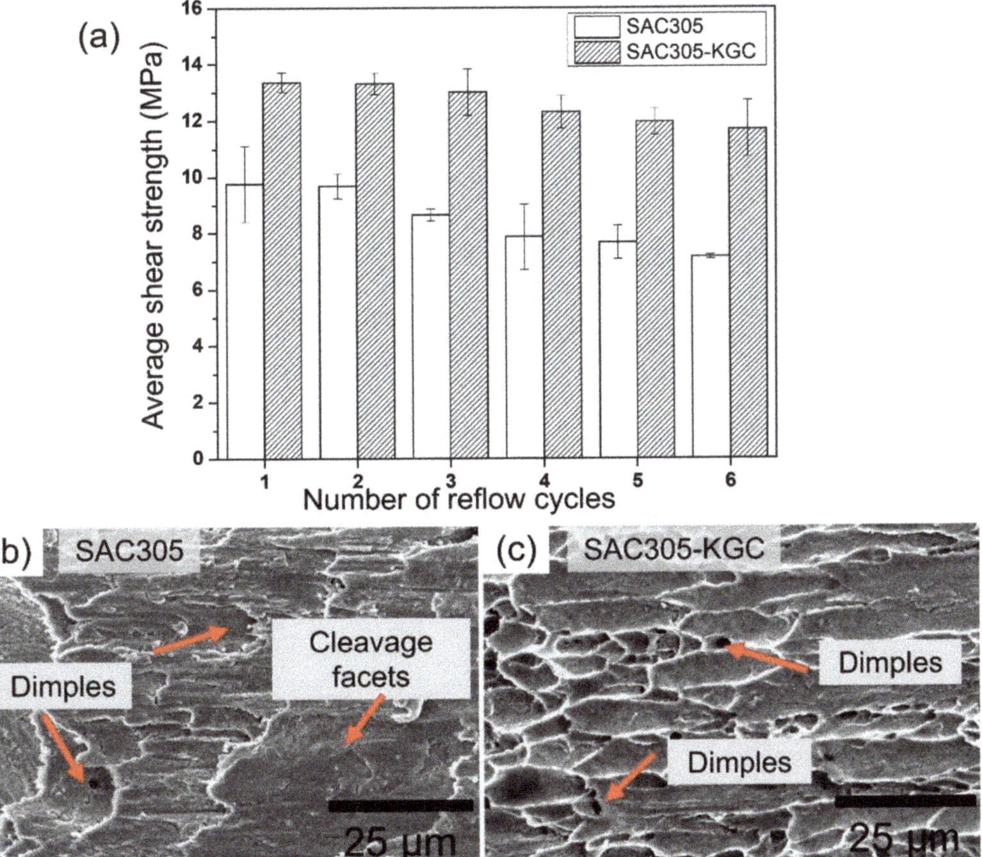

Figure 13. *Cont.*

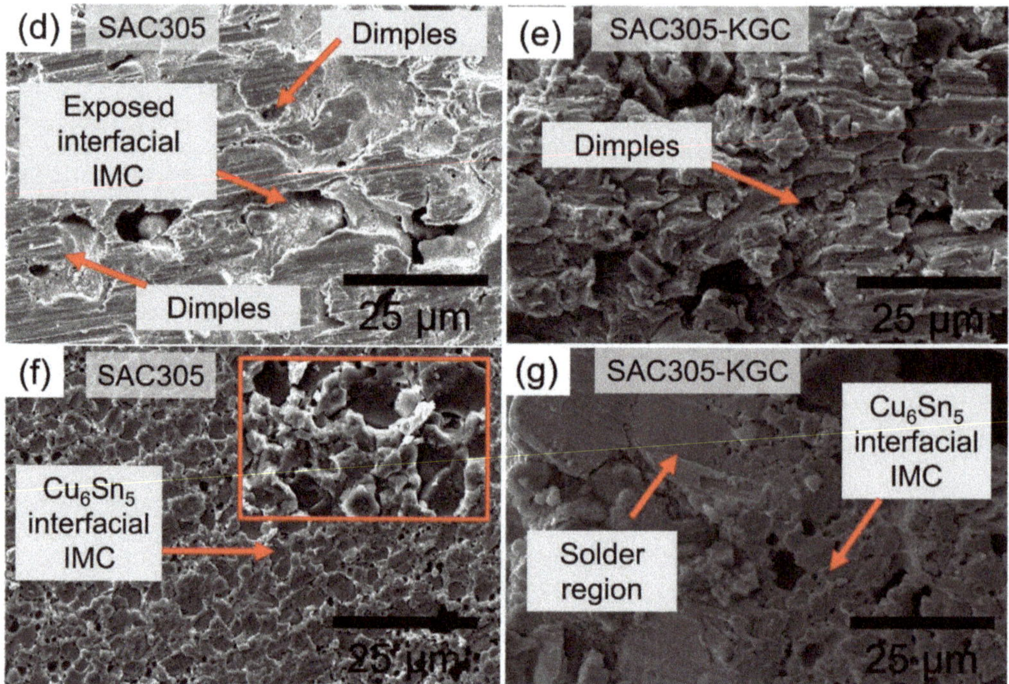

Figure 13. (**a**) Shear strength of SAC305 and SAC305-KGC subjected to multiple cycles of reflow soldering. SEM fracture surface of (**b**) SAC305 at 1st reflow, (**c**) SAC305-KGC at 1st reflow, (**d**) SAC305 at 3rd reflow, (**e**) SAC305-KGC at 3rd reflow, (**f**) SAC305 at 6th reflow, and (**g**) SAC305-KGC at 6th reflow cycle.

A comprehensive analysis of the failure mechanism in SAC305 and SAC305-KGC involved using a scanning electron microscope (SEM) to elucidate the failure modes during multiple reflow soldering. Figure 13b–g shows the fractography for SAC305 and SAC305-KGC solder joints during the firth, third, and sixth reflow cycles. During the first reflow cycle, SAC305 solder joints failed in the combination of brittle and ductile failure mode. The appearance of shallow shear dimples was corresponding to the ductile region as in Figure 13b. The cleavage fracture area indicates less energy was absorbed during the shear test, which corresponds to the brittle region [43]. Meanwhile, SAC305-KGC solder joints show ductile fracture mode during the first cycle of reflow soldering with the appearance of shear dimples as in Figure 13c. Then, after the sixth cycle of reflow soldering, SAC305 solder joints showed a prominent structure of Cu_6Sn_5 IMC, suggesting that the failure occurred along with the IMC in a brittle manner after the shearing indicated in Figure 13f. On the other hand, in SAC305-KGC solder joints, a combined fractured mode (brittle and ductile) was observed after the sixth cycle of reflow soldering, per Figure 13g.

4. Conclusions

The effects of the addition of kaolin geopolymer ceramic in SAC305 solder joints were elucidated via the microstructural analyses at the bulk solder and the interfacial layer. The addition of KGC in the SAC305 solder suppressed the growth of the IMC both at the primary and interfacial layers and improved the shear strength of the solder. It can therefore be concluded that:

1. It was observed that the KGC particles remained in contact with the grains of the interfacial Cu_6Sn_5 IMC during multiple reflow soldering, which decreased the maximum

average thickness of the IMC layer from ~12.6 μm (SAC305) to ~9.4 μm (SAC305-KGC). The scalloped interfacial IMC layer in SAC305-KGC became shorter and faceted after the sixth cycle of reflow. However, in SAC305, the elongated scalloped interfacial IMC layer grew longer into the solder's matrix. The growth rate constant calculated for SAC305-KGC was 0.37 $\mu m^2/s$, compared to SAC305, which is 0.81 $\mu m^2/s$.

2. During the in situ microstructure analysis, the primary Cu_6Sn_5 IMC in SAC305-KGC nucleated earlier at higher temperatures during the cooling stage. As a result, the maximum average growth rate achieved in the SAC305-KGC was 38 μm/s compared to SAC305, which is 49 μm/s. The lower growth rate resulted in shorter lengths of solidified primary Cu_6Sn_5 IMCs in SAC305-KGC.

3. It was also observed that after the third cycle, the primary Cu_6Sn_5 IMCs in SAC305-KGC did not fully melt during subsequent heating of the fourth, fifth, and sixth cycle of reflow soldering and grew with two growth directions, which differs from SAC305 where the primary only grows with one growth direction. The results obtained were likely related to the mechanism of adsorption of KGC particles on the surface of primary Cu_6Sn_5 IMCs during multiple reflow soldering.

4. The suppression of Cu_6Sn_5 IMC both as primary and interfacial layers in SAC305-KGC resulted in a reduction of ~13% of average shear strength after multiple reflow soldering. However, in SAC305, the average shear strength decreased by ~27% after multiple reflows soldering and experiencing the brittle fracture mode.

Author Contributions: Conceptualization, methodology and writing—original draft preparation, N.S.M.Z.; supervision and validation, M.A.A.M.S.; supervision, M.M.A.-B.A., data curation, N.I.M.N., K.N., H.Y.; investigation, M.I.I.R.; reviewing, validation and visualization, A.V.S., P.V., I.G.S. All authors have read and agreed to the published version of the manuscript.

Funding: This work was supported by ISIS Neutron and Muon Source Project under reference No: JPT.S (BPKI)2000/016/018/019(29) and COP26 Trilateral Research Initiative. The in-situ synchrotron X-ray radiography imaging was conducted using beamline BL20XU at SPring-8 synchrotron in Hyogo, Japan under the proposal number of 2019B1618 and supported by JSPS Grant-in-Aid for Scientific Research (S) (17H06155). Also this work was supported by Gheorghe Asachi Technical University of Iasi TUIASI Internal Grants Program (GI_Publications/2021), financed by the Romanian Government.

Institutional Review Board Statement: Not applicable.

Informed Consent Statement: Not applicable.

Data Availability Statement: Data is contained within the article.

Conflicts of Interest: The authors declare no conflict of interest.

References

1. Mohd Salleh, M.A.A.; McDonald, S.D.; Gourlay, C.M.; Yasuda, H.; Nogita, K. Suppression of Cu_6Sn_5 in TiO_2 reinforced solder joints after multiple reflow cycles. *Mater. Des.* **2016**, *108*, 418–428. [CrossRef]

2. Wang, H.; Hu, X.; Jiang, X.; Li, Y. Interfacial reaction and shear strength of ultrasonically-assisted Sn-Ag-Cu solder joint using composite flux. *J. Manuf. Processes* **2021**, *62*, 291–301. [CrossRef]

3. Zhang, Z.; Hu, X.; Jiang, X.; Li, Y. Influences of Mono-Ni(P) and Dual-Cu/Ni(P) Plating on the Interfacial Microstructure Evolution of Solder Joints. *Metall. Mater. Trans. A* **2019**, *50*, 480–492. [CrossRef]

4. Gao, H.; Wei, F.; Lin, C.; Shu, T.; Sui, Y.; Qi, J.; Zhang, X. Growth behavior of intermetallic compounds on Sn-10Bi-0.7Cu-0.15Co/Co interface under multiple reflows. *Mater. Lett.* **2019**, *252*, 92–95. [CrossRef]

5. Fix, A.; López, G.; Brauer, I.; Nüchter, W.; Mittemeijer, E. Microstructural development of Sn-Ag-Cu solder joints. *J. Electron. Mater.* **2005**, *34*, 137–142. [CrossRef]

6. Ma, H.R.; Kunwar, A.; Shang, S.Y.; Jiang, C.R.; Wang, Y.P.; Ma, H.T.; Zhao, N. Evolution behavior and growth kinetics of intermetallic compounds at Sn/Cu interface during multiple reflows. *Intermetallics* **2018**, *96*, 1–12. [CrossRef]

7. Chen, H.; Tsai, Y.-L.; Chang, Y.-T.; Wu, A.T. Effect of massive spalling on mechanical strength of solder joints in Pb-free solder reflowed on Co-based surface finishes. *J. Alloy. Compd.* **2016**, *671*, 100–108. [CrossRef]

8. Haseeb, A.S.M.A.; Arafat, M.M.; Johan, M.R. Stability of molybdenum nanoparticles in Sn–3.8Ag–0.7Cu solder during multiple reflow and their influence on interfacial intermetallic compounds. *Mater. Charact.* **2012**, *64*, 27–35. [CrossRef]

9. Tang, Y.; Luo, S.M.; Wang, K.Q.; Li, G.Y. Effect of Nano-TiO$_2$ particles on growth of interfacial Cu$_6$Sn$_5$ and Cu$_3$Sn layers in Sn–3.0Ag–0.5Cu–xTiO$_2$ solder joints. *J. Alloy. Compd.* **2016**, *684*. [CrossRef]

10. Ha, S.-S.; Jang, J.-K.; Ha, S.-O.; Yoon, J.-W.; Lee, H.-J.; Joo, J.-H.; Kim, Y.-H.; Jung, S.-B. Effect of multiple reflows on interfacial reaction and shear strength of Sn–Ag electroplated solder bumps for flip chip package. *Microelectron. Eng.* **2010**, *87*, 517–521. [CrossRef]

11. Wang, T.; Zhou, P.; Cao, F.; Kang, H.; Chen, Z.; Fu, Y.; Xiao, T.; Huang, W.; Yuan, Q. Growth behavior of Cu$_6$Sn$_5$ in Sn–6.5 Cu solders under DC considering trace Al: In situ observation. *Intermetallics* **2015**, *58*, 84–90. [CrossRef]

12. Pal, M.K.; Gergely, G.; Koncz-Horváth, D.; Gácsi, Z. Investigation of microstructure and wetting behavior of Sn–3.0Ag–0.5Cu (SAC305) lead-free solder with additions of 1.0 wt % SiC on copper substrate. *Intermetallics* **2021**, *128*, 106991. [CrossRef]

13. Vianco, P. Understanding the reliability of solder joints used in advanced structural and electronics applications: Part 2—Reliability performance: Factors that directly affect solder joint reliability, including fatigue and growth of intermetallic compound reaction layers, were studied. *Weld. J.* **2017**, *96*, 83s–94s.

14. Qiao, Y.; Ma, H.; Yu, F.; Zhao, N. Quasi-in-situ observation on diffusion anisotropy dominated asymmetrical growth of Cu-Sn IMCs under temperature gradient. *Acta Mater.* **2021**, *217*, 117168. [CrossRef]

15. Wang, H.; Hu, X.; Jiang, X. Effects of Ni modified MWCNTs on the microstructural evolution and shear strength of Sn-3.0Ag-0.5Cu composite solder joints. *Mater. Charact.* **2020**, *163*, 110287. [CrossRef]

16. Hu, X.; Xu, H.; Chen, W.; Jiang, X. Effects of ultrasonic treatment on mechanical properties and microstructure evolution of the Cu/SAC305 solder joints. *J. Manuf. Processes* **2021**, *64*, 648–654. [CrossRef]

17. Yahaya, M.Z.; Nazeri, M.F.M.; Kheawhom, S.; Illés, B.; Skwarek, A.; Mohamad, A.A. Microstructural analysis of Sn-3.0Ag-0.5Cu-TiO$_2$ composite solder alloy after selective electrochemical etching. *Mater. Res. Express* **2020**, *7*, 016583. [CrossRef]

18. Skwarek, A.; Ptak, P.; Górecki, K.; Hurtony, T.; Illés, B. Microstructure Influence of SACX0307-TiO$_2$ Composite Solder Joints on Thermal Properties of Power LED Assemblies. *Materials* **2020**, *13*, 1563. [CrossRef]

19. Wu, J.; Xue, S.; Wang, J.; Wu, M. Coupling effects of rare-earth Pr and Al$_2$O$_3$ nanoparticles on the microstructure and properties of Sn-0.3Ag-0.7Cu low-Ag solder. *J. Alloy. Compd.* **2019**, *784*, 471–487. [CrossRef]

20. Qu, M.; Cao, T.; Cui, Y.; Liu, F.; Jiao, Z. Effect of nano-ZnO particles on wettability, interfacial morphology and growth kinetics of Sn–3.0Ag–0.5Cu–xZnO composite solder. *J. Mater. Sci. Mater. Electron.* **2019**, *30*, 19214–19226. [CrossRef]

21. Mohd Nasir, S.S.; Yahaya, M.Z.; Erer, A.M.; Illés, B.; Mohamad, A.A. Effect of TiO$_2$ nanoparticles on the horizontal hardness properties of Sn-3.0Ag-0.5Cu-1.0TiO$_2$ composite solder. *Ceram. Int.* **2019**, *45*, 18563–18571. [CrossRef]

22. Chen, G.; Peng, H.; Silberschmidt, V.V.; Chan, Y.C.; Liu, C.; Wu, F. Performance of Sn–3.0Ag–0.5Cu composite solder with TiC reinforcement: Physical properties, solderability and microstructural evolution under isothermal ageing. *J. Alloy. Compd.* **2016**, *685*, 680–689. [CrossRef]

23. Gain, A.K.; Zhang, L. Nanosized samarium oxide (Sm$_2$O$_3$) particles suppressed the IMC phases and enhanced the shear strength of environmental-friendly Sn–Ag–Cu material. *Mater. Res. Express* **2019**, *6*, 066526. [CrossRef]

24. Tikale, S.; Narayan Prabhu, K. Effect of Multiple Reflow Cycles and Al$_2$O$_3$ Nanoparticles Reinforcement on Performance of SAC305 Lead-Free Solder Alloy. *J. Mater. Eng. Perform.* **2018**, *27*, 3102–3111. [CrossRef]

25. Li, Z.H.; Tang, Y.; Guo, Q.W.; Li, G.Y. Effects of CeO$_2$ nanoparticles addition on shear properties of low-silver Sn–0.3Ag–0.7Cu–xCeO$_2$ solder alloys. *J. Alloy. Compd.* **2019**, *789*, 150–162. [CrossRef]

26. Li, Z.H.; Tang, Y.; Guo, Q.W.; Li, G.Y. A diffusion model and growth kinetics of interfacial intermetallic compounds in Sn-0.3Ag-0.7Cu and Sn-0.3Ag-0.7Cu-0.5CeO$_2$ solder joints. *J. Alloy. Compd.* **2019**. [CrossRef]

27. Li, M.-l.; Zhang, L.; Jiang, N.; Zhang, L.; Zhong, S.-J. Materials modification of the lead-free solders incorporated with micro/nano-sized particles: A review. *Mater. Des.* **2021**, *197*, 109224. [CrossRef]

28. Zhang, P.; Xue, S.; Wang, J.; Xue, P.; Zhong, S.; Long, W. Effect of Nanoparticles Addition on the Microstructure and Properties of Lead-Free Solders: A Review. *Appl. Sci.* **2019**, *9*, 2044. [CrossRef]

29. Tang, Y.; Luo, S.; Li, G.; Yang, Z.; Hou, C. Ripening Growth Kinetics of Cu$_6$Sn$_5$ Grains in Sn-3.0Ag-0.5Cu-xTiO$_2$/Cu Solder Joints during the Reflow Process. *J. Electron. Packag.* **2017**, *140*. [CrossRef]

30. Zaimi, N.S.M.; Salleh, M.A.A.M.; Sandu, A.V.; Abdullah, M.M.A.B.; Saud, N.; Rahim, S.Z.A.; Vizureanu, P.; Said, R.M.; Ramli, M.I.I. Performance of Sn-3.0Ag-0.5Cu Composite Solder with Kaolin Geopolymer Ceramic Reinforcement on Microstructure and Mechanical Properties under Isothermal Ageing. *Materials* **2021**, *14*, 776. [CrossRef]

31. Liew, Y.M.; Heah, C.Y.; Li, L.; Jaya, N.A.; Abdullah, M.M.A.; Tan, S.J.; Hussin, K. Formation of one-part-mixing geopolymers and geopolymer ceramics from geopolymer powder. *Constr. Build. Mater.* **2017**, *156*, 9–18. [CrossRef]

32. Mohamad Zaimi, N.S.; Mohd Salleh, M.A.A.; Abdullah, M.M.A.B.; Ahmad, R.; Mostapha, M.; Yoriya, S.; Chaiprapa, J.; Zhang, G.; Harvey, D.M. Effect of kaolin geopolymer ceramic addition on the properties of Sn-3.0Ag-0.5Cu solder joint. *Mater. Today Commun.* **2020**, *25*, 101469. [CrossRef]

33. Muhd Amli, S.F.N.; Mohd Salleh, M.A.A.; Ramli, M.I.I.; Yasuda, H.; Chaiprapa, J.; Somidin, F.; Shayfull, Z.; Nogita, K. Origin of Primary Cu$_6$Sn$_5$ in Hypoeutectic Solder Alloys and a Method of Suppression to Improve Mechanical Properties. *J. Electron. Mater.* **2021**, *50*, 710–722. [CrossRef]

34. Hsiao, H.Y.; Hu, C.C.; Guo, M.Y.; Chen, C.; Tu, K.N. Inhibiting the consumption of Cu during multiple reflows of Pb-free solder on Cu. *Scr. Mater.* **2011**, *65*, 907–910. [CrossRef]

35. McDonald, S.; Nogita, K.; Read, J.; Ventura, T.; Nishimura, T. Influence of Composition on the Morphology of Primary Cu_6Sn_5 in Sn-4Cu Alloys. *J. Electron. Mater.* **2012**, *42*. [CrossRef]
36. Ma, H.T.; Qu, L.; Huang, M.L.; Gu, L.Y.; Zhao, N.; Wang, L. In-situ study on growth behavior of Ag_3Sn in Sn–3.5Ag/Cu soldering reaction by synchrotron radiation real-time imaging technology. *J. Alloy. Compd.* **2012**, *537*, 286–290. [CrossRef]
37. Mohd Salleh, M.A.A.; Gourlay, C.M.; Xian, J.W.; Belyakov, S.A.; Yasuda, H.; McDonald, S.D.; Nogita, K. In situ imaging of microstructure formation in electronic interconnections. *Sci. Rep.* **2017**, *7*, 40010. [CrossRef]
38. Gu, Y.; Zhao, X.; Li, Y.; Liu, Y.; Wang, Y.; Li, Z. Effect of nano-Fe_2O_3 additions on wettability and interfacial intermetallic growth of low-Ag content Sn–Ag–Cu solders on Cu substrates. *J. Alloy. Compd.* **2015**, *627*, 39–47. [CrossRef]
39. Ma, H.; Ma, H.; Kunwar, A.; Shang, S.; Wang, Y.; Chen, J.; Huang, M.; Zhao, N. Effect of initial Cu concentration on the IMC size and grain aspect ratio in Sn–xCu solders during multiple reflows. *J. Mater. Sci. Mater. Electron.* **2018**, *29*, 602–613. [CrossRef]
40. Liu, P.; Yao, P.; Liu, J. Effects of multiple reflows on interfacial reaction and shear strength of SnAgCu and SnPb solder joints with different PCB surface finishes. *J. Alloy. Compd.* **2009**, *470*, 188–194. [CrossRef]
41. Ramli, M.I.I.; Mohd Salleh, M.A.A.; Yasuda, H.; Chaiprapa, J.; Nogita, K. The effect of Bi on the microstructure, electrical, wettability and mechanical properties of Sn-0.7Cu-0.05Ni alloys for high strength soldering. *Mater. Des.* **2019**, *181*, 108281. [CrossRef]
42. Mohd Said, R.; Mohd Salleh, M.A.A.; Saud, N.; Ramli, M.I.I.; Yasuda, H.; Nogita, K. Microstructure and growth kinetic study in Sn–Cu transient liquid phase sintering solder paste. *J. Mater. Sci. Mater. Electron.* **2020**, *31*, 11077–11094. [CrossRef]
43. Barkia, B.; Courouau, J.L.; Perrin, E.; Lorentz, V.; Rivollier, M.; Robin, R.; Nicolas, L.; Cabet, C.; Auger, T. Investigation of crack propagation resistance of 304L, 316L and 316L(N) austenitic steels in liquid sodium. *J. Nucl. Mater.* **2018**, *507*, 15–23. [CrossRef]

materials

Article

Development of Ag–In Alloy Pastes by Mechanical Alloying for Die Attachment of High-Power Semiconductor Devices

Chin-Hao Tsai [1], Wei-Chen Huang [1] and Chengheng Robert Kao[1,2,*]

[1] Department of Materials Science and Engineering, National Taiwan University, Taipei 10617, Taiwan; f07527060@ntu.edu.tw (C.-H.T.); f08527014@ntu.edu.tw (W.-C.H.)
[2] Advanced Research Center for Green Materials Science & Technology, National Taiwan University, Taipei 10617, Taiwan
* Correspondence: crkao@ntu.edu.tw

Abstract: Sintered silver paste is widely used as the die-attachment material for power semiconductors. However, sintered silver joints encounter problems, such as severe coarsening of sintered pores and oxidation issues, in harsh high-temperature environments. These lead to the deterioration of the die-attachment joints. In this paper, a novel method of sintering silver joints is demonstrated, where silver–indium alloy paste is used to improve the reliability of sintered Ag joints. The silver–indium (Ag–In) alloy paste was fabricated through mechanical alloying using the ball-milling technique. The well-bonded sintered Ag–In alloy joints inhibited pore coarsening better than pure sintered Ag joints and significantly enhanced the mechanical properties at high operating temperatures. Lastly, an oxidation mechanism for the sintered joint was proposed, and strategies to prevent such high-temperature oxidation were discussed.

Keywords: Ag–In alloy pastes; mechanical alloying; power semiconductor packaging; die attachment; mechanical properties; oxidation mechanism

Citation: Tsai, C.-H.; Huang, W.-C.; Kao, C.R. Development of Ag–In Alloy Pastes by Mechanical Alloying for Die Attachment of High-Power Semiconductor Devices. *Materials* **2022**, *15*, 1397. https://doi.org/10.3390/ma15041397

Academic Editor: Kuo-Ning Chiang

Received: 20 December 2021
Accepted: 11 February 2022
Published: 14 February 2022

Publisher's Note: MDPI stays neutral with regard to jurisdictional claims in published maps and institutional affiliations.

1. Introduction

In the age of pursuing energy savings and reducing carbon emissions, the popularization of electric vehicles (EVs) has become the primary objective of next-generation transportation technology because the use of EVs can significantly reduce carbon emissions, which will diminish the greenhouse effect and global warming. Consequently, many countries advocate that internal combustion engine vehicles should be phased out and replaced by EVs or hybrid EVs within the next decade [1]. Therefore, it is important to accelerate the development of EVs and enhance the energy conversion efficiency of inverters, the power module responsible for switching between alternating current and direct current power in EVs.

Inverters need to withstand a considerable amount of current in EVs, which leads to a substantial amount of heat generation and a harsh operating temperature of above 200 °C [2,3]. Given that Si-based semiconductors become conductive and can fail at such high temperatures, wide-bandgap (WBG) semiconductors are better suited for use in inverter power chips because of their superior properties compared to Si-based semiconductors [4–6]. Silicon carbide (SiC) and gallium nitride (GaN) are two promising WBG semiconductors for application in EV power modules because they exhibit high breakdown voltage, high switching frequency, low switching losses, and low power conversion losses while operating above 200 °C [7–9]. Therefore, WBG semiconductor-based power chips increase the horsepower of EVs and simultaneously extend the operating duration owing to their low power consumption.

For power chips to function well, packaging materials with high-temperature reliability are essential for the die attachment, and they should provide power chips with excellent mechanical support and heat dissipation. However, the conventional Pb-free Sn-based packaging materials used in consumer electronics cannot withstand the high temperatures generated during power semiconductor operation due to their intrinsically low melting points and creep resistance [10–12]. Therefore, developing reliable die-attachment materials for high-temperature applications is crucial for EV power modules. Ag paste sintering is a die-attachment packaging technology that has been widely used in power modules because sintered silver joints possess outstanding mechanical reliability and heat dissipation properties at high operating temperatures. Nevertheless, sintered Ag joints have challenging Cu substrate oxidation problems that need to be resolved. It was demonstrated that the Cu oxide layers formed at operating temperatures above 200 °C deteriorated the mechanical strength of sintered Ag joints [13–17]. Furthermore, coarsening of sintered pores in high-temperature environments also resulted in reliability issues. To address the aforementioned issues, Lee and coworkers pointed out that a joint comprising Ag–In intermetallic compounds (IMCs) exhibited excellent properties compared to pure Ag joints [18,19]. Moreover, our previous studies have proven that the addition of In to sintered Ag joints has positive effects on the sintered joint [20–25]. The sintered Ag–In joint produced via the transient liquid phase (TLP) bonding reaction between the Ag paste and In foil significantly improved the mechanical properties and reduced oxidation problems of the sintered joints at high temperatures. However, the In foils used for TLP bonding are too soft and fragile, leading to handling difficulties during the fabrication process. The additional step of adding In foil also increases the production costs. Hence, the fabrication process for the sintered Ag–In joint using In foil is not fully compatible with large-scale manufacturing. Furthermore, the sintered Ag–In joint produced via the TLP reaction needed to undergo a long-term phase transformation from brittle Ag–In IMCs to a ductile Ag–In solid solution to achieve the optimal mechanical properties and compositional homogenization [24,25]. Consequently, despite the excellent properties of sintered Ag–In joints, it is not practical to use In foil in industrial applications because it cannot be mass-produced, which restricts the application of sintered Ag–In joints.

Herein, a novel method of directly sintering the Ag–In alloy pastes using a homogeneous solid solution is proposed. Through the employment of the Ag–In alloy paste, the additional step of adding In foil can be eliminated, thereby simplifying the fabrication process and bringing sintered Ag–In joints closer to practical industrial application. Such Ag–In alloy pastes can retain the benefits of In foils but at much-reduced manufacturing cost. A proper powder production method needs to be developed for the fabrication of homogeneous Ag–In alloy pastes. Powder production methods such as gas and water atomization are common in industrial applications because they can mass-produce metal and alloy powders [26,27]. However, gas and water atomization are not suitable for the early stages of research and development in Ag–In alloy powders because of the large quantities of raw material required. Therefore, the focus should be on testing the feasibility and optimizing the composition of the Ag–In alloy paste. Hence, this study uses mechanical alloying by the ball-milling technique to produce small quantities of Ag–In alloy powders [28]. In this way, various Ag–In alloy powders can be made and compared to determine the optimal composition.

In this study, a novel Ag–In alloy paste was developed by directly alloying In with Ag powder through the ball-milling process and mixing with an organic solvent. The Ag–In alloy paste was utilized to fabricate sintered joints by a hot-pressing process. In addition, we compared the bonding and sintering results of the fabricated Ag–In alloy joint to the pure Ag joint. Furthermore, the high-temperature reliability of both sintered joints was evaluated using high-temperature storage (HTS) and die shear tests, after which the microstructure and phase were investigated. Moreover, the correlation between the mechanical properties of both sintered joints during HTS was established. Lastly, the oxidation mechanism in the sintered joint and strategies to prevent oxidation were discussed.

2. Experimental

2.1. Ball-Milling Process for the Fabrication of Ag and Ag–In Alloy Pastes

Figure 1 shows a schematic of the Ag and Ag–In alloy paste fabrication processes. A high-energy planetary ball-milling machine (Pulverisette 7 Premium Line, Fritsch, Idar-Oberstein, Germany) with two atmosphere-controlled milling jars (Fritsch, Idar-Oberstein, Germany) was employed for the production of the Ag and Ag–In alloy powders. Initially, 10 g of Ag powder with an average particle size of 150 μm, 75 g of zirconium dioxide (ZrO_2) grinding balls, and 0.2 g of stearic acid were mixed into one of the ball-milling jars, as illustrated in Figure 1a. In the other ball-milling jar, 8 g of Ag powder and 2 g of In powder with an average particle size of 150 μm, 75 g of ZrO_2 grinding balls, and 0.2 g of stearic acid were mixed, as illustrated in Figure 1b. ZrO_2 with a diameter of 5 mm was chosen for the grinding balls because ZrO_2 ceramic is less prone to metal powder contamination during ball-milling. Stearic acid was added to inhibit the cold welding and agglomeration of ductile Ag and In powders during the high-energy ball-milling process. Furthermore, the atmosphere in both ball-milling jars was replaced with high-purity argon gas to prevent oxidation of the Ag and In powders. The milling speed in both jars was 600 rpm for 10 h, after which the jars were opened in a nitrogen glove box (Younme Technology Company, Taoyuan, Taiwan) with an oxygen concentration of less than 0.1 ppm. Then, the ball-milled Ag and Ag–In alloy powders were passed through a 500-mesh sieve. After sieving, Ag and Ag–In alloy powders with particle sizes less than 25 μm were homogeneously mixed with a polymer solvent to complete the preparation of the pastes.

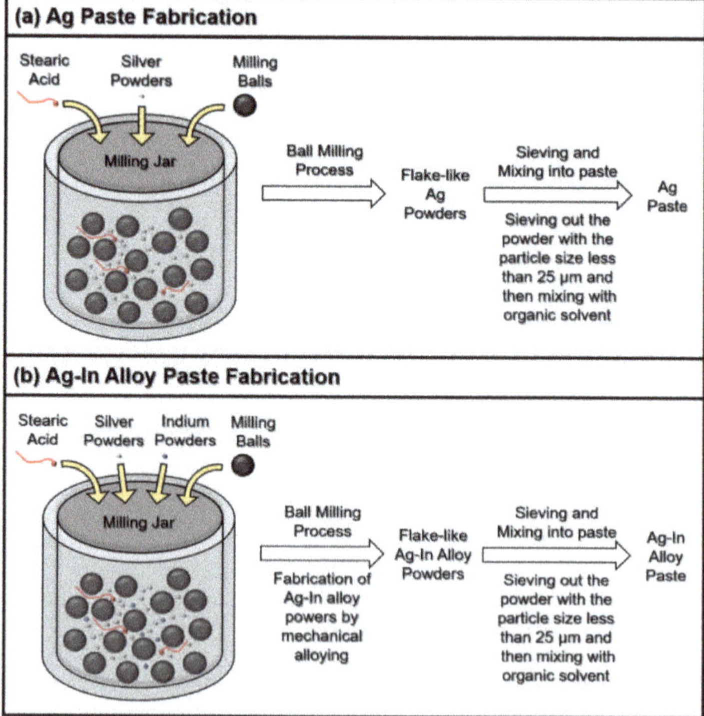

Figure 1. Schematic of the (**a**) Ag and (**b**) Ag–In alloy paste fabrication processes.

2.2. Experimental Bonding Procedure

Figure 2 shows a flowchart for the sample preparation process of the sintered Ag and Ag–In alloy joints. Cu (99.9% purity) substrates were chosen as top and bottom dies. The dimensions of top Cu substrates were 3 mm × 3 mm × 3 mm, and the dimensions of bottom Cu substrates were 10 mm × 10 mm × 3 mm (length × width × thickness). To remove the Cu oxide and create flat surfaces, the top and bottom Cu dies were metallographically polished using SiC abrasive sandpapers (#4000, Struers). Then, a 300 nm thick layer of Ag was sputtered onto the top and bottom Cu substrates as the surface finish layer, as shown in Figure 2a. During the stencil-printing and pre-drying processes shown in Figure 2b,c, respectively, the Ag and Ag–In alloy pastes produced by the ball-milling process were printed onto the bottom Ag-sputtered Cu dies using a 100 μm thick stencil. The pastes were then pre-cured on a hot plate at 130 °C for 25 min to volatilize the organic solvent. Finally, Ag-sputtered top dies were put on the pre-cured pastes, and the sintered Ag and Ag–In alloy joints were fabricated using a hot-pressing bonding machine. The bonding condition for the thermal-compressive bonding process is presented in Figure 3. The atmosphere in the chamber of the hot-pressing machine (Yongfa Technology Company, Taipei, Taiwan) was replaced with inert nitrogen during heating to prevent oxidation. The bonding temperature was set at 300 °C for 30 min with a compressive stress of 10 MPa or 20 MPa to promote bonding. After the bonding process, the bonded samples underwent the furnace cooling process. Finally, the as-bonded samples of sintered Ag and Ag–In alloy joints could be achieved after cooling.

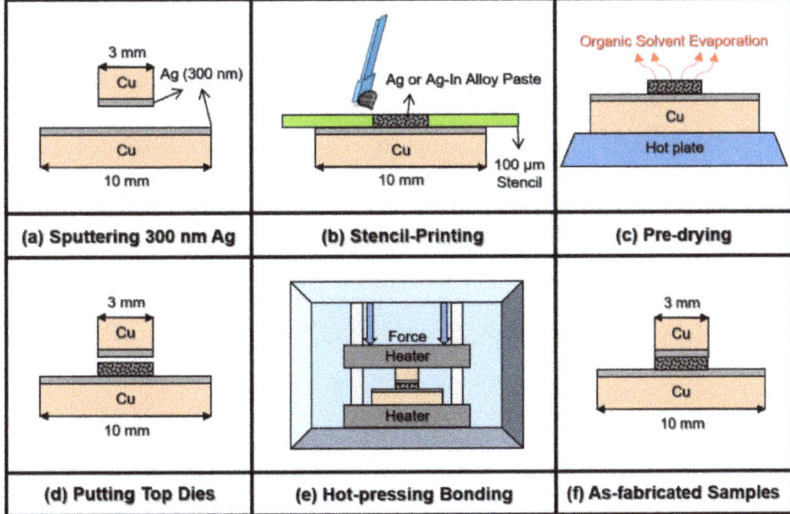

Figure 2. Flowchart of the sintered Ag and Ag–In alloy joint bonding processes. (**a**) Sputtering 300 nm Ag layer on Cu substrates as surface finish metallization, (**b**) printing 100 μm Ag or Ag-In alloy pastes in thickness on lower substrates, (**c**) pre-drying the pastes on a hot plate, (**d**) placing top dies on the pastes, (**e**) putting the assembly samples in a hot-pressing machine for the bonding process, and (**f**) obtaining as-fabricated samples after the bonding process.

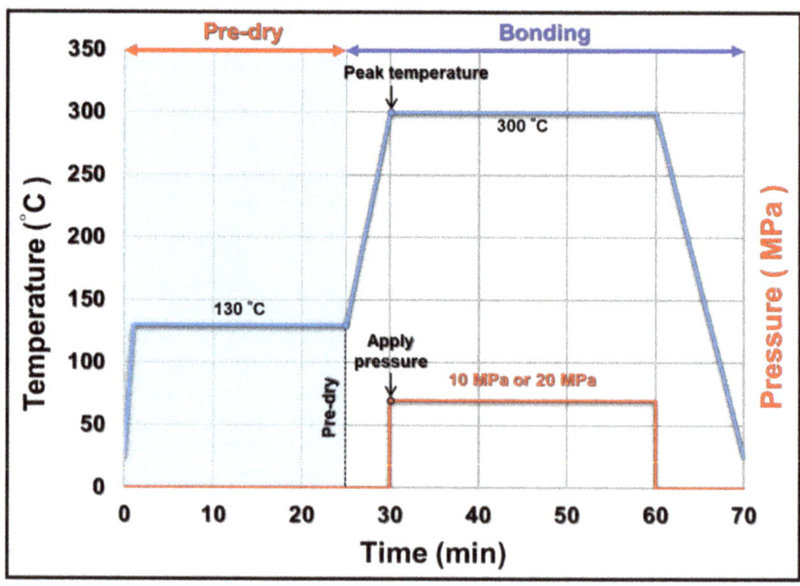

Figure 3. Temperature and pressure profiles for the sintering process.

2.3. Reliability Tests

HTS and die shear tests were performed to examine the reliability and mechanical properties of the sintered Ag and Ag–In alloy joints. The HTS test temperature was 300 °C in a furnace with an atmospheric environment to test the short- and long-term high-temperature reliability of both sintered joints. After the HTS test, die shear tests were performed using a shear tester (Condor Sigma Lite, XYZTEC, Panningen, The Netherlands) to measure the shear strengths and conduct the assessment for the mechanical reliability of both sintered joints. Figure 4 depicts the schematic diagram of the die shear test. The distance between the shear tool and the bottom die was 100 μm, and the speed for the die shear test was 500 μm/s.

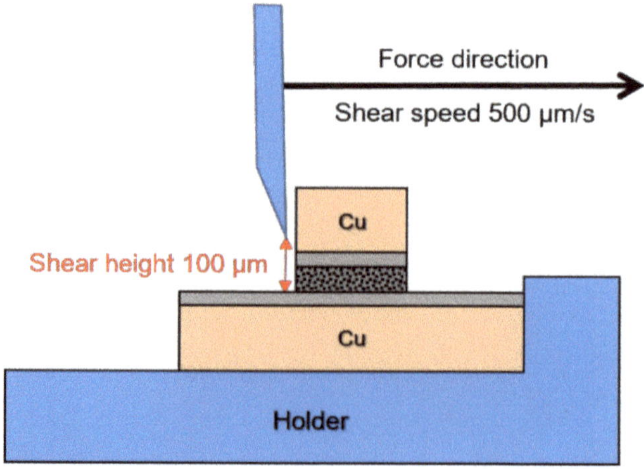

Figure 4. Schematic of the die shear test.

2.4. Phase Identification and Compositional Analysis

The morphology and microstructure of the ball-milled Ag and Ag–In alloy powders were characterized using a scanning electron microscope (SEM, Hitachi SU-5000, Hitachi, Tokyo, Japan) equipped with an energy-dispersive X-ray spectrometer (EDS, Bruker, Berlin, Germany). The phases of the Ag–In alloy powders were identified using X-ray diffraction (XRD, Rigaku TTRAX3, Tokyo, Japan) with Cu Kα radiation (λ = 0.15418 nm). A transmission electron microscope (TEM, FEI Tecnai G2 F20, FEI, Roanoke, VA, USA) was used to observe the crystal structure of the ball-milled powders. In addition, the size distributions of the powders were examined using a laser-diffraction particle size analyzer (Coulter LS230, Miami, FL, USA), with an examination range from 0.04 μm to 2000 μm. After the hot-pressing bonding process and the HTS test, bonded samples were mounted with epoxy resin and polished using abrasive sandpapers. For SEM observation, an artifact-free cross-section was created by using an ion-milling system equipped with an Ar^+ ion gun (Hitachi IM-4000, Hitachi, Tokyo, Japan). Moreover, an electron probe microanalyzer (EPMA, JEOL JXA-8530FPlus, JEOL, Tokyo, Japan) was utilized for elemental mapping and compositional analysis.

3. Results

3.1. Characterization of the Ag and Ag–In Alloy Powders after Ball-Milling

Figure 5 shows the characterization of the raw Ag, raw In, ball-milled Ag, and ball-milled Ag–In alloy powders. The morphology of both the ball-milled Ag and the Ag–In alloy powders consisted of flakes after milling, and both the ball-milled powders were refined compared to raw powders, as shown in Figure 5a–d. The ball-milled Ag–In alloy powders exhibited nanocrystalline structure from the grain contrast of the bright-field TEM image in Figure 5e. The generation of nanocrystalline domains was attributed to the constant high-energy impact on the Ag and In powders during the ball-milling process, causing the accumulation of numerous grain boundaries within the ball-milled powders. The crystal structure of Ag–In alloy powders was determined to be face-centered cubic (FCC) from the TEM diffraction pattern in Figure 5f. The diffraction ring of the Ag–In alloy powder was similar to that of pure Ag; only the lattice constants were different. On the basis of the TEM ring pattern, the lattice constant of the Ag–In alloy powder was calculated to be 0.414 nm, which was slightly higher than that of the Ag powders (0.409 nm) [29]. The difference in the lattice constant was due to lattice distortion in the Ag–In alloy powder caused by the solid solution of In in the Ag matrix.

The XRD patterns for the raw Ag and Ag–In alloy powders before and after the ball-milling process are shown in Figure 6. This measurement confirms that the crystal structure of the Ag–In alloy powder remained FCC after the ball-milling process. There was no change in the crystalline structure after ball-milling; only the XRD peak position shifted, which suggested a change in the lattice constant. The lattice constant of the Ag–In alloy powder was calculated to be 0.414 nm from the XRD pattern, which was consistent with the TEM analysis. Moreover, no peaks of Ag–In IMCs could be identified from the XRD pattern of the Ag–In alloy powder. This illustrates that a single phase of the Ag–In alloy powder with a homogeneous Ag-based Ag–In solid solution could be obtained via mechanical alloying. In other words, the solid solution and homogenization of the Ag–In alloy powder were achieved through sufficient diffusion between the pure Ag and In powders under such milling conditions. The SEM–EDS compositional analysis of the Ag–In alloy powder in Figure 5d is shown in Table 1. The phase of the Ag–In alloy powder was identified to be Ag-based Ag–In solid solution, labeled (Ag)–In. The elemental analytical result was consistent with the XRD result. TEM–EDS also indicated that the mean composition of the Ag–In alloy powder was 82.3 at.% Ag and 17.7 at.% In, which corresponds with the elemental compositions determined using SEM–EDS.

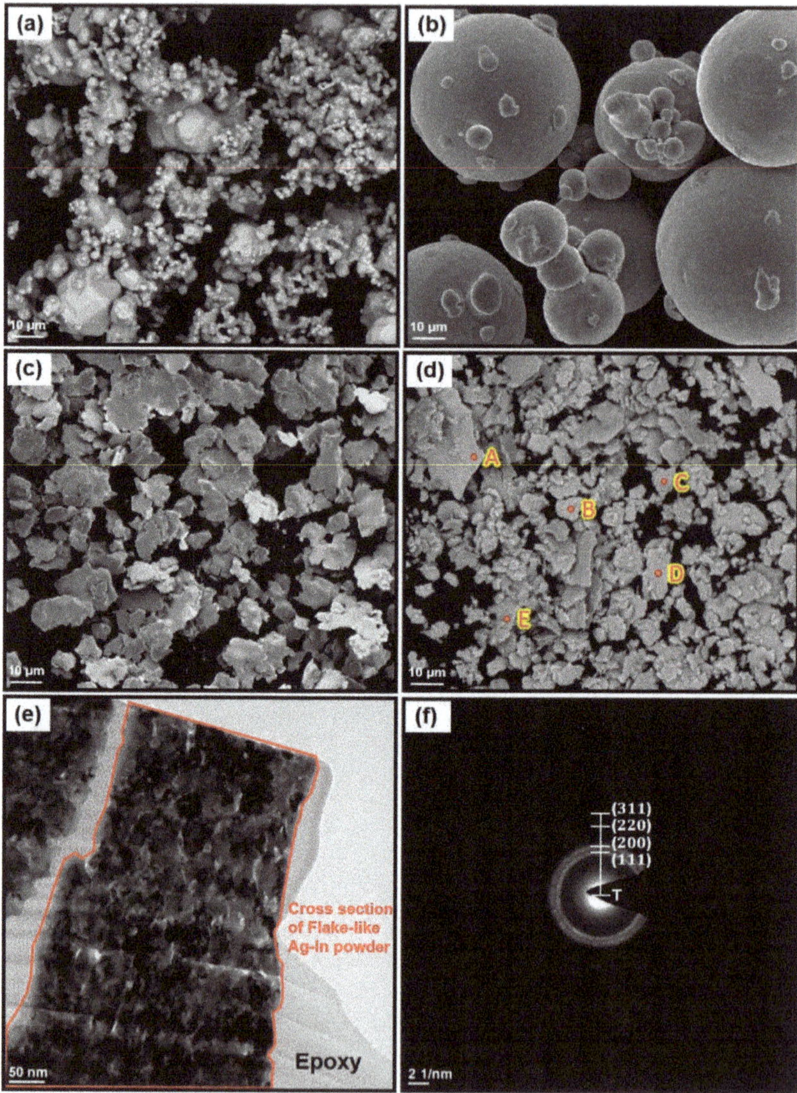

Figure 5. Raw and ball-milled powder characterizations. SEM images of (**a**) the raw Ag powder, (**b**) the raw In powder, (**c**) the ball-milled Ag powder, and (**d**) the ball-milled Ag–In alloy powder. (**e**) TEM images and (**f**) the corresponding TEM diffraction pattern for the ball-milled Ag–In alloy powder.

The particle size distributions (PSDs) for the Ag and Ag–In alloy powders after the ball-milling process are shown in Figure 7. In the PSD analysis, the value D_{50} indicates that 50 vol.% of the powder had a diameter less than this value. Herein, we regarded D_{50} as the average particle size and compared the D_{50} of powders before and after ball-milling. The D_{50} of the raw Ag powder was 28 μm, while the D_{50} of the ball-milled Ag and Ag–In alloy powders was 19 μm and 13 μm, respectively. Hence, both the ball-milled Ag and the Ag–In alloy powders were refined through the fracturing effect during the milling. The agglomeration and coarsening of ductile Ag and In powders caused by cold welding could be effectively inhibited by the addition of stearic acid, which was consistent with previous studies [30]. The Ag and Ag–In alloy powders were sieved with a 25 μm mesh after ball-milling. The resulting powders were mixed with the organic solvent to produce the Ag and Ag–In alloy pastes used for sintering and bonding applications.

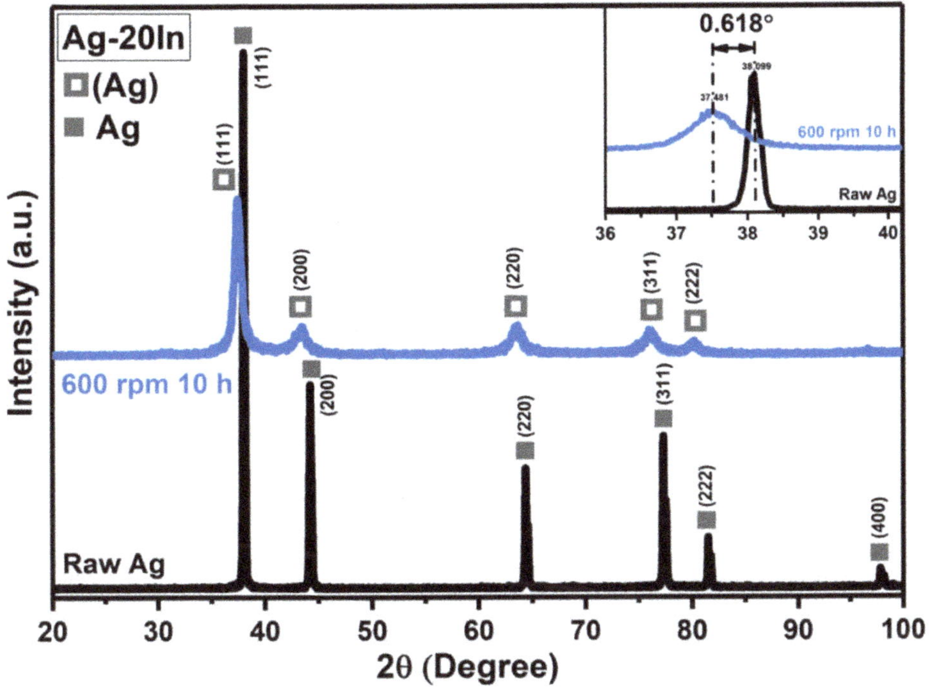

Figure 6. XRD patterns of the Ag–In alloy powders.

Table 1. Composition determined using EDS at the positions indicated in Figure 5d.

Position	Ag (at.%)	In (at.%)	Phase
A	81.1 ± 0.26	18.9 ± 0.26	(Ag)–In
B	81.2 ± 0.26	18.8 ± 0.26	(Ag)–In
C	81.9 ± 0.26	18.1 ± 0.26	(Ag)–In
D	81.8 ± 0.26	18.2 ± 0.26	(Ag)–In
E	81.7 ± 0.26	18.3 ± 0.26	(Ag)–In

3.2. Microstructure of the Sintered Ag Joints during HTS at 300 °C

Figure 8 shows SEM images of the microstructure cross-sections of the sintered Ag joints bonded at 10 MPa after HTS at 300 °C from 0 h to 2000 h. The porosity of the sintered joint was calculated using an image threshold in ImageJ software. The porosity was averaged across five representative positions at the sintered joint of one sample at the same magnification in SEM. Figure 8a shows the cross-sectional microstructure of the as-bonded Ag joint. After sintering the flake-like Ag particles with a compressive pressure of 10 MPa, a low porosity of 4.1% could be achieved in the as-bonded Ag joint, indicating a dense structure. Moreover, excellent wettability was observed at the interface between the 300 nm thick sputtered Ag layer and the sintered flake-like Ag particles. However, the porosity of the sintered Ag joint significantly increased to 9.3%, 11.3%, and 15.7% after the HTS for 100 h, 1000 h, and 2000 h, respectively, as shown in Figure 8b–d. The sintered pores in the Ag joint tended to coalesce via vacancy and grain boundary diffusion during the HTS to reduce the surface energy and achieve a more thermodynamically stable state. Nevertheless, the coarsening of the sintered pores after the HTS severely deteriorated the wettability of the Ag-sputtered Cu substrates, as shown in Figure 8d. Micro-voids were formed at the Ag/Cu wetting interface because of the accumulation and coarsening of the sintered pores in the high-temperature environment. The coarsening of the sintered pores and the poor wettability in the sintered Ag joint after the HTS led to severe stress concentration problems, which caused deterioration of the mechanical properties.

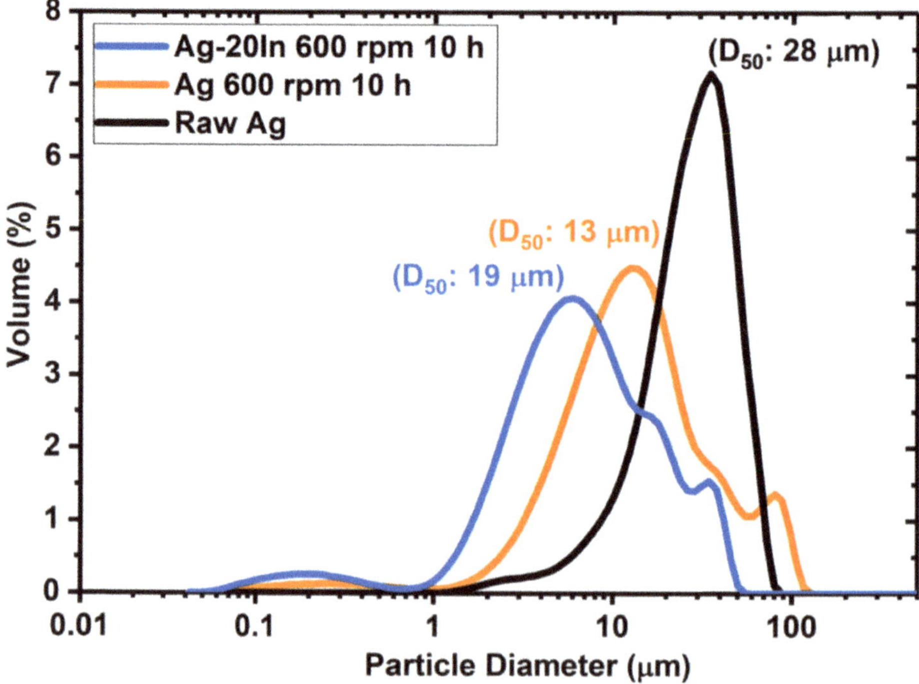

Figure 7. Particle size distributions for the Ag and Ag–In alloy powders after the ball-milling process.

Cu oxidation above 200 °C should have a considerable effect on the pure sintered Ag joint. Although it is not deeply understood yet, many researchers have discussed that the sintered Ag joint would suffer from Cu oxidation at high temperatures, which detrimentally affects the mechanical properties of sintered Ag joints. However, the Cu oxidation phenomenon was not observed in the sintered Ag joint during the HTS at 300 °C in this study. The main reason is the porosity of the as-bonded sintered Ag joint. There is a positive correlation between the porosity of the as-bonded sintered joint and the occurrence of such high-temperature oxidation, which is described by percolation theory. Specifically, the high porosity of the as-bonded sintered joint facilitates the oxidation process during HTS. Assuming that the pores are compact and uniformly distributed, three different theoretical porosity ranges can be calculated using percolation theory to describe whether external media (e.g., O_2) can invade the porous structure [31–33]. To elaborate, a theoretical porosity of less than 3% results in fully isolated and uniformly distributed pores in the sintered joint that are not exposed to external surfaces, isolating the sintered joint from environmental O_2. Meanwhile, isolated and interconnected pores coexist at the sintered joint in the theoretical porosity range from 3% to 20%, while fully interconnected pores exist to form a 3D porous network at the sintered joint for a theoretical porosity above 20%.

Hence, when the porosity of the sintered joint is beyond or close to this theoretical limit of 20%, the 3D porous network provides abundant pathways for O_2 to invade and penetrate the sintered joint.

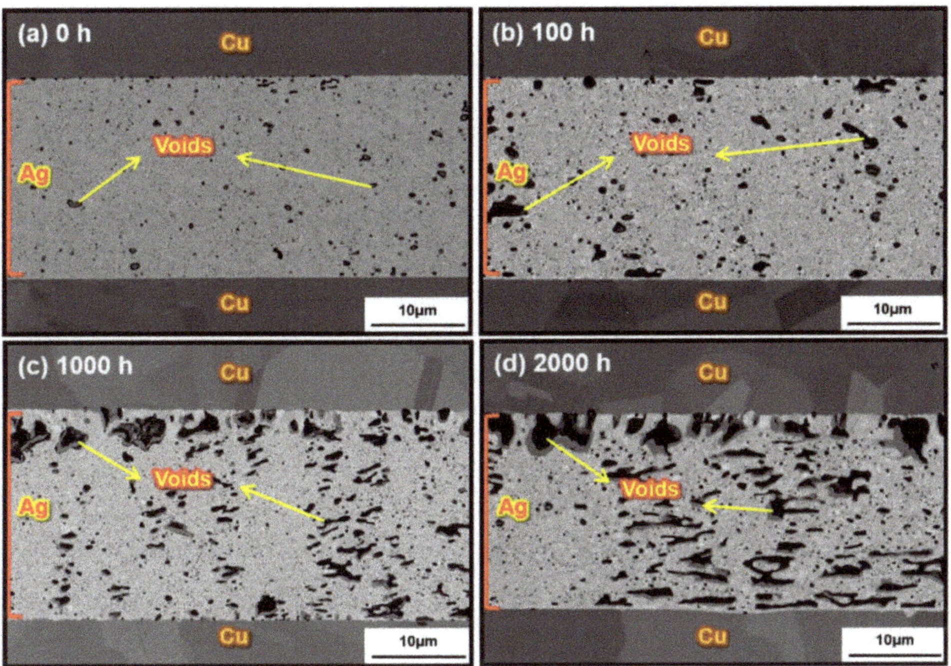

Figure 8. SEM images of the cross-section microstructure of the sintered Ag joints bonded at 10 MPa after HTS at 300 °C for (**a**) 0 h, (**b**) 100 h, (**c**) 1000 h, and (**d**) 2000 h.

In previous studies, a low-pressure or pressureless Ag sintering process was attempted to reduce the production cost. However, the high porosity occurring from low-pressure bonding resulted in interconnected pores in the as-bonded sintered Ag joint; therefore, O_2 could easily penetrate the edges of the sintered Ag joint during the HTS. As a result, Cu diffusing from the substrate reacted with O_2 to form a cuprous oxide (Cu_2O) layer near the interface between the sintered Ag joint and the Cu substrate owing to the high affinity of Cu and O_2. By contrast, a relatively high bonding pressure was utilized for the flake-like Ag sintering in this study. As a result, a low porosity of 4.1% was achieved in the as-bonded sintered Ag joint due to densification during the sintering process, and most of the pores in the sintered joint were unconnected and isolated. Consequently, O_2 was effectively prevented from invading the sintered joint, reacting with Cu, and forming the Cu_2O layer. Although the porosity of the sintered Ag joint significantly increased during the HTS and oxidation should occur, the tendency of the pores to coalesce instead of maintaining a uniform distribution caused most of the pores to remain isolated despite the relatively high porosity. However, although the oxidation issue during the HTS could be effectively addressed by increasing the applied pressure, the sintered pore coarsening phenomenon in the sintered Ag joint would still cause degradation of the mechanical properties.

3.3. Microstructure and Phase Identification of the Sintered Ag–In Alloy Joints during HTS at 300 °C

Figure 9 depicts SEM images of the cross-sectional microstructure of the sintered Ag–In alloy joint bonded at a bonding pressure of 10 MPa during the 2000 h HTS test at 300 °C. The microstructures of the sintered Ag–In alloy joint bonded at 10 MPa were quite different in the side and the central regions of the joint at the early stage of the HTS. Figure 9a,e,i,m indicate the observation positions for the joint after different HTS periods: as-bonded and after 50, 100, and 2000 h, respectively. SEM images of the side (Figure 9b,f,j,n) and central regions (Figure 9c,g,k,o) are shown for the joints. In the same bonding condition, the average porosity of the as-bonded Ag–In alloy joint in Figure 9b,c was 18.9%, which was much higher than that of the as-bonded Ag joint in Figure 8a. This implied that the sinterability of Ag–In alloy powders was worse than that of pure Ag powders. Additionally, micro-voids could be observed at a few regions of the wetting interfaces, as shown in Figure 9d, indicating that the wettability of the as-bonded Ag–In alloy joint was not as good as the as-bonded Ag joint. The composition of the as-bonded Ag–In alloy joint in the sintering zone was also determined to be the homogeneous Ag-based Ag–In solid solution phase from EDS analysis, labeled (Ag)–In in Figure 9b–d.

The porosity of the Ag–In alloy joint did not increase as dramatically as that of the pure Ag joint after the HTS at 300 °C for 50 h and 100 h, as shown in Figure 9f,g and Figure 9j,k, respectively. The porosity of the sintered Ag–In alloy joint was 17.6% and 17.1% after the HTS for 50 h and 100 h, respectively, which was similar to the porosity of the as-bonded Ag–In alloy joint. In other words, the sintered pore coarsening phenomenon was not evident in the Ag–In alloy joint at high temperatures, in contrast to the sintered Ag joint. The most likely reason for the inhibition of the pore coarsening in the Ag–In alloy joint during the HTS was the solute drag effect that restricted the migration of the Ag grain boundaries [34–36]. According to the sintering theory, the grain growth is accompanied by the coarsening of sintered pores, and this process is highly dependent on grain boundary diffusion. However, the indium element as the solute alloying in the Ag–In alloy joint would prohibit the movement of grain boundaries. Consequently, the sintered pores could not coalesce via grain boundary diffusion. Hence, the pore coarsening phenomenon during the HTS could be effectively inhibited by adding In. However, severe oxidation issues still occurred in the Ag–In alloy joint bonded at 10 MPa because the porosity of the as-bonded Ag–In alloy joint was close to the percolated porosity limit of 20%.

The EPMA elemental mapping of the Ag–In alloy joint after HTS at 300 °C for 100 h is shown in Figure 10. The elemental content of the area indicated in Figure 10a is listed in Table 2. After the HTS for 50 h and 100 h, a lamellar Cu_2O layer could be identified between the Ag–In alloy joint and the substrate near the side regions of the joint in Figure 9f,j. In addition, we observed that the Cu_2O layer converged and ended near the central region of the joint, as shown in Figure 9g,k. This revealed that O_2 entered the Ag–In alloy joint through the interconnected pores from the edges of the joint. At the beginning of the HTS, the O_2 concentration was too low to form a lamellar Cu_2O layer near the central region of the joint. Instead, Cu_2O clusters with dark-gray contrast were formed around the sintered pores in the central region of the joint, as shown in Figure 9g,h. In_2O_3 clusters with light-gray contrast were also found around the sintered pores. However, In_2O_3 clusters were identified around the pores in all sintered regions of the joint after the HTS, which was not the case for the Cu_2O clusters. Furthermore, owing to the interdiffusion of Ag and Cu at the wetting interface, the 300 nm thick sputtered Ag was transformed into the Ag-based Ag–Cu solid solution phase, labeled (Ag)–Cu in Figure 9f,h,j,l,n,p. After the long-term HTS for 2000 h, adequate O_2 could reach the central region of the Ag–In alloy joint and form a continuous lamellar Cu_2O layer. The formation and growth of the thick Cu_2O layers generated cracks at the interface between the continuous Cu_2O layer and the substrate, which negatively affected the mechanical properties.

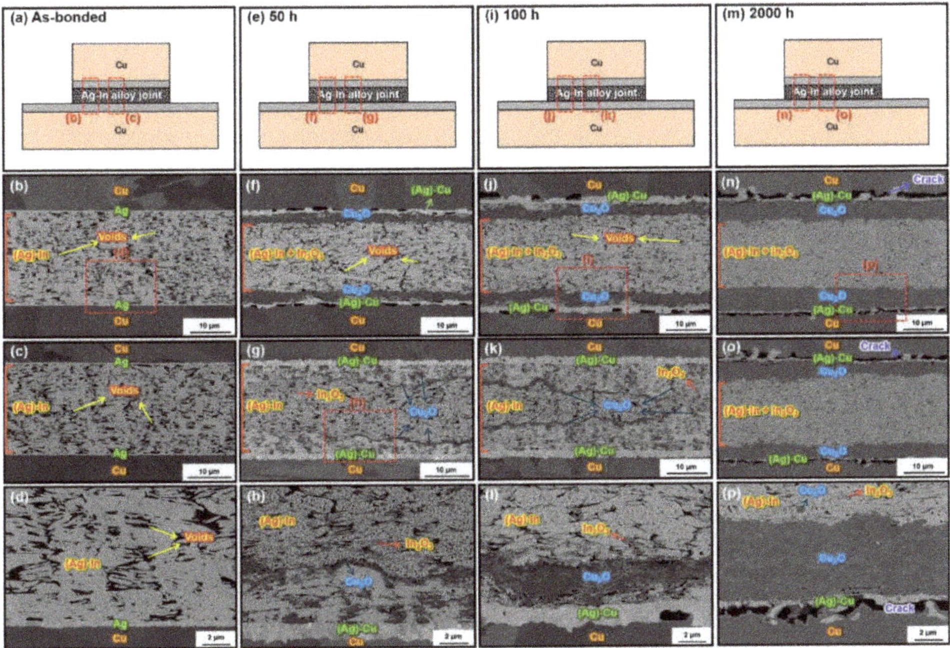

Figure 9. SEM images of the cross-section microstructure of the sintered Ag–In alloy joints bonded at 10 MPa after HTS at 300 °C. (**a,e,i,m**) Schematics of the cross-section of the Ag–In alloy joints as-bonded and after HTS for 50 h, 100 h, and 2000 h, respectively. The dashed boxes in (**a,e,i**), and (**m**) indicate the regions near the edge shown in (**b,f,j**), and (**n**), as well as the regions away from the edge shown in (**c,g,k**), and (**o**), respectively. Zoomed-in images of the regions indicated in (**b,g,j**) and (**n**) are shown in (**d,h,l**), and (**p**), respectively.

Figure 11 shows SEM images of the cross-sectional microstructure of the sintered Ag–In alloy joint bonded at a bonding pressure of 20 MPa during the 2000 h HTS test at 300 °C. As shown in Figure 11a, the porosity of this joint was dramatically lower at 10.6%. Compared to the as-bonded Ag–In alloy joint bonded at 10 MPa, not only did the porosity decrease but the wettability also increased at the higher bonding pressure. Although interconnected pores still existed at this porosity according to the predictions of the percolation theory, the relatively low porosity could still reduce the penetration of O_2. After 50 h to 2000 h of HTS, as shown in Figure 11b–e, no continuous Cu_2O layer was found in the sintered joint, which indicated that the amount of O_2 entering the joint bonded at 20 MPa was significantly reduced compared to the joint bonded at 10 MPa. Nevertheless, clusters of Cu_2O and In_2O_3 could still be observed around the sintered pores after the HTS; interconnected pores were still present in the as-bonded Ag–In alloy joint, which provided pathways for O_2 penetration. Although the Cu_2O and In_2O_3 clusters were formed in the Ag–In alloy joint, the oxide clusters around the sintered pores had a negligible effect on the mechanical properties of the joints compared to the lamellar oxides and cracks. More importantly, the pores in the sintered Ag–In alloy joint bonded at 20 MPa did not coarsen, which was consistent with the results at 10 MPa.

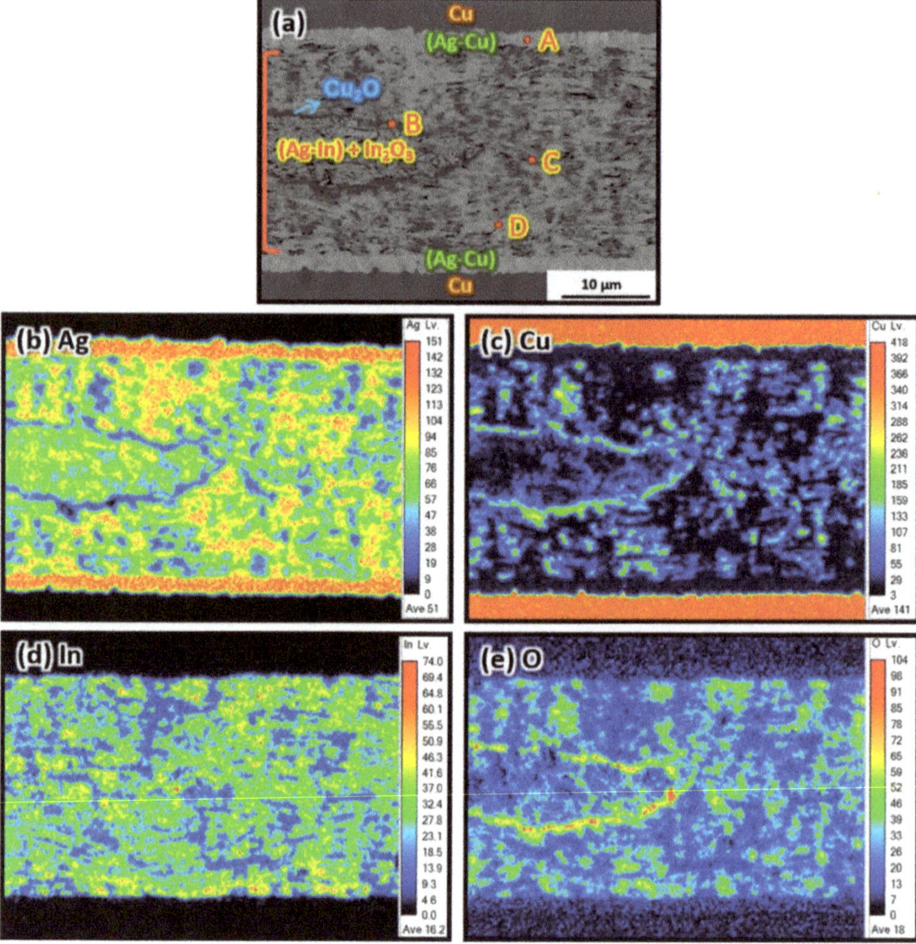

Figure 10. (**a**) Cross-section microstructure of the middle of the Ag–In alloy joint bonded at 10 MPa after HTS at 300 °C for 100 h. (**b–e**) Elemental distributions of (**a**).

Table 2. Composition determined using EPMA at the positions indicated in Figure 10a.

Position	Ag (at.%)	In (at.%)	Cu (at.%)	O (at.%)	Phase
A	92.6	0	7.4	0	(Ag)–Cu
B	7.8	2.9	58.2	30.1	Cu_2O
C	0.9	38.2	1.2	59.7	In_2O_3
D	86.7	9.4	2.7	1.2	(Ag)–In

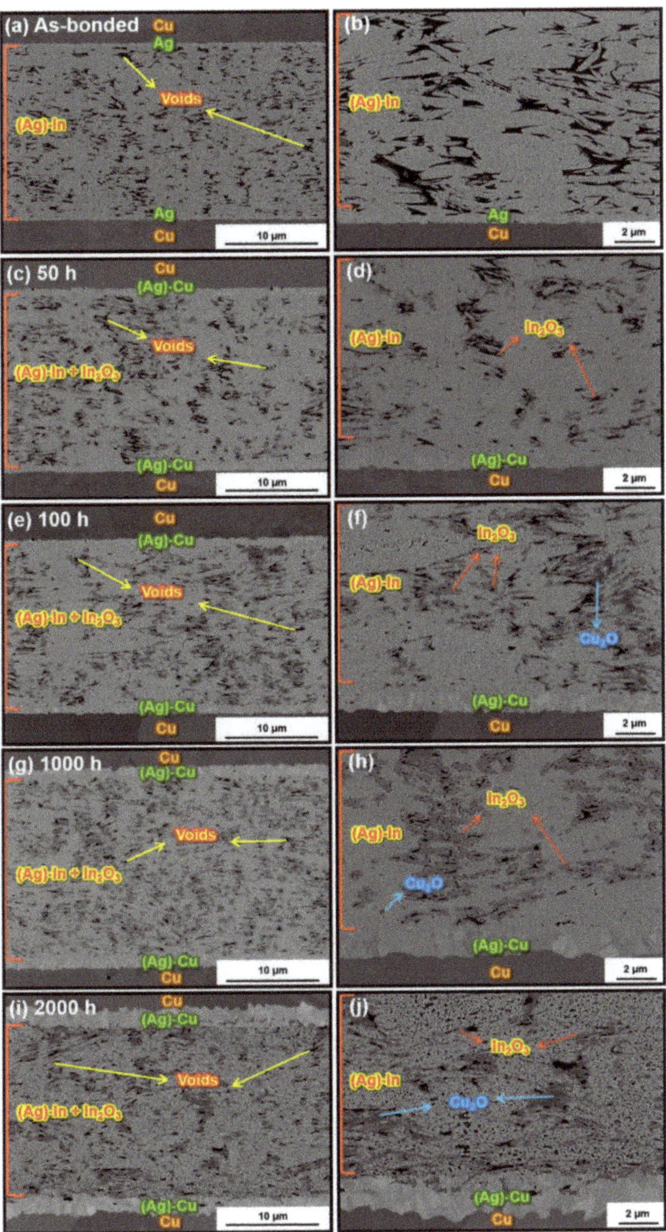

Figure 11. SEM images of the cross-section microstructure of the sintered Ag–In alloy joints bonded at 20 MPa after HTS at 300 °C for the (**a**) as-bonded joints and sintered joints after HTS for (**c**) 50 h, (**e**) 100 h, (**g**) 1000 h, and (**i**) 2000 h. Magnified images of (**a,c,e,g**), and (**i**) are shown in (**b,d,f,h**), and (**j**), respectively.

3.4. Mechanical Properties of the Sintered Ag and Ag–In Alloy Joints during HTS at 300 °C

The shear strength variations of the sintered Ag and Ag–In alloy joints with the HTS treatment time are illustrated in Figure 12. Each data point was averaged over four sintered joints. The gray dashed line in Figure 12 depicts the change in shear strength of the sintered Ag joint at a bonding pressure of 10 MPa during the HTS. The mean shear strength of the as-bonded Ag joint bonded at 10 MPa was 73 MPa, which was higher than that of the as-bonded Ag–In alloy joint with the same bonding condition due to the lower porosity and the better wettability. However, the shear strength of the Ag joint bonded at 10 MPa dramatically decreased after the HTS. The deterioration of mechanical properties was attributed to the sintered pore coarsening and the correspondingly poor wettability, which gave rise to stress concentration that significantly deteriorated the mechanical reliability of the sintered Ag joints. The green dashed line in Figure 12 depicts the change in shear strength of the sintered Ag joints at the bonding pressure of 20 MPa during the HTS. The shear strength of the as-bonded Ag joint bonded at 20 MPa was 106 MPa. This value was better than that of the as-bonded Ag joint bonded at 10 MPa, demonstrating that the mechanical properties can be improved by increasing the bonding pressure. The reinforcement of the sintered Ag joint was attributed to the lower porosity and better wettability. Nevertheless, although the as-bonded sintered Ag joint was strengthened by the higher pressure, the mechanical strength was still compromised during the HTS due to the coarsening of the sintered pores and the wettability issues.

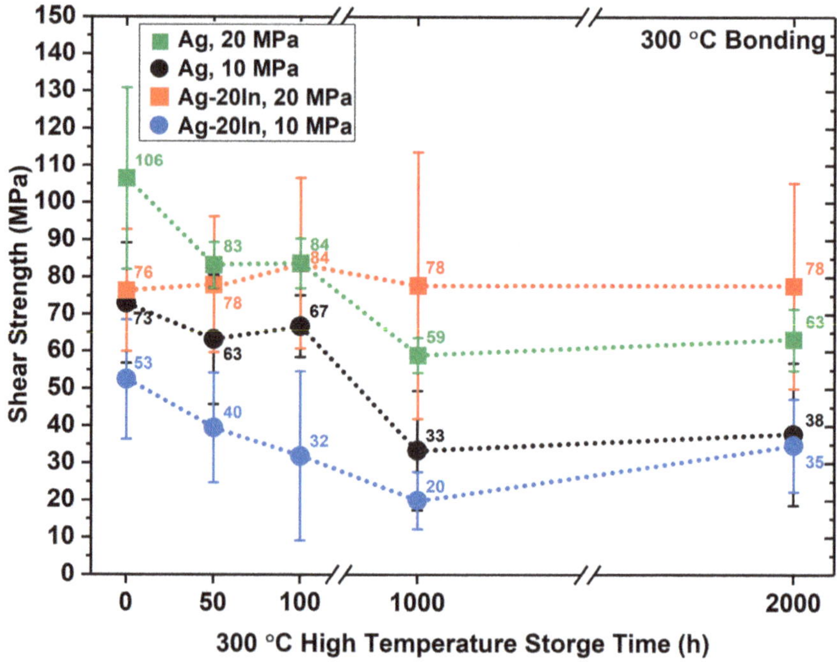

Figure 12. Variation of the shear strength for sintered Ag and Ag–In alloy joints with the thermal aging time at 300 °C.

The blue dashed line in Figure 12 illustrates the change in shear strength of the sintered Ag–In alloy joints bonded at 10 MPa during the HTS. The shear strength of the as-bonded Ag–In alloy joints bonded at 10 MPa was 53 MPa, which was slightly lower than that of the as-bonded Ag joints with the same bonding conditions owing to the higher porosity. The mechanical properties of the Ag–In alloy joints bonded at 10 MPa were found to significantly deteriorate during the 2000 h HTS. The reason for the decrease in shear strength of the Ag–In alloy joints was different from that of the sintered Ag joints. The deterioration of the Ag–In alloy joint shear strength was not due to the coarsening of the sintered pores but rather from the formation of continuous Cu_2O layers and cracks.

The red dashed line in Figure 12 illustrates the change in shear strength of the sintered Ag–In alloy joints bonded at 20 MPa during the HTS. The shear strength of the as-bonded Ag–In alloy joints bonded at 20 MPa was better than that of the as-bonded Ag–In alloy joints bonded at 10 MPa owing to the lower porosity at the higher bonding pressure. The mechanical properties of the Ag–In alloy joints bonded at 20 MPa were found to exhibit excellent shear strengths and stable mechanical properties during the 2000 h HTS. The low porosity of the Ag–In alloy joints achieved at a bonding pressure of 20 MPa was sufficient to prevent a large amount of external O_2 from entering the interior of the sintered joint and forming fragile Cu_2O layers and cracks. Furthermore, the addition of In was proven to help prevent sintered pore coarsening. Consequently, the Ag–In alloy joint with the optimal bonding condition was the most mechanically reliable at high temperatures because it simultaneously overcame the oxidation and pore coarsening issues.

The surface fracture analyses of the sintered Ag joints are shown in Figure 13. Large quantities of micro-dimples were formed on the fracture surfaces of the as-bonded Ag joints, indicating a ductile deformation of the joint, as shown in Figure 13b,c. Figure 13d depicts the schematic drawing of the fracture failure pattern of the as-bonded Ag joint; the failure was along the sintering zone in the sintered Ag joint after the die shear test. Although a few dimples could still be observed on the fracture surfaces after the HTS for 100 h and 2000 h, as shown in Figure 13f,g,j,k, the density of dimples on the fracture surfaces gradually decreased as the HTS time increased. This was due to sintered pore coarsening, which led to stress concentration in the sintered Ag joint. Hence, cracks originated from these large pores when subjected to external forces, resulting in the deterioration of the shear strength, a tendency to cleave, and a decrease in the density of dimples on the fracture surfaces. Figure 13h,l show schematics of the fracture failure pattern of the Ag joint after the HTS for 100 h and 2000 h, respectively. Although the sintered pores dramatically coarsened after the HTS, cracks still propagated along the sintering zone in the Ag joint.

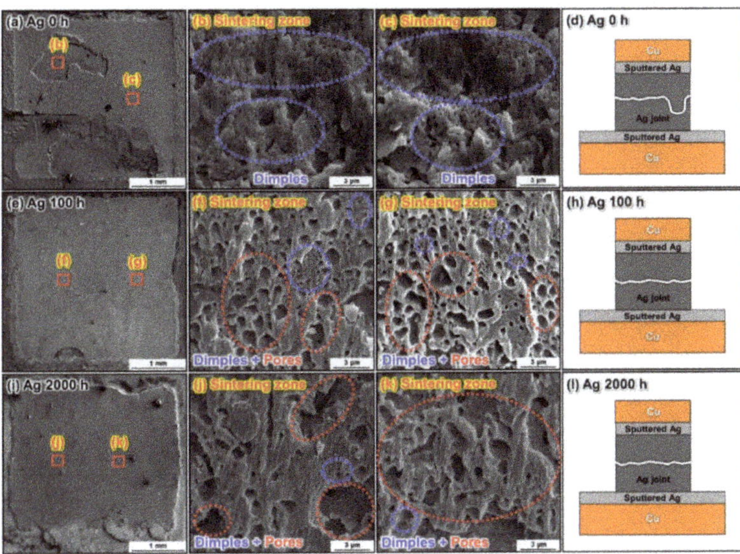

Figure 13. Surface fracture analysis of the sintered Ag joints bonded at 10 MPa. Fracture planes of the sintered Ag joints (**a**) as-bonded and after (**e**) 100 h and (**i**) 2000 h of HTS at 300 °C. (**b,c**), (**f,g**), and (**j,k**) are magnified images of the fracture plane areas marked in (**a,e**), and (**i**), respectively. Schematic of the crack propagation in the sintered Ag joints (**d**) as-bonded and after (**h**) 100 h and (**l**) 2000 h of HTS at 300 °C.

The fracture surface analysis of the sintered Ag–In alloy joints bonded at 10 MPa and 20 MPa is shown in Figures 14 and 15, respectively. Ductile deformation characteristics could be identified on the fracture surfaces of the as-bonded Ag–In alloy joints, as shown in Figures 14a–c and 15a–c. The composition was determined to be (Ag)–In on the fracture surfaces of both as-bonded Ag–In alloy joints. However, Cu surfaces were observed on the fracture surfaces of the Ag–In alloy joint bonded at 10 MPa after the HTS for 100 h, as shown in Figure 14e–g. At the early stages of the HTS, the Cu_2O layers mainly formed near the edges of the sintered joints. Hence, the Cu surfaces were near the edges of the fracture plane, which was consistent with the cross-sectional SEM analysis. This indicated that cracks tended to propagate along the fragile Cu_2O layers near the edges, rather than along the sintered joints, as illustrated in Figure 14h. After 2000 h of HTS, continuous Cu_2O layers were formed throughout the sintered joint, and the cracks only propagated along the Cu_2O layers, as shown in Figure 14i–l. By contrast, a Cu layer could not be located on the fracture surfaces of the Ag–In alloy joint bonded at 20 MPa after the HTS for 100 h and 2000 h, as shown in Figure 15e–l. The compositions of the fracture surfaces were mainly the (Ag)–In solid solution. This again demonstrated that the oxide clusters had little impact on the mechanical properties, and that the continuous oxide layer was not formed in the Ag–In alloy joint bonded at 20 MPa after the HTS due to the low porosity. Moreover, micro-dimples could be identified on all the fracture planes of the Ag–In alloy joint after the HTS, indicating that the Ag–In alloy joint maintained excellent ductility at high temperatures.

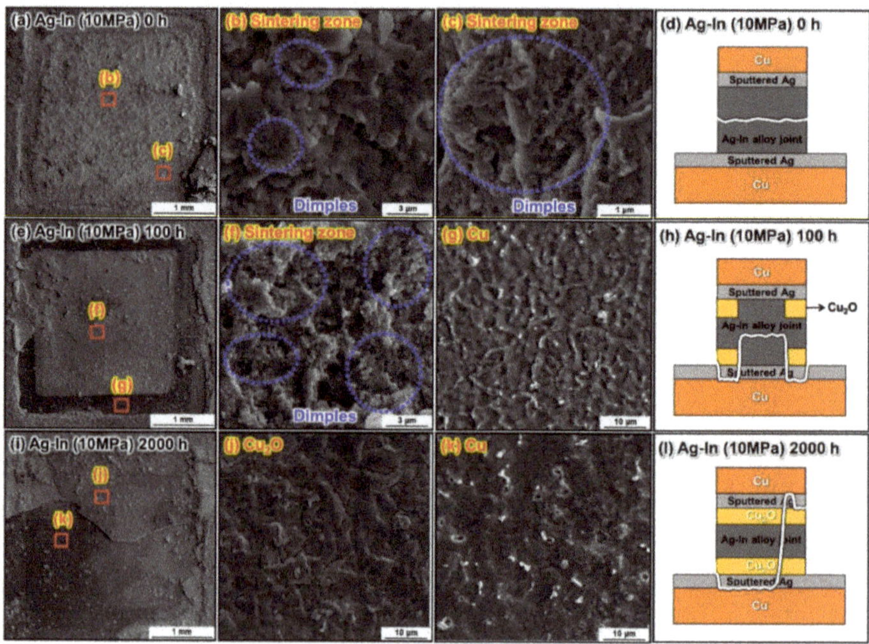

Figure 14. Surface fracture analysis of the sintered Ag–In alloy joints bonded at 10 MPa. Fracture planes of the Ag–In alloy joints (**a**) as-bonded and after (**e**) 100 h and (**i**) 2000 h of HTS at 300 °C. (**b**,**c**), (**f**,**g**), and (**j**,**k**) are magnified images of the fracture plane areas marked in (**a**,**e**), and (**i**), respectively. Schematic of the crack propagation in the sintered Ag–In alloy joints bonded at 10 MPa (**d**) as-bonded and after (**h**) 100 h and (**l**) 2000 h of HTS at 300 °C.

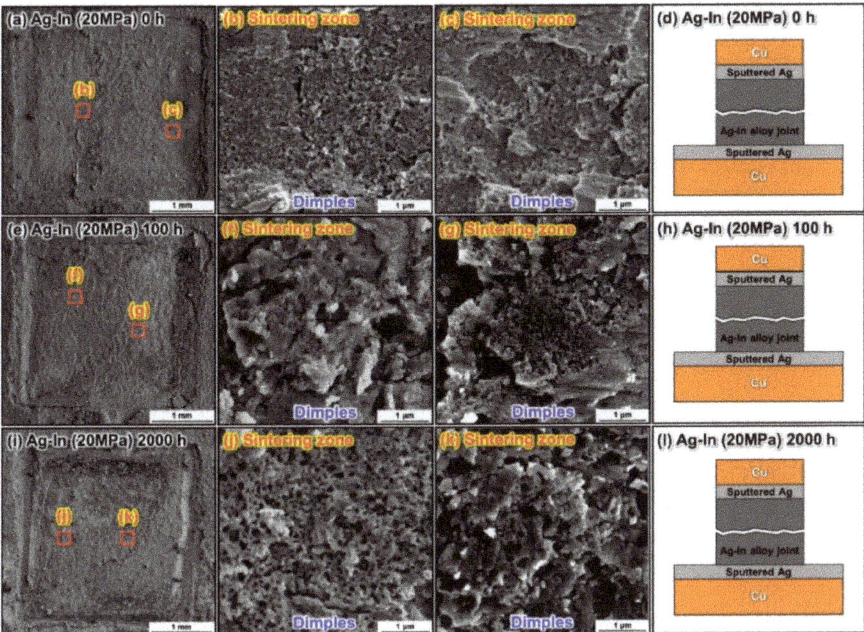

Figure 15. Surface fracture analysis of the sintered Ag–In alloy joints bonded at 20 MPa. Fracture planes of the Ag–In alloy joints (**a**) as-bonded and after (**e**) 100 h and (**i**) 2000 h of HTS at 300 °C. (**b**,**c**), (**f**,**g**), and (**j**,**k**) are magnified images of the fracture plane areas marked in (**a**,**e**), and (**i**), respectively. Schematic of the crack propagation in the sintered Ag–In alloy joints bonded at 20 MPa (**d**) as-bonded and after (**h**) 100 h and (**l**) 2000 h of HTS at 300 °C.

4. Discussion

4.1. Oxidation Mechanism of the Ag–In Alloy Joint Bonded at 10 MPa

Two types of Cu_2O and one type of In_2O_3 were observed in the Ag–In alloy joint bonded at 10 MPa after the HTS, including a Cu_2O layer, Cu_2O clusters, and In_2O_3 clusters, as shown in Figure 9. The porosity of the Ag–In alloy joint for this bonding condition was not sufficiently low to completely block external O_2; therefore, O_2 could still attack the sintered Ag–In alloy joint through interconnected pores and react with Cu and In. This resulted in the formation of a Cu_2O layer, which led to severe reliability issues.

Herein, a mechanism for the oxidation process in the sintered Ag–In joint was established, as illustrated in Figure 16. The diffusion of four elements (Ag, Cu, In, and O) in and around the sintered Ag–In alloy joint needs to be considered during the HTS. Initially, O_2 from the environment entered the Ag–In alloy joint via the interconnected pores at the edges of the joints; the high porosity favors the penetration of O_2. Consequently, the concentration of O_2 near the edges would be higher than that near the middle of the joint at the beginning of the HTS. Therefore, continuous Cu_2O layers were observed between the Ag–In alloy joint and the Ag-sputtered Cu substrate near both edges of the sintered joints, as shown in Figure 16b. Cu entering the Ag–In alloy joint through Cu–Ag interdiffusion from the substrate contacts O_2 and immediately reacts to form the Cu_2O layer near the wetting interfaces at both edges of the joint. From the EPMA analysis, it was found that the 300 nm thick sputtered Ag layer was transformed into a Ag-rich Ag–Cu solid solution phase, labeled (Ag)–Cu, which was the evidence for the Ag–Cu interdiffusion during the HTS. In addition, near the middle region of the Ag–In alloy joint at the early stages of the HTS, the convergence of the Cu_2O layer can be seen, as shown in Figure 16b. This indicates that the external O_2 was not sufficient to form a Cu_2O layer in the middle region of the joint at this stage. Therefore, Cu from the substrate needed to diffuse a relatively long distance before it formed the Cu_2O layer near the middle region of the join, as illustrated in Figure 16b. Cu_2O also existed in the form of clusters located around the sintered pores in the middle region of the joint. The Cu_2O clusters were mainly formed there in the early stages of the HTS because O_2 that had reached the middle region via the interconnected pores was not sufficient to form a continuous Cu_2O layer. However, Cu_2O clusters could not be found near the edges of the joint; the Cu_2O layer caused the Cu diffusing from the substrates to continue to react with O_2, forming a thick layer. In addition to the two types of Cu_2O contained in the joint, In_2O_3 clusters were found in all sintered regions of the joint after the HTS. In_2O_3 clusters precipitated around the sintered pores in the Ag–In alloy joint because of the O_2, which entered the joint through the interconnected pores, similar to the case for the Cu_2O clusters. Additionally, mixed In_2O_3 and Cu_2O clusters precipitated in the middle region of the joint because O_2 reacted with both In from the joint and Cu diffusing from the substrates. After the long-term HTS in Figure 16c, continuous Cu_2O layers and cracks were formed at both interfaces between the Ag–In alloy joint and the substrate. At this stage of the HTS, a considerable amount of O_2 diffused into the middle of the joint over a long period, causing the Cu_2O layers near the edges to merge and form continuous Cu_2O layers. Having developed a complete understanding of the oxidation mechanism in the sintered joint, it is now important to find solutions for overcoming oxidation issues.

4.2. Strategies to Overcome Oxidation and Enhance Mechanical Properties of Sintered Joints at High Temperatures

Sintered nano- and micro-Ag materials are promising for die attachment of power chips. In recent years, many studies have focused on pursuing low-pressure or pressureless bonding processes for Ag sintering to achieve cost reductions. However, the porosity of the sintered joint would unavoidably increase when a low-pressure or pressureless bonding was attempted. Consequently, this would significantly increase the risk of high-temperature oxidation, which affects the mechanical properties of the sintered joint. Therefore, developing a strategy for overcoming oxidation problems is a critical and urgent issue for sintered joints. Meanwhile, it is also crucial to inhibit the sintered pore coarsening phenomenon at high temperatures because the pore coarsening also decreases mechanical strength. Alloying the Ag joint has become the primary strategy to combat this issue in pure sintered Ag joints.

In this study, we found a positive correlation, described by percolation theory, between the porosity of the as-bonded sintered joint and the occurrence of such high-temperature oxidation. In other words, the sintering porosity of the as-bonded sintered joint is an important reliability evaluation metric for high-temperature oxidation. According to the predictions of the percolation theory, high-temperature oxidation can be completely prevented when the porosity is below approximately 3%. Hence, reducing the porosity of the as-bonded sintered joint as much as possible helps prevent high-temperature oxidation. Previous studies have proven that the TLP bonding reaction can help prevent oxidation because the introduction of low-melting-point liquid metals during bonding can fill the gaps between the nano- and micro-Ag particles and form isolated pores. This not only resolves the oxidation problem but also addresses the sintered pore coarsening phenomenon. However, the TLP bonding method is not favored in industrial applications. Hence, in this study, the sintering behavior of the Ag–In alloy paste developed by ball-milling was investigated. The results demonstrated that increasing the bonding pressure is another effective method to lower the porosity. The low-porosity Ag–In alloy joint with the optimized bonding condition exhibited excellent reliability because it simultaneously avoided the oxidation and sintered pore coarsening problems. Consequently, the Ag–In alloy joint could retain its outstanding mechanical properties at high operating temperatures. In addition to reducing the porosity, enhancing the heat dissipation of power chip modules is another way to avoid oxidation. If the heat dissipation is sufficient to keep the working temperature from exceeding 200 °C, then the oxidation problem can also be avoided because the driving force to form the Cu_2O layer in the sintered joints is insufficient. Lastly, another potential strategy is to introduce a surface finish as a diffusion barrier to prevent the diffusion of Cu from the substrate at high temperatures.

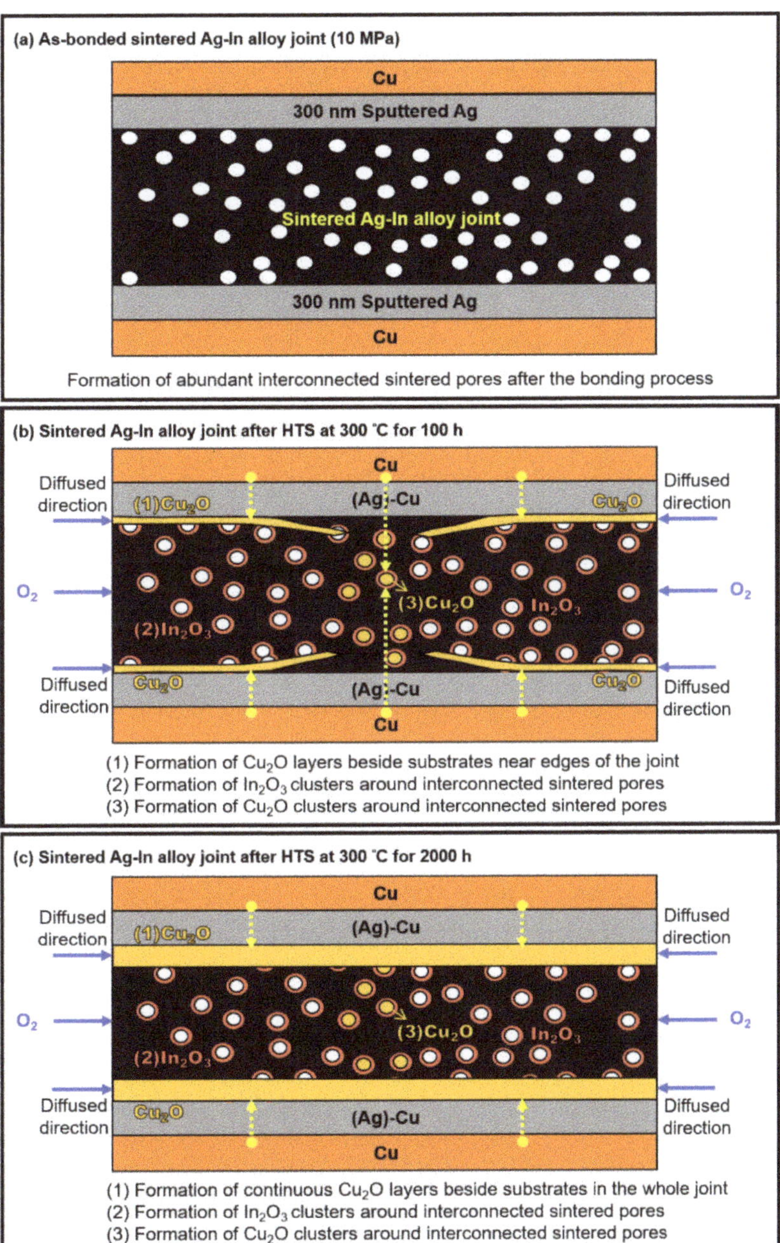

Figure 16. Proposed mechanism for the oxidation process of the sintered Ag–In alloy joints at a bonding pressure of 10 MPa. Schematic drawings for the cross-section of the sintered Ag-In alloy joints after the HTS at 300 °C for (**a**) 0 h, (**b**) 100 h, and (**c**) 2000 h.

5. Conclusions

In this study, pure Ag and In powders were refined to make flake-like Ag and Ag–In alloy powders using a high-energy ball-milling process. Homogeneous Ag–In alloy powders in solid solution could be obtained via mechanical alloying. Moreover, Ag and Ag–In alloy pastes containing particle sizes less than 25 μm were synthesized for power chip attachment after sieving and mixing with an organic solvent. The graphical summary of the sintered Ag joints and Ag-In alloy joints is exhibited in Figure 17. The Ag joint bonded at 10 MPa exhibited excellent mechanical properties. In comparison to low-pressure or pressureless Ag bonding in previous studies, the Ag joint bonded at an optimized pressure significantly reduced the porosity, preventing the joint from being oxidized at high temperatures. However, the well-bonded Ag joint with low porosity still could not avoid the sintered pore coarsening issue in a high-temperature application environment, which significantly deteriorated the mechanical properties. Conversely, the as-bonded Ag–In alloy joint bonded at 10 MPa exhibited outstanding mechanical properties with a shear strength of 53 MPa as well. Sintered pore coarsening was inhibited in the Ag–In alloy joint during the HTS. However, the Ag–In alloy joint bonded at 10 MPa suffered from oxidation problems because it had relatively high porosity and interconnected pores, which provided pathways for O_2 to enter the joint from the edges and deteriorated the sintered joint. Increasing the bonding pressure to 20 MPa significantly reduced the porosity of the Ag–In alloy joint, blocking external O_2 at high temperatures. Therefore, the robust Ag–In alloy joint possessed the most stable shear strength and reliable mechanical properties by maintaining high shear strength of above 76 MPa without deterioration in a high-temperature environment because it inhibited the sintered pore coarsening and oxidation issues simultaneously. Lastly, the mechanism for the oxidation process in sintered joints was established in this study and used to formulate strategies to prevent such oxidation. In conclusion, a novel ball-milling method was employed in this study to produce the Ag and Ag–In alloy pastes for die attachment through mechanical alloying, and a promising and reliable Ag–In alloy material was developed for power electronic packaging. It is considered that the materials and technologies developed in this study have great potential for practical application in power-chip attachment and can meet the urgent requirements of future industries for highly reliable power semiconductor packaging materials for EVs or aerospace technology.

Properties	Sintered Ag Joint (10 MPa)	Sintered Ag Joint (20 MPa)	Ag-In Alloy Joint (10 MPa)	Ag-In Alloy Joint (20 MPa)
Cu Oxidation Issue	No (porosity « percolation limit)		Yes (porosity ≈ percolation limit)	No (porosity < percolation limit)
Interfacial Wettability	Excellent with Ag Metallization			
Pores Coarsening Issue	Yes	Yes	No	No
Mechanical Properties	Unstable	Unstable	Unstable	Stable
High Temperature Reliability	Fair	Fair	Poor	Excellent

Figure 17. Graphical summary of the benefits of robust Ag–In alloy joints bonded with 20 MPa compared to other joints in 300 °C high-temperature applications.

Author Contributions: Conceptualization, C.-H.T., W.-C.H. and C.R.K.; methodology, C.-H.T., W.-C.H. and C.R.K.; validation, C.-H.T., W.-C.H.; investigation, C.-H.T. and W.-C.H.; data curation, C.-H.T. and W.-C.H.; writing—original draft preparation, C.-H.T.; writing—review and editing, C.-H.T.; supervision, C.R.K.; project administration, C.R.K.; funding acquisition, C.R.K. All authors have read and agreed to the published version of the manuscript.

Funding: This research was funded by the Advanced Research Center for Green Materials Science and Technology from the Featured Area Research Center Program within the framework of the Higher Education Sprout Project by the Ministry of Education in Taiwan (under Grant 110L9096) and the Ministry of Science and Technology in Taiwan (under Grant MOST 110–2223–E–002–002 and MOST 110–2634–F–002–043).

Institutional Review Board Statement: Not applicable.

Informed Consent Statement: Not applicable.

Data Availability Statement: Not applicable.

Acknowledgments: The authors would like to thank Jing-Yao Chang at Industrial Technology Research Institute in Taiwan for great assistance in the assembly process, as well as Ya-Yun Yang and Ching-Yen Lin from the Precious Instrument Center of College of Science at National Taiwan University for assistance with the ion milling.

Conflicts of Interest: The authors declare no conflict of interest.

References

1. Senecal, P.K.; Leach, F. Diversity in transportation: Why a mix of propulsion technologies is the way forward for the future fleet. *Results Eng.* **2019**, *4*, 100060. [CrossRef]
2. Woo, D.R.M.; Yuan, H.H.; Li, J.A.J.; Ling, H.S.; Bum, L.J.; Songbai, Z. High power SiC inverter module packaging solutions for junction temperature over 220C. In Proceedings of the 2014 IEEE 16th Electronics Packaging Technology Conference (EPTC), Singapore, 3–5 December 2014; pp. 31–35.
3. Buttay, C.; Planson, D.; Allard, B.; Bergogne, D.; Bevilacqua, P.; Joubert, C.; Lazar, M.; Martin, C.; Morel, H.; Tournier, D.; et al. State of the art of high temperature power electronics. *Mater. Sci. Eng. B* **2011**, *176*, 283–288. [CrossRef]
4. Millan, J.; Godignon, P.; Perpiñà, X.; Perez-Tomas, A.; Rebollo, J. A survey of wide bandgap power semiconductor devices. *IEEE Trans. Power Electron.* **2014**, *29*, 2155–2163. [CrossRef]
5. Shenai, K.; Dudley, M.; Davis, R.F. Current status and emerging trends in wide bandgap (WBG) semiconductor power switching devices. *ECS J. Solid State Sci. Technol.* **2013**, *2*, N3055–N3063. [CrossRef]
6. Matallana, A.; Ibarra, E.; López, I.; Andreu, J.; Garate, J.I.; Jordà, X.; Rebollo, J. Power module electronics in HEV/EV applications: New trends in wide-bandgap semiconductor technologies and design aspects. *Renew. Sustain. Energy Rev.* **2019**, *113*, 109264. [CrossRef]
7. Roccaforte, F.; Fiorenza, P.; Greco, G.; Nigro, R.L.; Giannazzo, F.; Iucolano, F.; Saggio, M. Emerging trends in wide band gap semiconductors (SiC and GaN) technology for power devices. *Microelectron. Eng.* **2018**, *187–188*, 66–77. [CrossRef]
8. Milligan, J.W.; Sheppard, S.; Pribble, W.; Wu, Y.-F.; Muller, G.; Palmour, J.W. SiC and gan wide bandgap device technology overview. In Proceedings of the 2007 IEEE Radar Conference, Waltham, MA, USA, 17–20 April 2007; pp. 960–964.
9. Shen, Z.J.; Omura, I. Power semiconductor devices for hybrid, electric, and fuel cell vehicles. *Proc. IEEE* **2007**, *95*, 778–789. [CrossRef]
10. Moon, K.-W.; Boettinger, W.J.; Kattner, U.R.; Biancaniello, F.S.; Handwerker, C.A. Experimental and thermodynamic assessment of Sn-Ag-Cu solder alloys. *J. Electron. Mater.* **2000**, *29*, 1122–1136. [CrossRef]
11. Zeng, G.; Xue, S.; Zhang, L.; Gao, L.; Dai, W.; Luo, J. A review on the interfacial intermetallic compounds between Sn–Ag–Cu based solders and substrates. *J. Mater. Sci. Mater. Electron.* **2010**, *21*, 421–440. [CrossRef]
12. Kotadia, H.R.; Howes, P.D.; Mannan, S.H. A review: On the development of low melting temperature Pb-free solders. *Microelectron. Reliab.* **2014**, *54*, 1253–1273. [CrossRef]
13. Chua, S.; Siow, K. Microstructural studies and bonding strength of pressureless sintered nano-silver joints on silver, direct bond copper (DBC) and copper substrates aged at 300 °C. *J. Alloys Compd.* **2016**, *687*, 486–498. [CrossRef]
14. Yang, F.; Zhu, W.; Wu, W.; Ji, H.; Hang, C.; Li, M. Microstructural evolution and degradation mechanism of SiC–Cu chip attachment using sintered nano-Ag paste during high-temperature ageing. *J. Alloys Compd.* **2020**, *846*, 156442. [CrossRef]
15. Zhang, Z.; Chen, C.; Yang, Y.; Zhang, H.; Kim, D.; Sugahara, T.; Nagao, S.; Suganuma, K. Low-temperature and pressureless sinter joining of Cu with micron/submicron Ag particle paste in air. *J. Alloys Compd.* **2019**, *780*, 435–442. [CrossRef]
16. Zhao, S.-Y.; Li, X.; Mei, Y.-H.; Lu, G.-Q. Study on high temperature bonding reliability of sintered nano-silver joint on bare copper plate. *Microelectron. Reliab.* **2015**, *55*, 2524–2531. [CrossRef]
17. Zhang, H.; Nagao, S.; Suganuma, K.; Albrecht, H.-J.; Wilke, K. Thermostable Ag die-attach structure for high-temperature power devices. *J. Mater. Sci. Mater. Electron.* **2016**, *27*, 1337–1344. [CrossRef]
18. Chuang, R.W.; Lee, C.C. Silver-indium joints produced at low temperature for high temperature devices. *IEEE Trans. Compon. Packag. Technol.* **2002**, *25*, 453–458. [CrossRef]

19. Lee, C.C.; So, W.W. High temperature silver-indium joints manufactured at low temperature. *Thin Solid Film* **2000**, *366*, 196–201. [CrossRef]

20. Yang, C.A.; Kao, C.R. Study of Sintered Nano-Silver Die Attachment Materials Doped with Indium. In Proceedings of the 2016 IEEE 66th Electronic Components and Technology Conference (ECTC), Las Vegas, NV, USA, 31 May–3 June 2016; pp. 2468–2474.

21. Yang, C.A.; Kao, C.R.; Nishikawa, H. Development of Die Attachment Technology for Power IC Module by Introducing Indium into Sintered Nano-Silver Joint. In Proceedings of the 2017 IEEE 67th Electronic Components and Technology Conference (ECTC), Orlando, FL, USA, 30 May–2 June 2017; pp. 1974–1980.

22. Yang, C.A.; Kao, C.R.; Nishikawa, H.; Lee, C.C. High Reliability Sintered Silver-Indium Bonding with Anti-Oxidation Property for High Temperature Applications. In Proceedings of the 2018 IEEE 68th Electronic Components and Technology Conference (ECTC), San Diego, CA, USA, 29 May–1 June 2018; pp. 1993–1999.

23. Tsai, C.-H.; Huang, W.-C.; Kao, C.R.; Chew, L.M.; Schmitt, W.; Nishikawa, H. Sintered Micro-Silver Paste Doped with Indium for Die Attachment Applications of Power ICs. In Proceedings of the 2020 IEEE 70th Electronic Components and Technology Conference (ECTC), Orlando, FL, USA, 3–30 June 2020; pp. 1430–1435.

24. Yang, C.A.; Yang, S.; Liu, X.; Nishikawa, H.; Kao, C.R. Enhancement of nano-silver chip attachment by using transient liquid phase reaction with indium. *J. Alloys Compd.* **2018**, *762*, 586–597. [CrossRef]

25. Tsai, C.-H.; Huang, W.-C.; Chew, L.M.; Schmitt, W.; Li, J.; Nishikawa, H.; Kao, C.R. Low-pressure micro-silver sintering with the addition of indium for high-temperature power chips attachment. *J. Mater. Res. Technol.* **2021**, *15*, 4541–4553. [CrossRef]

26. Lagutkin, S.; Achelis, L.; Sheikhaliev, S.; Uhlenwinkel, V.; Srivastava, V. Atomization process for metal powder. *Mater. Sci. Eng. A* **2004**, *383*, 1–6. [CrossRef]

27. Antony, L.V.M.; Reddy, R.G. Processes for production of high-purity metal powders. *JOM* **2003**, *55*, 14–18. [CrossRef]

28. Canakci, A.; Varol, T. A novel method for the production of metal powders without conventional atomization process. *J. Clean. Prod.* **2015**, *99*, 312–319. [CrossRef]

29. Foiles, S.M.; Baskes, M.I.; Daw, M.S. Embedded-atom-method functions for the fcc metals Cu, Ag, Au, Ni, Pd, Pt, and their alloys. *Phys. Rev. B* **1986**, *33*, 7983–7991. [CrossRef] [PubMed]

30. Li, Q.; Liu, S.; Li, S.; Guo, W.; Wu, C. Preparation of micro-size flake silver powder by planetary ball mill. *J. Mater. Sci. Mater. Electron.* **2016**, *27*, 452–457. [CrossRef]

31. Kováčik, J. Correlation between Young's modulus and porosity in porous materials. *J. Mater. Sci. Lett.* **1999**, *18*, 1007–1010. [CrossRef]

32. Bertei, A.; Choi, H.-W.; Pharoah, J.G.; Nicolella, C. Percolating behavior of sintered random packings of spheres. *Powder Technol.* **2012**, *231*, 44–53. [CrossRef]

33. Lee, S.-M.; Kang, S.-J.L. Theoretical analysis of liquid-phase sintering: Pore filling theory. *Acta Mater.* **1998**, *46*, 3191–3202. [CrossRef]

34. Drolet, J.P.; Galibois, A. The impurity-drag effect on grain growth. *Acta Met.* **1968**, *16*, 1387–1399. [CrossRef]

35. Mendelev, M.I.; Srolovitz, D.J. Impurity effects on grain boundary migration. *Model. Simul. Mater. Sci. Eng.* **2002**, *10*, R79–R109. [CrossRef]

36. Heo, T.W.; Bhattacharyya, S.; Chen, L.-Q. A phase field study of strain energy effects on solute–grain boundary interactions. *Acta Mater.* **2011**, *59*, 7800–7815. [CrossRef]

Article

Exploring Dielectric Constant and Dissipation Factor of LTCC Using Machine Learning

Yu-chen Liu [1,2], Tzu-Yu Liu [3], Tien-Heng Huang [3], Kuo-Chuang Chiu [3] and Shih-kang Lin [1,2,4,5,*]

1   Department of Materials Science and Engineering, National Cheng Kung University, Tainan 70101, Taiwan; 10809071@gs.ncku.edu.tw

2   Hierarchical Green-Energy Materials (Hi-GEM) Research Center, National Cheng Kung University, Tainan 70101, Taiwan

3   Material and Chemical Research Laboratories, Industrial Technology Research Institute, Hsinchu 31040, Taiwan; jill.t.y.liu@itri.org.tw (T.-Y.L.); TomHuang@itri.org.tw (T.-H.H.); ckc@itri.org.tw (K.-C.C.)

4   Core Facility Center, National Cheng Kung University, Tainan 70101, Taiwan

5   Program on Smart and Sustainable Manufacturing, Academy of Innovative Semiconductor and Sustainable Manufacturing, National Cheng Kung University, Tainan 70101, Taiwan

*   Correspondence: linsk@mail.ncku.edu.tw

Abstract: Low-temperature co-fired ceramics (LTCCs) have been attracting attention due to rapid advances in wireless telecommunications. Low-dielectric-constant (Dk) and low-dissipation-factor (Df) LTCCs enable a low propagation delay and high signal quality. However, the wide ranges of glass, ceramic filler compositions, and processing features in fabricating LTCC make property modulating difficult via experimental trial-and-error approaches. In this study, we explored Dk and Df values of LTCCs using a machine learning method with a Gaussian kernel ridge regression model. A principal component analysis and k-means methods were initially performed to visually analyze data clustering and to reduce the dimension complexity. Model assessments, by using a five-fold cross-validation, residual analysis, and randomized test, suggest that the proposed Dk and Df models had some predictive ability, that the model selection was appropriate, and that the fittings were not just numerical due to a rather small data set. A cross-plot analysis and property contour plot were performed for the purpose of exploring potential LTCCs for real applications with Dk and Df values less than 10 and 2×10^{-3}, respectively, at an operating frequency of 1 GHz. The proposed machine learning models can potentially be utilized to accelerate the design of technology-related LTCC systems.

Keywords: low-temperature co-fired ceramics (LTCCs); dielectric constant; dissipation factor; machine learning

1. Introduction

Low-temperature co-fired ceramics (LTCCs) have been attracting attention over recent decades due to rapid advances in wireless telecommunications, including the 5th generation (5G) tactile internet and the Internet of Things (IoT) [1]. LTCCs have characteristics that require sintering at temperatures of less than 1000 °C in order to be co-fired with electrode materials, such as Cu (melting point: 1083 °C), Ag (melting point: 961 °C), or Au (melting point: 1061 °C) [2]. LTCC devices provide a solution for integrating passive components, e.g., capacitors and resistors, with these electrodes into a three-dimensional module at the same time. A more recent review article from Sebastian et al. suggested that new LTCCs with ultra-low sintering temperatures (e.g., <700 °C) are becoming popular when pursing applications at a much lower temperature [3]. In recent years, the development of millimeter wave (mmWave) systems with typical frequencies above 24 GHz has led to performance benefits in 5G systems [4]. In the meantime, the increasing operating frequency from the current 4G systems at 3.5 GHz requires more reliable LTCC devices with a low dielectric constant (Dk), low dissipation factor (Df), and comparable mechanical strengths [1]. Wang et al. had pointed out that Dk and Df values should be lower than 10 and 2×10^{-3}, respectively, for real applications at high frequency operation [5]. Ohsato et al. recently reviewed the current status and prospects of LTCC applications in microwave and mmWave telecommunications [6]. These low Dk and Df values enable a low propagation delay and high signal quality in 5G systems.

LTCC fabrications are mostly based on glass-ceramic (GC) and glass/ceramic composite (GCC) in order to lower the sintering temperature [7]. Al_2O_3-based glass/ceramic composites are extensively used due to their good electrical and physical properties [8]. CaO-B_2O_3-SiO_2-Al_2O_3/Al_2O_3 composites have been reported to be promising materials due to their low firing temperature and low dielectric loss [9]. B_2O_3-SiO_2-Al_2O_3 glass and ZnO-B_2O_3-SiO_2 glass/Al_2O_3 composites have been shown to exhibit a low dielectric loss and good mechanical and thermal performance [10]. La_2O_3-B_2O_3-CaO-P_2O_5 glass/cordierite [5] or La_2O_3-B_2O_3-CaO glass/$LaBO_3$ composites [11] have recently been reported as potential candidates for real LTCC applications. Sebastian and Jantunen had made a thorough review on the material selections, fabrication methods, and properties of LTCCs [12]. The review suggests that LTCC fabrications involve modulating a wide range of glass, ceramic composition, and sintering conditions to meet the desired physical properties. The modulation is mainly carried out via experimental trial-and-error processes, and thus is time consuming and economically unfeasible. The wide ranges of input parameters make optimization even more difficult.

Theoretical modeling is an effective way to guide an experimental design. While there have been no existing models to properly simulate Df values, intrinsic Dk can be simulated via the density functional theory (DFT) in terms of polarizability. Peng et al. performed DFT calculation and the classical Clausius–Mossotti equation to model the dielectric constant of $Li_2(Mg_{1-x}Ni_x)SiO_4$ (x = 0.00–0.10) [13] and $(Zn_{1-x}Ni_x)_3B_2O_6$ (x = 0.00–0.07) ceramics [14]. However, microstructure and processing features make extrinsic contributions which are not easily simulated. Therefore, a more reliable simulation method is required. Recently, machine learning methods are considered powerful tools to predict material properties which do not have existing physical models, e.g., effective charges in electromigration [15], permittivity of microwave dielectric ceramics [16], and dielectric constants of crystals [17]. Morgan and Jacobs had made a thorough review of recent applications of machine learning methods used in the field of materials science [18]. In this study, we thus employed the machine learning method to explore Dk and Df properties of LTCC. We used the glass phase content, ceramic filler content, and GC content as well as the processing features (e.g., calcination temperature and time) to fit the experimentally-determined Dk and Df data. Gaussian kernel ridge regression was used as it is powerful for interpolating data points which has fewer hyperparameters than typical deep learning method (e.g., neural network). Fitting to fewer hyperparameters is beneficial to a small-scale data set. We assessed the model using a cross-validation, randomized test, and cross-plot analysis. The results suggest that the proposed models had a reasonable predictive ability. We explored the composition and processing feature spaces to find potential LTCC systems with low Dk and Df values. The proposed models may serve as a quick guideline for new LTCC material design in future technology-related systems. To the best of the author's knowledge, this is the first paper to use a machine learning method to explore Dk and Df of LTCC systems.

2. Methods

2.1. Data Set

The data set used in this study was provided by the Industrial Technology Research Institute (ITRI), Hsinchu, Taiwan. We refer to the database in this study as "ITRI-LTCC database." The database is focused on exploring potential LTCC systems with low Dk and Df values that can be applied in real 5G applications using various GC, GCC, and fabrication methods. The database consists of glass phase content, ceramic filler content, GC content, processing parameters (e.g., calcination temperature and time), and Dk

and *Df* measurements at various operating frequencies. The glass phases comprise two categories—(1) commercial glass and (2) and glass self-fabricated by ITRI. Commercial glasses consist of MgO-Al_2O_3-SiO_2 glass (MASG), CaO-B_2O_3-SiO_2 glass (CBSG), borosilicate glass (BG), and borosilicate glass + filler (BGF). For CBSG, the database includes two different compositions—one with higher SiO_2 content and the other with higher B_2O_3 content. For the convenience of reference in the following discussion, we refer to the CBSG with higher SiO_2 content as "CBSG-S," and the one with higher B_2O_3 content as "CBSG-B." Self-fabricated glass comprises mainly of MgO-Al_2O_3-SiO_2-based glass (MAS), with a few data points for Li_2O-Al_2O_3-SiO_2-based glass and Al_2O_3-SiO_2-based glass. MAS consists of different oxide compositions, i.e., different compositions of MgO, Al_2O_3, SiO_2, and other relevant oxides. The ceramic filler and GC in the database are alumina and cordierite, respectively. A given self-fabricated glass might be chosen with a given mole percent and added into a glass or GCC comprising commercial glass, ceramic filler, and/or GC in order to tailor the microstructure and the corresponding properties. For example, 2 mol% MAS was added in a GCC composed of 30 wt.% alumina, 20 wt.% cordierite, and 50 wt.% CBSG-S. Processing parameters consist of three stages of calcination at various temperatures and times. We used *T1*, *T2*, and *T3* throughout this article to represent the first to third calcination temperatures, while used *time_1*, *time_2*, and *time_3* to represent the first to third calcination times, respectively (e.g., *T1* and *time_1* represent the first stage calcination temperature and time). *Dk* and *Df* values were measured at four different operating frequencies in the database, including 0.1, 1, 10, and 11 GHz, but mainly at 1 GHz. In the present study, the data measured at 1 GHz was pulled out to become an initial data set, which had a total number of 116 data points. Table 1 shows the feature information of these 116 data points. A principle component analysis (PCA) and k-means method from the scikit-learn library [19] were used for analyzing the data clustering in the initial data set. The proposed machining learning models for *Dk* and *Df* were thus developed on one of the clusters (see Section 3.1) with 63 data points. The proposed machine learning models are referred to as "the proposed models" in the following discussion.

Table 1. The feature information of the initial data set (number of data points = 116).

Feature	Maximum	Minimum	Average	Standard Deviation	Unit	Category
Dielectric constant (*Dk*) at 1 GHz	7.8	2.71	5.33	1	-	Target feature
Dissipation factor (*Df*) at 1 GHz	16.7×10^{-3}	0.07×10^{-3}	2.7×10^{-3}	3.4×10^{-3}	-	Target feature
Al_2O_3 (Alumina)	50	0	8.5	16.1	wt.%	Ceramic filler
$Mg_2Al_4Si_5O_{18}$ (Cordierite)	72.9	0	9.9	19.8	wt.%	Glass-ceramic
Borosilicate glass + filler (BGF)	100	0	3.7	18.4	wt.%	Glass phase
MgO-Al_2O_3-SiO_2 glass (MASG)	100	0	7.7	26.7	wt.%	Glass phase
CaO-B_2O_3-SiO_2 glass (high SiO_2, CBSG-S)	100	0	25.8	31.6	wt.%	Glass phase
Borosilicate glass (BG)	55	0	6.7	14.4	wt.%	Glass phase
CaO-B_2O_3-SiO_2 glass (high B_2O_3, CBSG-B)	100	0	26.5	39.6	wt.%	Glass phase
MgO-Al_2O_3-SiO_2-based ceramic (MAS)	100	0	5.6	15.6	mol%	Glass phase
First stage calcination temperature (*T1*)	1650	27	373.7	386.0	°C	Processing parameter
Second stage calcination temperature (*T2*)	750	27	599.3	263.8	°C	Processing parameter
Third stage calcination temperature (*T3*)	1200	27	852.0	85.6	°C	Processing parameter
First stage calcination time (*time_1*)	3	0	2.7	0.7	h	Processing parameter
Second stage calcination time (*time_2*)	2	0	1.6	0.8	h	Processing parameter
Third stage calcination time (*time_3*)	2	0	1.3	0.8	h	Processing parameter

Table 2 shows the feature information of the 63 data points after the PCA and k-means analysis used in developing the proposed models. The input features consisted of the glass phase content, GC (i.e., cordierite) content, ceramic filler (i.e., alumina) content, and the processing parameters. The glass phases were MASG, CBSG-S, CBSG-B, BG, and MAS. MAS had a constant composition of MgO, Al_2O_3, SiO_2, and other relevant oxides. *T1* and *time_1* was constant at 750 °C and 3 h, respectively, while *T2, T3, time_2* and *time_3* remained variables. Note that, in Table 2, we defined two further features labeled as the *X* stage calcination reaction product (*TX_R*) and *X* stage calcination temperature and time product (*TX_time*), where *X* = second (2) or third (3). *TX_R* was defined as the calcination time multiplied by $\exp\left(\frac{-1}{\text{calcination temperature}}\right)$, while *TX_time* was defined as the calcination time multiplied by the calcination temperature. We found that these features showed a better model fit, for which the details are discussed in Section 3.1.

Table 2. The feature information of the data set used in the proposed models (number of data points = 63).

Feature	Maximum	Minimum	Average	Standard Deviation	Unit	Category
Dielectric constant (*Dk*) at 1 GHz	7.8	2.79	5.3	0.97	-	Target feature
Dissipation factor (*Df*) at 1 GHz	17.6×10^{-3}	0.34×10^{-3}	2.74×10^{-3}	3.57×10^{-3}	-	Target feature
Al_2O_3 (Alumina)	50	0	20.16	19.6	wt.%	Ceramic filler
$Mg_2Al_4Si_5O_{18}$ (Cordierite)	70	0	14.29	21.27	wt.%	Glass-ceramic
MgO-Al_2O_3-SiO_2 glass (MASG)	100	0	7.94	27.03	wt.%	Glass phase
CaO-B_2O_3-SiO_2 glass (high SiO_2, CBSG-S)	100	0	37.6	30.61	wt.%	Glass phase
Borosilicate glass (BG)	55	0	7.72	14.88	wt.%	Glass phase
CaO-B_2O_3-SiO_2 glass (high B_2O_3, CBSG-B)	75	0	7.54	19.6	wt.%	Glass phase
MgO-Al_2O_3-SiO_2-based ceramic (MAS)	100	2	8.49	20.54	mol%	Glass phase
Second stage calcination temperature (*T2*)	760	650	711.1	42.5	°C	Processing parameter
Third stage calcination temperature (*T3*)	1200	27	848.8	116.0	°C	Processing parameter
Second stage calcination time (*time_2*)	2	0.5	1.93	0.32	h	Processing parameter
Third stage calcination time (*time_3*)	2	0	1.60	0.71	h	Processing parameter
Second stage calcination reaction product (*T2_R*)	2	0.5	1.93	0.32	h/K	Processing parameter
Third stage calcination reaction product (*T3_R*)	2	0	1.6	0.71	h/K	Processing parameter
Second stage calcination temperature and time product (*T2_time*)	2400	0	1382.54	613.73	°C × h	Processing parameter
Third stage calcination temperature and time product (*T3_time*)	1500	375	1368.65	237.04	°C × h	Processing parameter

2.2. Machine Learning Modeling

The machine learning model used in this work was the Gaussian kernel ridge regression (GKRR). The GKRR model uses the radial basis function kernel, where a hyperparameter γ represents the length scale between two given features. The Gaussian kernel has the form shown in Equation (1):

$$k_{ij} = \exp\left(-\gamma \| x_i - x_j \|^2\right) \tag{1}$$

where x_i and x_j are given features vectors for LTCC i and j. k_{ij} ranges from 0, which occurs when all the LTCCs j are infinitely far from LTCC i as measured by the kernel, to 1, which occurs when all the LTCCs j are infinitely close to LTCC i as measured by the kernel. The ridge regression uses a hyperparameter α as the coefficient of the L2-norm to penalize the fitting coefficients. The cost function Φ thus has the matrix form shown in Equation (2):

$$\Phi = ||Y - K\beta||_2^2 + \alpha\beta^T K\beta \tag{2}$$

where Y is the target feature, K is the Gaussian kernel, and β are the kernel regression coefficients. The model analysis and exploration were primarily performed with the MAterials Simulation Toolkit for Machine Learning (MAST-ML, version 3.x, University of Wisconsin-Madison Computational Materials Group, Madison, WI, USA.) [20], an open-source Python package with scikit-learn [19] library to automate machine learning workflows and model assessments. The hyperparameters (α, γ) of the GKRR model were optimized using a genetic algorithm (GA) with the five-fold cross validation (CV) root-mean-square error (RMSE) as the scoring metric. Table 3 shows the optimized hyperparameters of Dk and Df models.

Table 3. The optimized hyperparameters of Dk and Df models.

Model	α	γ
Dk model	0.0012	0.3530
Df model	0.0420	3.2551

2.3. Model Assessment

A five-fold CV was used to assess the predictive ability of the model. The five-fold CV randomly partitioned the data set into five folds and took four of them as the training sub-dataset to build a GKRR model, with the remaining fold was used as the validation sub-data set. The CV process was repeated five times in one iteration, with each of the folds used exactly once as the validation dataset. The five-fold CV was repeated for 20 iterations. The 20 results were averaged to yield one prediction for each data point, and the average RMSE of these predictions was called the five-fold CV RMSE. The error bar for the five-fold CV RMSE was the standard deviation of the distribution of the RMSE values from the 20 iterations. The R^2 score of the model was calculated, for which the method can be found elsewhere [15]. A randomized test was used to determine whether the model fitted to physical correlations that are not real. A randomized test involves randomly associating each Dk or Df with a given feature vector, but not the correct one. This gives a new data set that is exactly like the original one in terms of the actual values, where all the physical associations of the features and Dk or Df were removed. A five-fold CV was then performed for these randomized data to show the RMSE and R^2 score. The interpolative quality of the model was examined using a cross-plot analysis, which shows how well the model predicted the target against a given variable with all the other variables held constant. A pair plot was generated using Python Library Seaborn [21].

3. Results and Discussion

3.1. Model Development

We intended to use the initial data set with 116 data points (see Table 1) to develop machine learning models for Dk and Df; however, we failed to develop decent models. In developing the model for Dk, the best five-fold CV RMSE was 0.64; the CV RMSE over the standard deviation of the data set (RMSE/σ) was 0.12, and the R^2 was only 0.28. The optimal input feature vector was found to be *T1*, *T2*, *T3*, alumina, cordierite, MASG, CBSG-S, CBSG-B, BG, BGF, and MAS. Two issues were raised: (1) The model did not capture the calcination time; (2) in the initial data set, because MgO, Al_2O_3, SiO_2, and the other relevant oxides content were not constant in MAS, the oxide composition should also be considered as part of the input features. Nevertheless, the model did not capture the oxide composition either. Similar circumstances were found when developing the Df model. We suspected that there was data bias due to clustering originated from the unbalancing weight of the limited data sampling. To examine the clustering issue, we performed the PCA and k-means method and show the cluster plot in Figure 1. There were two groups that were apparent in the cluster plot, i.e., the green group and the red group. Red points were found much closer to the clustering center than the green one. After careful examination of the data set, we found that the red group shared the same oxide composition in MAS, while the green group did not. In other words, the oxide composition in MAS could be held constant when using the red group data as the training data set. On the other hand, the first stage of the calcination temperature and time (i.e., $T1$ = 750 °C, $time\_1$ = 3 h), as well as the BGF content (i.e., BGF = 0 wt.%) were also constant in the red group. Figure 2 shows a pair plot of the glass, ceramic filler, and GC content distribution in the red group data. The scattered points in the plot show how the data sampled along the composition spaces. We thus pulled out the red group data with a total number of 63 data points to build Dk and Df models (see Table 2 for the feature information). By using a cluster plot analysis, feature dimensions and complexity were reduced although we paid the price of losing some data. Nevertheless, it was still better than having models that were not consistent with the actual experimental fabrication process.

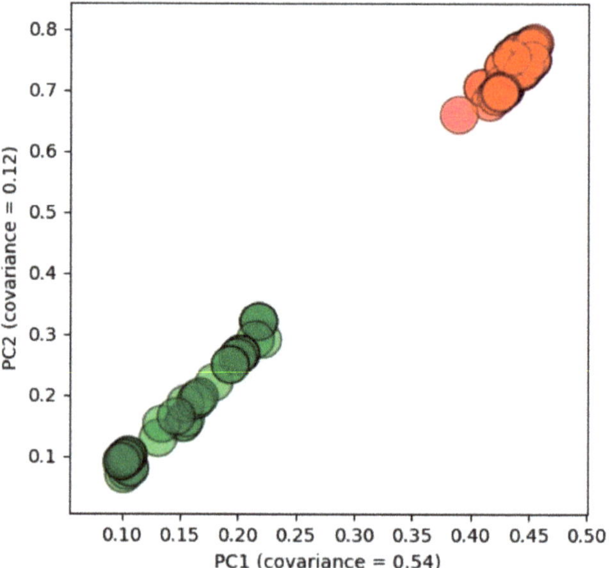

Figure 1. Cluster plot of the initial data set.

When developing Dk and Df models using the red group data, we firstly explored the optimal input feature vector. Figure 3 shows the five-fold CV RMSE vs. the top 12 input features, where the plot was sorted based on the CV RMSE values. The x-axis shows the input feature used in building the model. We showed only the processing parameters in the x-axis label because the glass, ceramic filler, and GC content (i.e., alumina, cordierite, MASG, CBSG-S, CBSG-B, BG, and MAS) remained the same in each feature vector. For instance, "T3_R" in Figure 3a means we used $T3\_R$, alumina, cordierite, MASG, CBSG-S, CBSG-B, BG, and MAS as the input for building Dk model. This input feature showed the lowest CV RMSE for Dk model. "T2_time, T3_time" in Figure 3b means we used $T2\_time$, $T3\_time$, alumina, cordierite, MASG, CBSG-S, CBSG-B, BG, and MAS as the input for building Df model and that this input feature showed the lowest CV RMSE for Df model. Note that, in Figure 3a, even though "T2_R, T3_R" showed the third-lowest CV RMSE, the difference in the RMSE between the lowest one was only 0.01. On the other hand, using the "T2_R, T3_R" feature set would be more consistent with actual experimental conditions. We felt that the model was actually capturing the real processing parameters, but due to the limited sampling of the data set, a numerical fitting issue occurred. Therefore, we manually selected the *T2_R*, *T3_R*, alumina, cordierite, MASG, CBSG-S, CBSG-B, BG, and MAS as the input for building *Dk* model. We did our best to explore as many processing parameter combinations as possible to see if the models could be further improved. For instance, we attempted to multiply the *T2* term by the *T3* term to see if coupling between the processing temperatures occurred. Nevertheless, it turned out that these CV RMSEs were not further improved. Overall, the designed features of *T2_R*, *T3_R*, *T2_time*, and *T3_*time worked better than using *T2*, *T3*, *time_2*, and *time_3* directly. This may suggest some underlying physics in terms of *Dk* and *Df* function that works against the processing parameters although the physics is not easily examined directly using a machine learning model. Using "*T2_R*, *T3_R*" or "*T2_time*, *T3_time*" also helped to reduce the number of feature dimensions and helped avoid the overfitting issue. In the meantime, the models built on these features were consistent with the real processing circumstances.

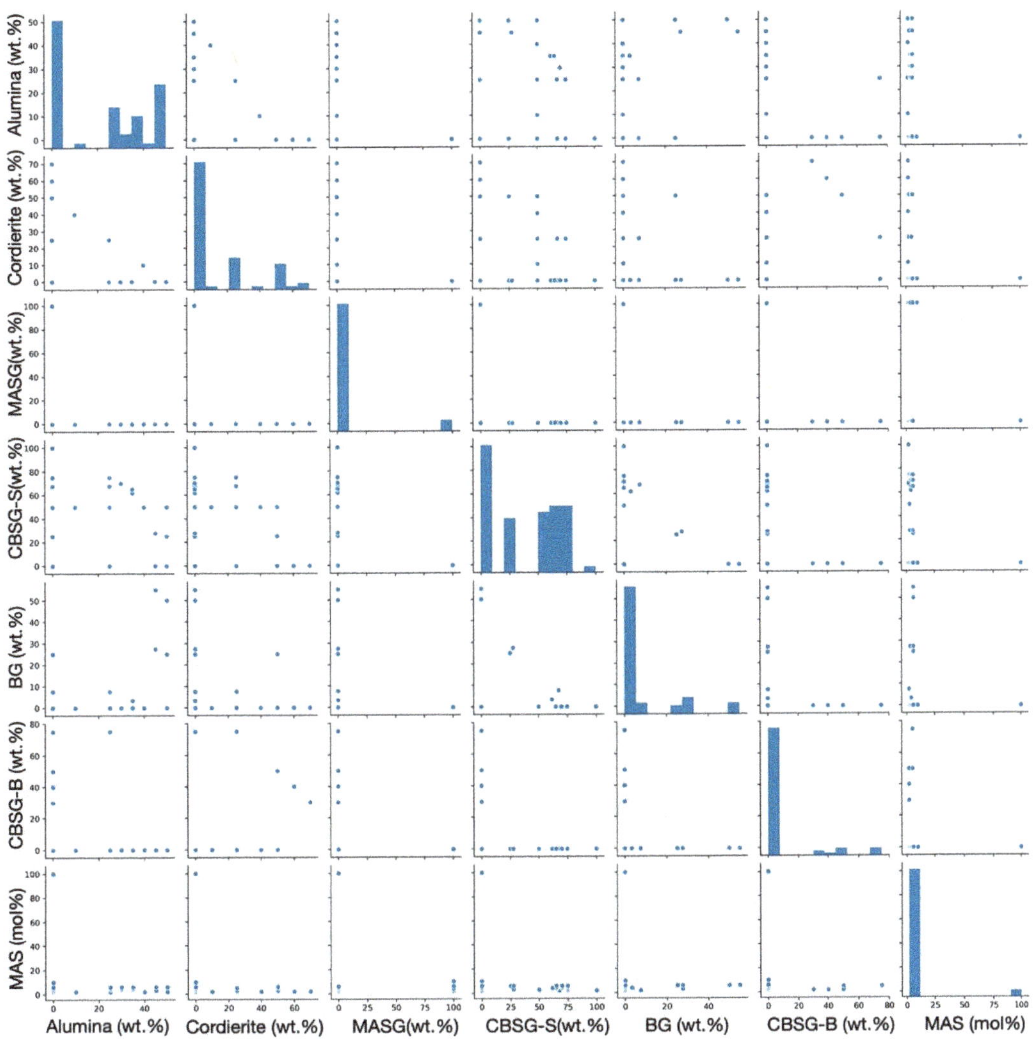

Figure 2. Pair plot of the input glass phase, ceramic filler, and GC content distribution.

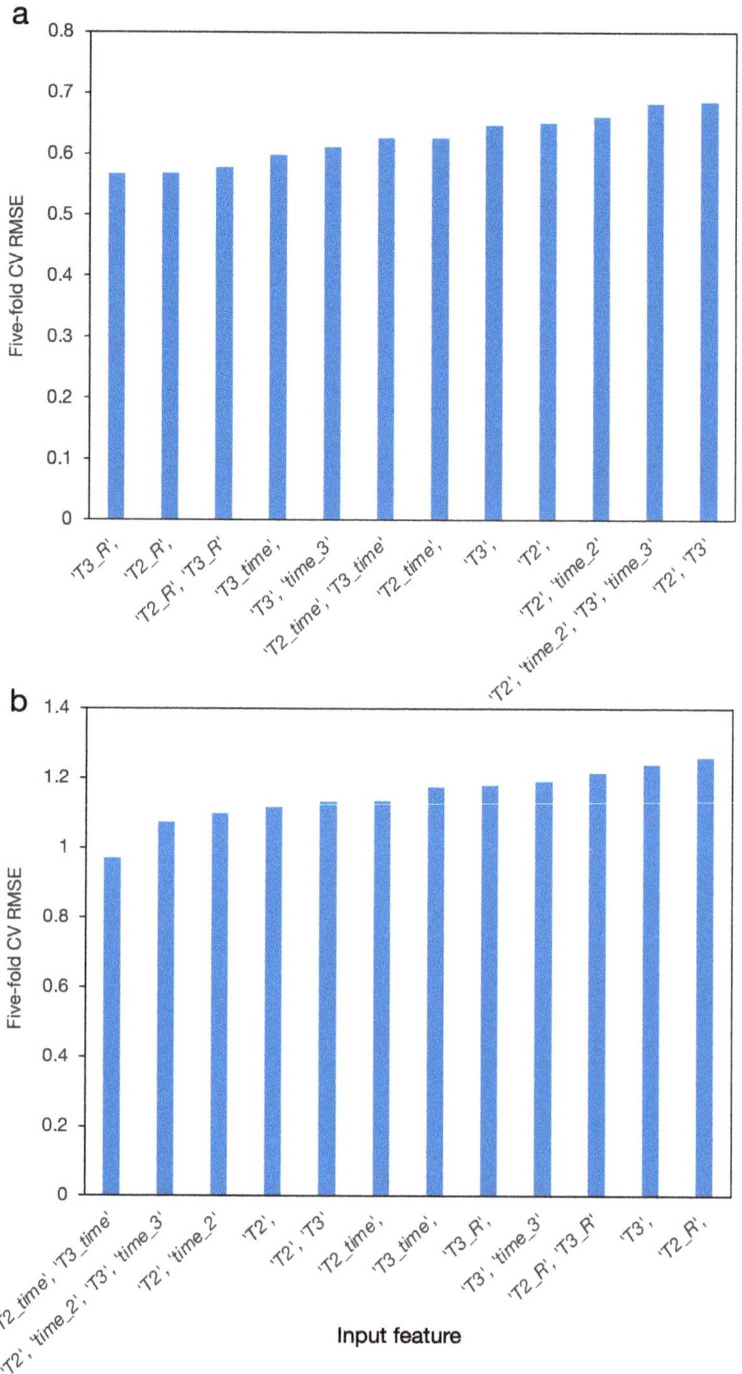

Figure 3. Five-fold CV RMSE vs. input feature to the model built for (**a**) *Dk* and (**b**) *Df*. The x-axis was sorted based on the CV RMSE values.

3.2. Model Assessment

Figure 4a,b show the parity plots of the full-fit for the proposed *Dk* and *Df* models fitted to the optimal features, respectively. The RMSE was 0.37, the RMSE/σ was 0.38, and the R^2 was 0.82 for *Dk* model. The RMSE was 0.39×10^{-3}, the RMSE/σ was 0.11, and the R^2 was 0.99 for *Df* model. Figure 4c,d show the parity plots of the five-fold CV for the proposed *Dk* and *Df* models, respectively. The CV RMSE was 0.59, the CV RMSE /σ was 0.61, and the R^2 was 0.57 for the *Dk* model. The CV RMSE was 1.12×10^{-3}, the CV RMSE/σ was 0.31, and the R^2 was 0.91 for the *Df* model. Overall, the RMSE/σ values were all less than one, which suggests that our models captured the complex *Dk* and *Df* properties by providing information only for the glass, ceramic filler, GC content, and the processing parameters. Figure 4e,f show the residual plots of the five-fold CV, for which the results showed an approximately normal distribution. This suggests that the choice of the GKRR model was appropriate.

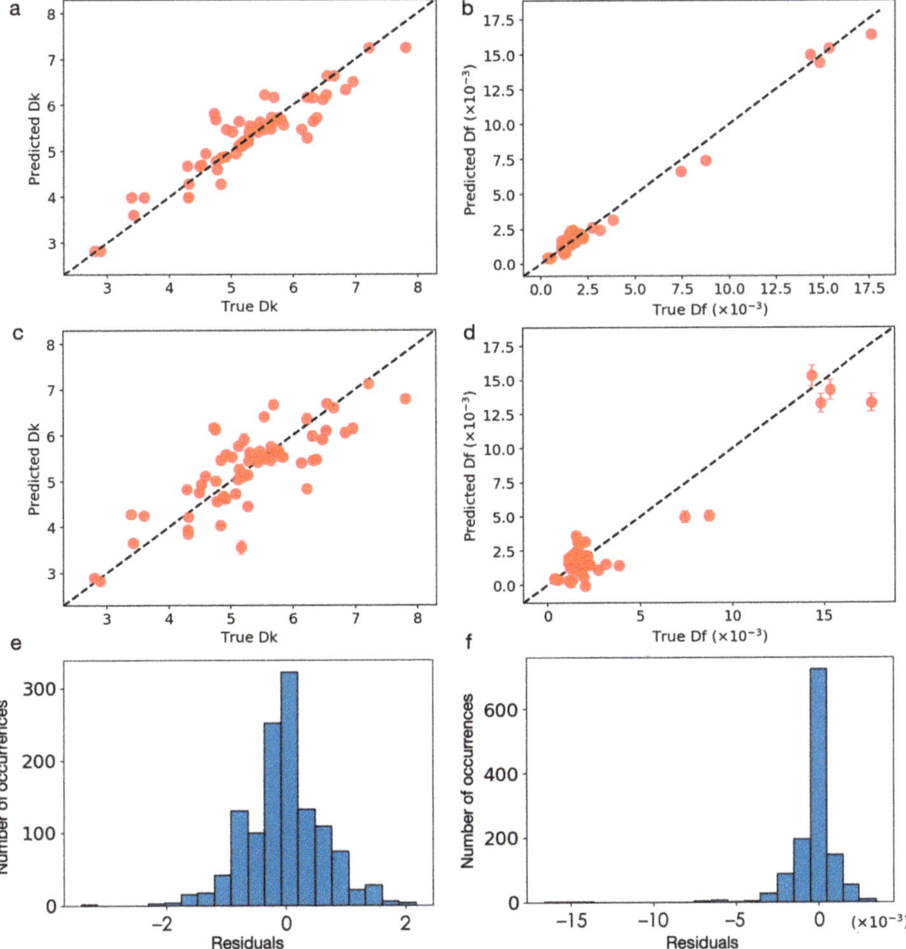

Figure 4. Parity plot of the (**a**,**b**) full-fit and (**c**,**d**) five-fold CV for (**a**,**c**) *Dk* and (**b**,**d**) *Df* models. (**a**) RMSE = 0.37, RMSE/σ = 0.38, R^2 = 0.82. (**b**) RMSE=0.39×10^{-3}, RMSE/σ = 0.11, R^2 = 0.99. (**c**) CV RMSE = 0.59, CV RMSE/σ = 0.61, R^2 = 0.57. (**d**) CV RMSE=1.12×10^{-3}, CV RMSE/σ = 0.31, R^2 = 0.91. The residual plot of the five-fold CV for (**e**) *Dk* and (**f**) *Df* model.

Due to a rather small scale of the data set with limited sampling (see Figure 2), it is conceivable that the model may not fit the correct physical correlations. To investigate this, we performed a randomized test [15]. Figure 5a,b shows the parity plots of the randomized tests for *Dk* and *Df* models, respectively. The RMSE was 1.19, the RMSE/σ was 1.22, and the R^2 was −1.58 for *Dk* model in the randomized test. The RMSE was 4.33×10^{-3}; the RMSE/σ was 1.21, and the R^2 was −0.94 for *Df* model in randomized test. All models for the randomized test were significantly worse than the models for the original data fits. The results suggest that the models for the original data were physically meaningful.

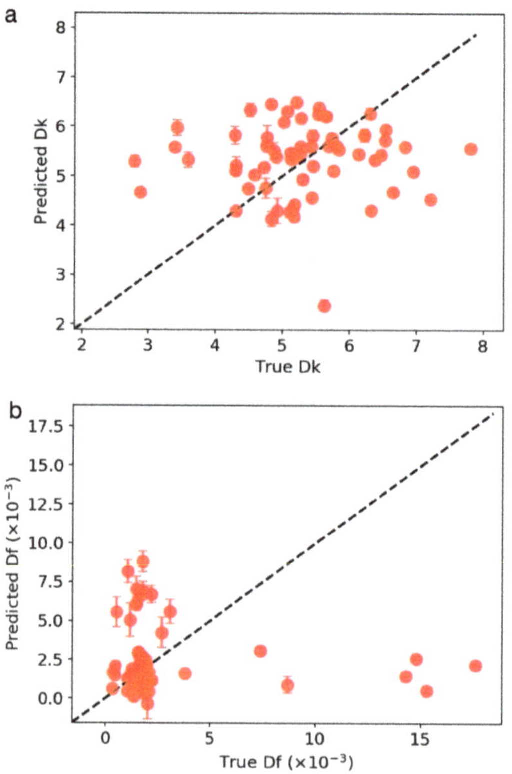

Figure 5. Parity plot of the randomized test for (**a**) *Dk* and (**b**) *Df* models. (**a**) RMSE = 1.19, RMSE/σ = 1.22, R^2 = −1.58. (**b**) RMSE=4.33×10^{-3}, RMSE/σ = 1.21, R^2 = −0.94.

3.3. Dk and Df Exploration

Because the data set was rather small and sampling was less homogeneous, it would be difficult to obtain an accurate prediction from extrapolation. Nevertheless, exploring potential LTCC candidates within the composition range of the data set as an interpolation would be still beneficial, especially in the relatively uniform composition spaces shown in Figure 2, e.g., in the alumina, cordierite, and CBSG-S spaces. To see how *Dk* and *Df* evolved along these feature spaces within the data set range, we performed a cross-plot analysis. We chose *T2* = 700 °C, *time_2* = 2 h, *T3* = 850 °C, *time_3* = 2 h (i.e., *T2_R*, and *T3_R* (*T2_time*, and *T3*_time) were set at 1.9979, and 1.9982 h/K (1400, and 1700 °C × h)), CBSG-S = 50 wt.%, MAS = 2 mol%, and all the other glass phases = 0 wt.% for the analysis because a series of LTCCs was fabricated at this range in the data set. Figure 6 shows the cross plot of *Dk* and *Df* against alumina and cordierite content. We found that *Dk* increased when the alumina content increased, as shown in Figure 6a. *Df* in general increased when the alumina content was increased until 40 wt.% and then decreased when alumina content was higher than

40 wt.%, as shown in Figure 6b. A small hump of *Df* increase at the alumina content of 0 to 5 wt.% was also found. Prediction for *Df* was slightly higher than the measured data but the difference was minor. Overall, the trend between the real measurements and the machine learning prediction agreed well with each other. From the cross plots, one would expect to obtain low *Dk* and *Df* values at a low (high) alumina (cordierite) content.

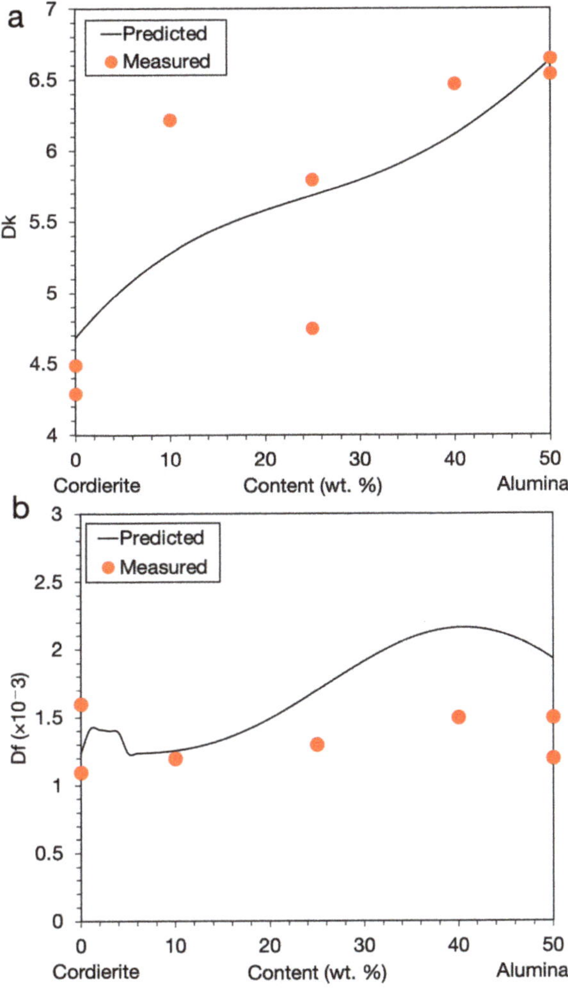

Figure 6. Cross plot of (**a**) *Dk* and (**b**) *Df* against the alumina and cordierite content. MAS was 2 mol%, *T2_R*, *T3_R* (*T2_time*, *T3*_time) were at 1.9979, 1.9982 h/K (1400, 1700 °C × h), and CBSG-S was 50 wt.%. All the other glass phase contents were zero.

As the goal for the proposed models is to explore potential LTCCs which have both low *Dk* and *Df* values to control signal delay and energy loss, a property contour plot would be visually useful to define the region fulfilling the given criteria. We followed the criteria provided in Ref [5], i.e., the notion that *Dk* and *Df* should be lower than 10 and 2×10^{-3}, respectively. We therefore set these criteria for the property contour plot and explored potential candidates within the alumina, cordierite, and CBSG-S content. MAS was again chosen at 2 mol%, and *T2_R*, *T3_R* (*T2_time*, *T3*_time) were set at 1.9979, 1.9982 h/K (1400, 1700 °C × h), respectively. All the other glass phase contents were set at zero. Figure 7 shows the property contour plot. The blue and red regions represent the LTCCs that fulfilled one, and fulfilled both criteria, respectively. This contour plot reveals that if one chooses the compositions in the red region and calcinates them in three different stages, i.e., (1) 750 °C for 3 h, (2) 700 °C for 2h, and then (3) 850 °C for 2 h, these LTCCs will be likely to have Dk and Df values less than 10 and 2×10^{-3}, respectively, at an operating frequency of 1 GHz. The plot provides a quick guideline for developing potential LTCCs with low Dk and Df values, as well as for saving both time and cost. Once more data become available, the proposed models could be further improved and extended to more complex systems.

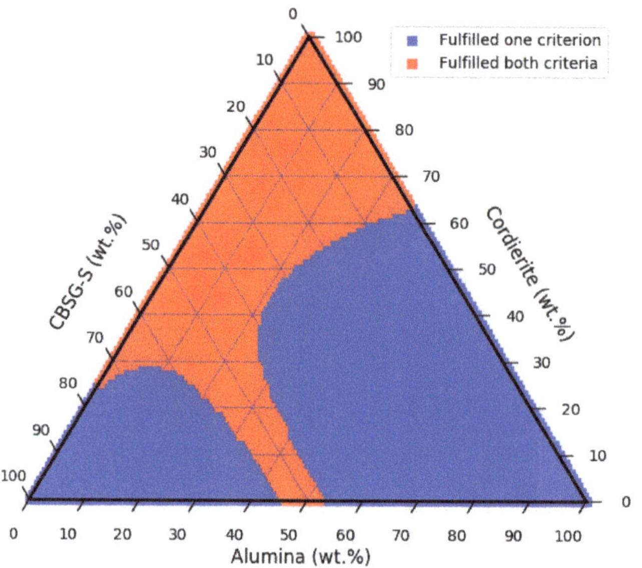

Figure 7. Property contour plot of LTCCs composed of CBSG-S, cordierite, and alumina. The MAS glass was 2 mol%, and the first to third calcination temperatures (time) were 750 °C, 700 °C, and 850 °C (3 h, 2 h, and 2 h), respectively. All the other glass phase contents were zero. The operating frequency was set at 1 GHz.

4. Conclusions

In this paper, we built machine learning models for predicting Dk and Df, and explored potential LTCCs with low Dk and Df values. Data at an operating frequency of 1 GHz were pulled out from the ITRI-LTCC database to build models. PCA and k-means methods were initially performed to visually analyze data clustering and to reduce the dimension complexity that inherently caused the model to fail. In optimizing the input features, we found that using the calcination reaction product (i.e., $T2\_R$ and $T3\_R$), as well as the calcination temperature and time product (i.e., $T2\_time$ and $T3\_time$), led to a better model performance (i.e., a lower five-fold CV RMSE) as opposed to using temperature and time separately (i.e., $T2$, $T3$, $time\_2$ and $time\_3$) for building Dk and Df models, respectively. The five-fold CV RMSE was 0.59, the CV RMSE/σ was 0.61, and the R^2 was 0.57 for the Dk model. The CV RMSE was 1.12×10^{-3}, the CV RMSE/σ was 0.31, and the R^2 was 0.91 for the Df model. CV results suggest that the proposed models captured the complex Dk and Df properties. Randomized test showed a worse model performance than that for the original data fits. It suggests that the proposed models were not only numerical due to the rather small data set, but were physically meaningful. Cross-plot analysis showed that the machine learning prediction agreed well with the real measurements. Cross-plot analysis suggests that the proposed models had the potential to predict Dk and Df within the input feature ranges as an interpolation. A property contour plot was built to explore LTCCs for real applications with *Dk* and *Df* values less than 10 and 2×10^{-3}, respectively, at an operating frequency of 1 GHz. Explorative models were obtained in the current work, and the models can be further improved as new data become available in the future. The proposed machine learning models can potentially be utilized to accelerate the design of LTCCs used in fifth-generation telecommunications.

Author Contributions: Methodology, Y.-c.L.; Software, Y.-c.L.; Validation, Y.-c.L. and T.-Y.L.; Formal Analysis, Y.-c.L.; Investigation, Y.-c.L. and T.-Y.L.; Writing—Original Draft Preparation, Y.-c.L.; Writing—Review & Editing, Y.-c.L., T.-Y.L. and S.-k.L.; Project Administration, T.-Y.L., T.-H.H., K.-C.C. and S.-k.L. All authors have read and agreed to the published version of the manuscript.

Funding: This research was funded by the Ministry of Science and Technology, Taiwan with project number of 109-3111-8-006-001, 110-2622-8-006-017-SB, 110-2634-F-006-017, 110-2636-E-006-016 and 110-2222-E-006-008; by the Ministry of Education, Taiwan, in Higher Education Sprout Project; by the Ministry of Economic Affairs, Taiwan, and ITRI/MCL, Taiwan with project number of L301AR4700.

Institutional Review Board Statement: Not applicable.

Informed Consent Statement: Not applicable.

Data Availability Statement: The ITRI-LTCC database is not publicly available and therefore this data is not included in any of the shared files. Requests for the ITRI-LTCC database should be directly sent to Tzu-Yu Liu at jill.t.y.liu@itri.org.tw.

Acknowledgments: Y.-c.L. and S.-k.L. gratefully acknowledge the financial supports from the Ministry of Science and Technology (MOST) in Taiwan (109-3111-8-006-001, 110-2622-8-006-017-SB, 110-2636-E-006-016 and 110-2222-E-006-008). This work was also partially supported by the Hierarchical Green-Energy Materials (Hi-GEM) Research Center, from The Featured Areas Research Center Program within the framework of the Higher Education Sprout Project by the Ministry of Education (MOE) and MOST (110-2634-F-006 -017) in Taiwan. T.-Y.L., T.-H.H. and K.-C.C. gratefully acknowledge funding from the Ministry of Economic Affairs (MEA), and ITRI/MCL of Taiwan. (Project No. L301AR4700).

Conflicts of Interest: The authors declare no conflict of interest.

References

1. Marley, P.M.; Tormey, E.S.; Yang, Y.; Gleason, C. Low-K LTCC Dielectrics: Novel High-Q Materials for 5G Applications. In Proceedings of the 2019 IEEE MTT-S International Microwave Workshop Series on Advanced Materials and Processes for RF and THz Applications (IMWS-AMP), Bochum, Germany, 16–18 July 2019; pp. 88–90.
2. Sebastian, M.T. Chapter Twelve—Low Temperature Cofired Ceramics. In *Dielectric Materials for Wireless Communication*; Sebastian, M.T., Ed.; Elsevier: Amsterdam, The Netherlands, 2008; pp. 445–512. [CrossRef]
3. Sebastian, M.T.; Wang, H.; Jantunen, H. Low temperature co-fired ceramics with ultra-low sintering temperature: A review. *Curr. Opin. Solid State Mater. Sci.* **2016**, *20*, 151–170. [CrossRef]
4. Mahon, S. The 5G Effect on RF Filter Technologies. *IEEE Trans. Semicond. Manuf.* **2017**, *30*, 494–499. [CrossRef]
5. Wang, F.; Zhang, W.; Chen, X.; Mao, H. Synthesis and characterization of low CTE value La$_2$O$_3$-B$_2$O$_3$-CaO-P$_2$O$_5$ glass/cordierite composites for LTCC application. *Ceram. Int.* **2019**, *45*, 7203–7209. [CrossRef]
6. Ohsato, H.; Varghese, J.; Vahera, T.; Kim, J.S.; Sebastian, M.T.; Jantunen, H.; Iwata, M. Micro/Millimeter-Wave Dielectric Indialite/Cordierite Glass-Ceramics Applied as LTCC and Direct Casting Substrates: Current Status and Prospects. *J. Korean Ceram. Soc* **2019**, *56*, 526–533. [CrossRef]
7. Zhou, J. Towards rational design of low-temperature co-fired ceramic (LTCC) materials. *J. Adv. Ceram.* **2012**, *1*, 89–99. [CrossRef]
8. Yu, M.; Zhang, J.; Li, X.; Liang, H.; Zhong, H.; Li, Y.; Duan, Y.; Jiang, D.L.; Liu, X.; Huang, Z. Optimization of the tape casting process for development of high performance alumina ceramics. *Ceram. Int.* **2015**, *41*, 14845–14853. [CrossRef]
9. Ren, L.; Zhou, H.; Li, X.; Xie, W.; Luo, X. Synthesis and characteristics of borosilicate-based glass–ceramics with different SiO$_2$ and Na2O contents. *J. Alloys Compd.* **2015**, *646*, 780–786. [CrossRef]
10. Shang, Y.; Zhong, C.; Xiong, H.; Li, X.; Li, H.; Jian, X. Ultralow-permittivity glass/Al2O3 composite for LTCC applications. *Ceram. Int.* **2019**, *45*, 13711–13718. [CrossRef]
11. Wang, F.; Zhang, W.; Chen, X.; Mao, H.; Liu, Z.; Bai, S. Low temperature sintering and characterization of La2O3-B2O3-CaO glass-ceramic/LaBO3 composites for LTCC application. *J. Eur. Ceram. Soc.* **2020**, *40*, 2382–2389. [CrossRef]
12. Sebastian, M.T.; Jantunen, H. Low loss dielectric materials for LTCC applications: A review. *Int. Mater. Rev.* **2008**, *53*, 57–90. [CrossRef]

13. Peng, R.; Su, H.; An, D.; Lu, Y.; Tao, Z.; Chen, D.; Shi, L.; Li, Y. The sintering and dielectric properties modification of Li2MgSiO4 ceramic with Ni2+-ion doping based on calculation and experiment. *J. Mater. Res. Technol.* **2020**, *9*, 1344–1356. [CrossRef]

14. Peng, R.; Su, H.; Li, Y.; Lu, Y.; Yu, C.; Shi, L.; Chen, D.; Liao, B. Microstructure and microwave dielectric properties of Ni doped zinc borate ceramics for LTCC applications. *J. Alloys Compd.* **2021**, *868*, 159006. [CrossRef]

15. Liu, Y.-c.; Afflerbach, B.; Jacobs, R.; Lin, S.-k.; Morgan, D. Exploring effective charge in electromigration using machine learning. *MRS Commun.* **2019**, *9*, 567–575. [CrossRef]

16. Qin, J.; Liu, Z.; Ma, M.; Li, Y. Machine learning approaches for permittivity prediction and rational design of microwave dielectric ceramics. *J. Mater.* **2021**. [CrossRef]

17. Morita, K.; Davies, D.W.; Butler, K.T.; Walsh, A. Modeling the dielectric constants of crystals using machine learning. *J. Chem. Phys.* **2020**, *153*, 024503. [CrossRef] [PubMed]

18. Morgan, D.; Jacobs, R. Opportunities and challenges for machine learning in materials science. *Annu. Rev. Mater. Res.* **2020**, *50*, 71–103. [CrossRef]

19. Pedregosa, F.; Varoquaux, G.; Gramfort, A.; Michel, V.; Thirion, B.; Grisel, O.; Blondel, M.; Prettenhofer, P.; Weiss, R.; Dubourg, V.; et al. Scikit-learn: Machine Learning in Python. *J. Mach. Learn.* **2011**, *12*, 2825–2830.

20. Jacobs, R.; Mayeshiba, T.; Afflerbach, B.; Miles, L.; Williams, M.; Turner, M.; Finkel, R.; Morgan, D. The Materials Simulation Toolkit for Machine learning (MAST-ML): An automated open source toolkit to accelerate data-driven materials research. *Comput. Mater. Sci.* **2020**, *176*, 109544. [CrossRef]

21. Waskom, M.L. Seaborn: Statistical data visualization. *J. Open Source Softw.* **2021**, *6*, 3021. [CrossRef]

Article

Investigation of Adhesive's Material in Hermetic MEMS Package for Interfacial Crack between the Silver Epoxy and the Metal Lid during the Precondition Test

Mei-Ling Wu * and **Jia-Shen Lan**

Department of Mechanical and Electro-Mechanical Engineering, National Sun Yat-sen University, 70 Lien-Hai Rd., Kaohsiung 804201, Taiwan; d023020004@student.nsysu.edu.tw
* Correspondence: meiling@mail.nsysu.edu.tw

Abstract: A hermetic Micro-Electro-Mechanical Systems (MEMS) package with a metal lid is investigated to prevent lid-off failure and improve its reliability during the precondition test. While the MEMS package benefits from miniaturization and low cost, a hermetic version is highly sensitive to internal pressure caused by moisture penetration and the reflow process, thus affecting its reliability. In this research, the finite element method is applied to analyze the contact stress between the metal lid and the silver epoxy by applying the cohesive zone model (CZM). Moreover, the red dye penetration test is applied, revealing a microcrack at the metal lid/silver epoxy interface. Further analyses indicate that the crack is caused by internal pressure. According to the experimental testing and simulation results, the silver epoxy material, the curing process, the metal lid geometry, and the bonding layer contact area can enhance the bonding strength between the metal lid and the substrate.

Keywords: Micro-Electro-Mechanical Systems; moisture sensitivity level test; reflow process; finite element method; cohesive zone model; bonding strength; precondition test

check for **updates**

Citation: Wu, M.-L.; Lan, J.-S. Investigation of Adhesive's Material in Hermetic MEMS Package for Interfacial Crack between the Silver Epoxy and the Metal Lid during the Precondition Test. *Materials* **2021**, *14*, 5626. https://doi.org/10.3390/ma14195626

Academic Editor: Arkadiusz Żak

Received: 10 August 2021
Accepted: 23 September 2021
Published: 27 September 2021

Publisher's Note: MDPI stays neutral with regard to jurisdictional claims in published maps and institutional affiliations.

1. Introduction

The MEMS package is widely used in automotive, medical, and consumer electronic devices for measuring the mechanical, thermal, optical, and magnetic phenomena. Therefore, it needs to be highly reliable to maintain its functionality and safety for several years. A metal lid is included in the MEMS package to provide protection from the external environment, while also reducing the cost, weight, and size, thus improving production efficiency. While the hermetic MEMS package offers considerable benefits, it also causes reliability issues for the lid bonding technology and hermeticity. Hsu et al. [1] investigated the characteristics of polymeric materials in the CMOS image sensor (CIS). They found that the hygroscopic swelling of polymer material is induced by absorbing the moisture in humid environments, which weakens the interfacial strength and causes delamination failure. To improve performance and reliability in RF-MEMS applications, Jeong et al. [2] developed novel wafer-level hermetic package technology. As a part of their study, low-temperature bonding technology was applied through gold/tin eutectic solder at the peripheral edge. The results revealed that thermal cycling, high-temperature storage, high-humidity storage, and a pressure cooker test failed to induce failure. Zhang et al. [3] investigated the bonding strength of a nanosilver sintered hermetic cavity with copper and silicon lids. Their results showed that the copper lid suffered delamination in the bonding layer, whereas the silicon lid exhibited great bonding quality. Farisi et al. [4] developed a low-temperature wafer-level hermetic packaging technology based on thermal compression bonding. Their analyses revealed that the bonding shear strength of the newly proposed technology exceeded 100 MPa and its leak rate was below 1.67×10^{-15} Pa·m^3·s^{-1}. Huang et al. [5] also developed low-cost and low-temperature hermetic technology based on a eutectic PbSn solder and Cr/Ni/Cu bonding pad. The bonding strengths of glass–glass, silicon–glass, and silicon–silicon pairs were measured at 4.5, 7, and 5.3 MPa, respectively. Jang et al. [6] proposed a diffusion-based governing equation to investigate the effects of polymer seal diffusion properties and geometries on the MEMS package performance. The numerical results revealed that both factors affected the lag time. Premachandran et al. [7] developed a wafer-level vacuum package with a wafer cap under the vacuum (1 mTorr). The package performance measured up to standard, evaluated via shear test and reliability tests. Jiang et al. [8] used a laser-assisted bonding method for a cavity-based package with a liquid crystal polymer (LCP). Their results showed that both silicon and glass substrates had high bonding quality. They also measured shear strength in the 20.8–26.1 MPa range, depending on the bonding assembly (glass–glass, silicon–glass, silicon–silicon, and silicon–package). Sandvand et al. [9] analyzed the bonding material stress in the MEMS pressure sensor for the glass-frit bonding process by conducting a finite element analysis. The authors observed microcracks at the outer perimeter of the glass-frit material due to the high stress levels induced by the thermal cycling test. As can be seen from the above, the bonding strength and the hermeticity of the MEMS package with the vacuum cavity have been thoroughly investigated. Nonetheless, the MEMS package reliability needs to be improved further for its greater use in automotive, medical, and consumer applications.

In the present research, the reliability of a hermetic MEMS package with a metal lid is evaluated through the precondition test. The hermetic MEMS package adopted for this purpose comprises of a ceramic substrate, two dies, and a metal lid. The metal lid and the ceramic substrate are bonded together with silver epoxy under atmospheric pressure, as the aim is to reduce cost and improve the fabrication process efficiency. However, as moisture inside the MEMS package cavity is a potential risk, during the reliability test, the aim is to prevent the lid-off and improve the bonding strength between the metal lid and the ceramic. Thus, in the analyses, focus is given to the curing process, the silver epoxy material, the metal lid geometry, and the bonding layer contact area.

2. Fabrication Process

The hermetic MEMS package with the metal lid used in this study was fabricated as shown in Figure 1. As can be seen from the diagram, once the stacked dies were bonded to the ceramic substrate with silver epoxy, the lid attachment and the precondition test were conducted.

(a) Lid attachment: The metal lid is attached to the ceramic substrate with silver epoxy, which is applied between the metal lid and the ceramic substrate of the peripheral MEMS package. To fully cure silver epoxy, it is exposed to the 175 °C temperature for 2 h. Subsequent evaluations confirm that silver epoxy fully adheres with the metal lid and the ceramic substrate.

(b) Moisture sensitivity level (MSL) 1 test: The MSL 1 test is carried out to determine the sensitivity level of the hermetic MEMS package under humid conditions. For this purpose, the hermetic MEMS package is exposed to high humidity and high temperature (85 °C/85% RH). When the moisture penetrates into the MEMS package cavity via silver epoxy, it weakens the metal lid/silver epoxy and the ceramic substrate/silver epoxy bonding strength.

(c) Reflow process: During the reflow process, the moisture concentration inside the MEMS package causes damage to the metal lid/silver epoxy interface. When the MEMS package is exposed to the maximum temperature of 265 °C for three cycles, vapor pressure and thermal pressure are induced by the residual moisture in the hermetic cavity. Furthermore, thermal stress is generated at the metal lid/silver epoxy and the ceramic substrate/silver epoxy interface due to the coefficient of thermal expansion (CTE) mismatch.

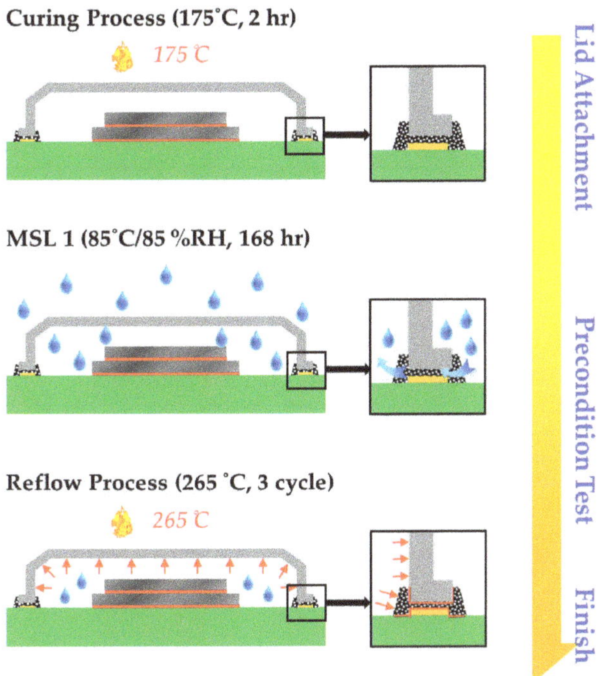

Figure 1. The fabrication process of MEMS package with metal lid.

3. Root Cause

As shown in Figure 2, the precondition test results in an interfacial crack on the exterior of the hermetic MEMS package. The crack extends from the exterior along the metal lid/silver epoxy interface. This causes a phenomenon known as "lid-off" indicating that the metal lid is separated from the ceramic substrate. In the MEMS package, lid-off failure occurs because upward force is applied on the metal lid. To determine its root cause in the precondition test, the experimental design shown in Figure 3 was adopted in this study. The shear test and the red ink penetration test were performed to record the results, which are denoted as Result A (only reflow), Result B (only MSL-1), and Result C (reflow and MSL-1).

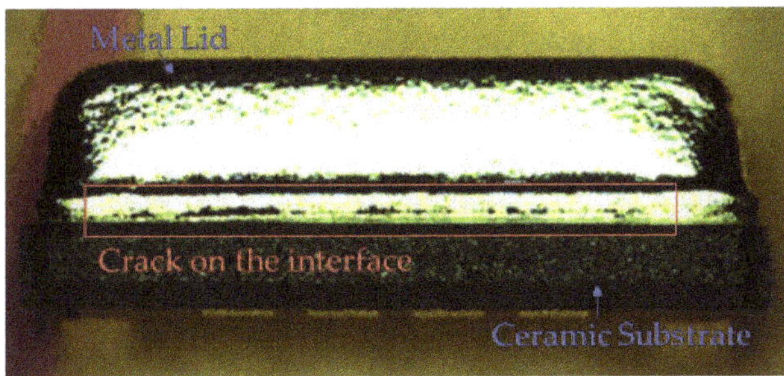

Figure 2. The crack occurred between the metal lid and the silver epoxy in the hermetic MEMS package.

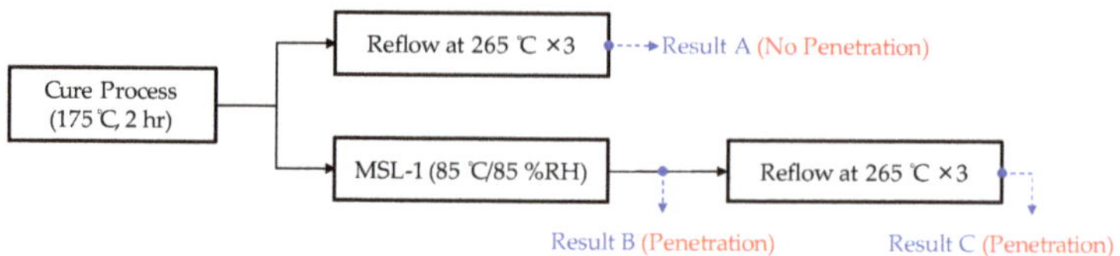

Figure 3. The process flow of investigating root cause.

The shear test and the red dye penetration test results are presented in Figure 4. As can be seen from Result A (only reflow), the red ink is located at the outside of the MEMS package, indicating that no cracks have occurred at the interface or in the silver epoxy under the reflow process. When the MEMS package is exposed to high humidity, the moisture penetrates into the silver epoxy. As the residual moisture weakens the metal lid/silver epoxy bonding strength, the red ink penetrates inside the silver epoxy and the package, as indicated by both Result B (only MSL-1) and Result C (MSL-1 and reflow). The maximum shear force also decreases as a result of moisture penetration. Result C further reveals that the metal lid has separated from the silver epoxy, as the moisture inside the MEMS package cavity vaporizes and generates vapor pressure during the reflow process. Hence, the vapor pressure and the thermal pressure have a potential to cause the lid-off.

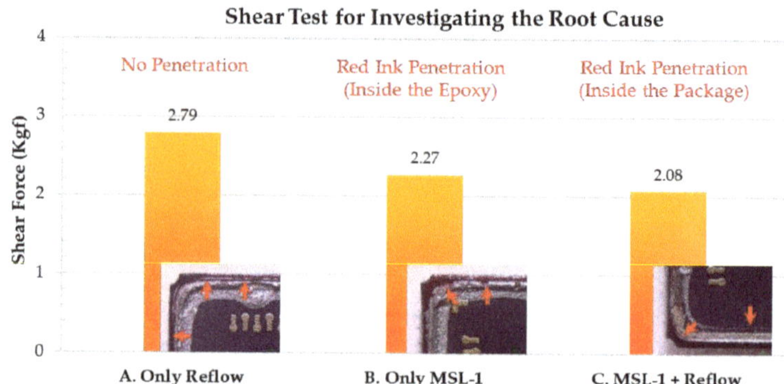

Figure 4. The shear test and red dye penetration test for investigating the root cause.

4. The Shear Test in the Different Manufacture Condition

In the shear test, the thrust force is applied on the bottom side of the metal lid to remove it from the MEMS package. During this process, the maximum shear force is measured to determine the shear strength of both the metal lid and the silver epoxy. To investigate the influence of the internal pressure on the likelihood of lid-off failure, a hermetic MEMS package with vent hole was designed, as shown in Figure 5. The vent hole was drilled at the corner and the top of the metal lid, allowing the internal pressure to be released during the reflow process. To analyze the maximum shear force under different manufacturing conditions, the hermetic MEMS package with a vent hole was compared to that without a vent hole, as shown in Figures 6 and 7. As can be seen from Figure 6, the maximum shear force of the hermetic MEMS package without a vent hole after the precondition test (1.39 kgf) is lower than that measured for the hermetic MEMS package without a vent hole before the precondition test (3.18 kgf). However, the maximum shear force of hermetic MEMS package with a vent hole measured before the precondition test is similar to that obtained after the test. These results indicate that the internal pressure is a critical factor for lid-off failure under the reflow process.

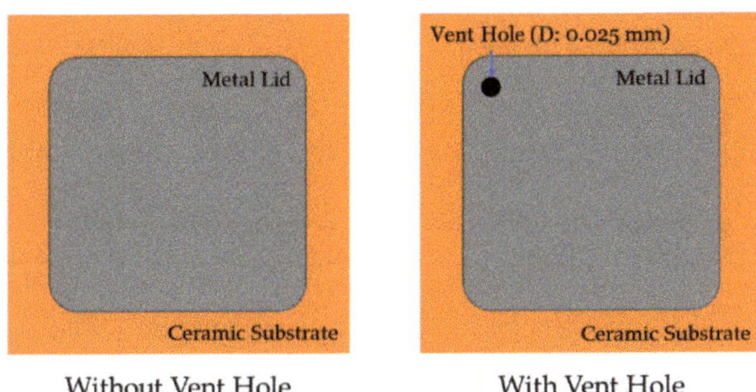

Figure 5. The schematic diagram of the hermetic MEMS package with a vent hole and without a vent hole.

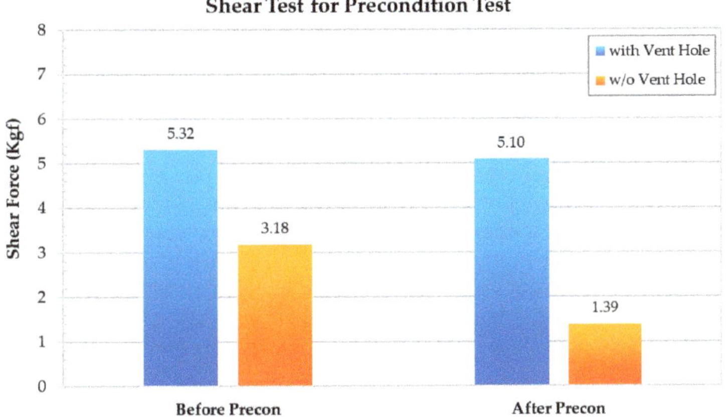

Figure 6. The shear test before and after the precondition test comparison.

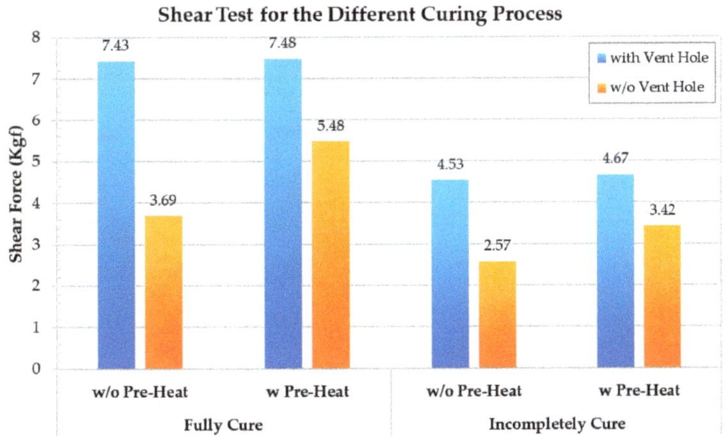

Figure 7. The shear test for discussing the degree of curing and the pre-heat condition.

To improve the maximum shear force of the hermetic MEMS package, additional tests were performed while controlling for the degree of curing and the pre-heat conditions, as these factors affect the material characteristics and the interfacial contact strength of the silver epoxy. To obtain fully cured and incompletely cured epoxy, the following conditions were respectively applied: 175 °C/1 h + 190 °C/1 h and 175 °C/1 h. As fully cured silver epoxy is harder and has a higher Young's modulus, its maximum shear force is higher than that of the partially cured epoxy. Our analyses further indicate that when the fully cured epoxy is used in the hermetic MEMS package with a vent hole and the pre-heat (110 °C/0.5 h) step is performed, the maximum shear force increases by about 60% relative to the partially cured epoxy. In addition, when the fully cured epoxy is used in the hermetic MEMS package without a vent hole, the maximum shear force increases by about 48.5% after pre-heating. By observing the experimental testing results, the pre-heat does not have an effect on the hermetic MEMS package with a vent hole. The pre-heat condition can relieve the internal pressure applied on the metal lid without a vent hole in the curing process.

5. Finite Element Method

The metal lid detaches from the ceramic substrate because of internal pressure during the reflow process. To analyze the stress and the deformation of the hermetic MEMS package with a metal lid under the reflow process, a finite element model was adopted by using ANSYS APDL. Specifically, the CZM method was used to calculate the contact stress at the metal lid/silver epoxy interface, which were denoted as contact and target elements. The MEMS package structure comprised of stacked dies, a die attach, a ceramic substrate, a metal ring, a silver epoxy, and lid metal, as shown in Figure 8. For modeling this structure, a two-dimensional finite element model with quadratic elements was established and was matched with scanning electron microscope (SEM) cross-section images. The material properties of the finite element model are presented in Table 1. During modeling, internal pressure was applied on the inside surface of the metal lid to simulate air pressure and vapor pressure in the cavity under the reflow process.

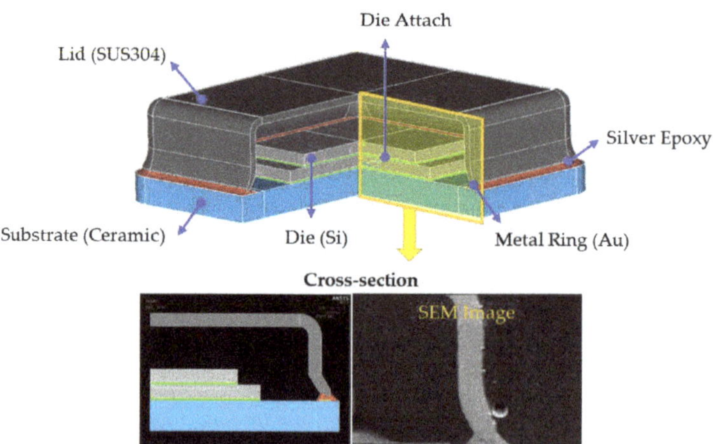

Figure 8. The finite element model of the hermetic MEMS package with metal lid.

Table 1. The material properties of the finite element model.

	E (MPa)	ν	CTE (ppm/K)
Die	131×10^3	0.27	2.8
Lid	190×10^3	0.30	16.3
Substrate	310×10^3	0.30	7.1
Au	77.2×10^3	0.42	14.4
Die Attach	7.1×10^3 @25 °C 0.6×10^3 @260 °C	0.30	18 < 175 °C 35 > 175 °C
Silver Epoxy	3900@25 °C 2000@150 °C 300@250 °C	0.30	40 < 120 °C 150 > 120 °C

The internal pressure inside the cavity can be obtained by using the ideal gas equation, as the following equations:

$$PV = nRT \tag{1}$$

$$P_{Internal} = P_{Air} + P_{Vapor} \tag{2}$$

$$P_{Air\_265°C} = P_{Air\_25°C} \cdot \frac{T_{265°C}}{T_{25°C}} \tag{3}$$

$$
\begin{aligned}
P_{Vapor\_265°C} &= P_{85°C/85\%RH} \cdot \frac{T_{265°C}}{T_{85°C}} \\
&= (0.85 \cdot P_{85°C,SAT}) \cdot \frac{T_{265°C}}{T_{85°C}}
\end{aligned}
\tag{4}
$$

where P is the pressure, V is the volume, n is the number of moles of gas, R is the idea gas constant ($8.317\ J \cdot mol^{-1} \cdot K^{-1}$), T is the absolute temperature, $P_{Internal}$ is the internal pressure, P_{Air} is the air pressure, P_{Vapor} is the vapor pressure, $P_{Air\_25°C}$ is the air pressure at 25 °C, $P_{85°C,SAT}$ is the saturated vapor pressure at 85 °C, and RH is the relative humidity.

The findings pertaining to the hermetic MEMS package with and without a vent hole were once again contrasted to investigate the contact stress and SEM observations, as shown in Figure 9. For evaluating the crack location, contact stress was defined as normal interface stress. In the hermetic MEMS package without a vent hole, the highest contact stress was located at the bottom of the metal lid. The fracture occurred at the same location during experimental testing. In the hermetic MEMS package with a vent hole, the contact stress was negligible and no fracture could be observed on the SEM images.

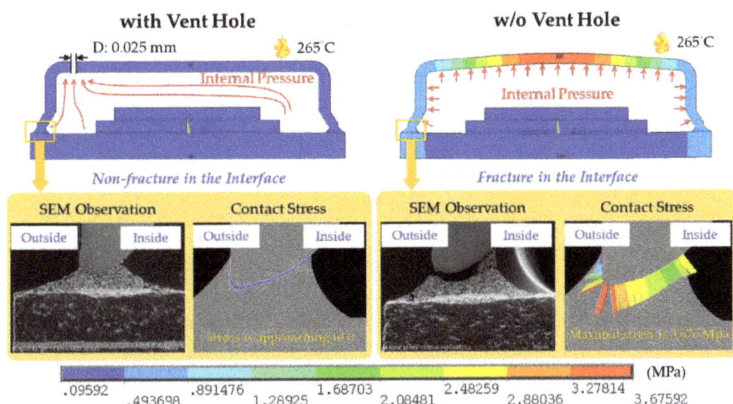

Figure 9. The contact stress (simulation result) and the SEM observation (experimental testing result) of the hermetic MEMS package with a vent hole and without a vent hole.

6. Optimization

6.1. One Factor Design

To decrease the contact stress at the metal lid/silver epoxy interface, one factor design was performed, considering lid thickness, connecting angle, epoxy height, lid height, lid size, and substrate height as factors, as presented in Figure 10. These design factors were chosen to evaluate the contact area effect, the lid geometry effect, and the material property effect, as indicated in Table 2. The lid thickness, the connecting angle, and the epoxy height are considered to exhibit the contact area effect since these factors are related to the contact interface area. The lid geometry not only affects the lid size but also has an influence on the cavity volume. Therefore, a lid of greater size would have higher contact stress due to withstanding higher internal pressure.

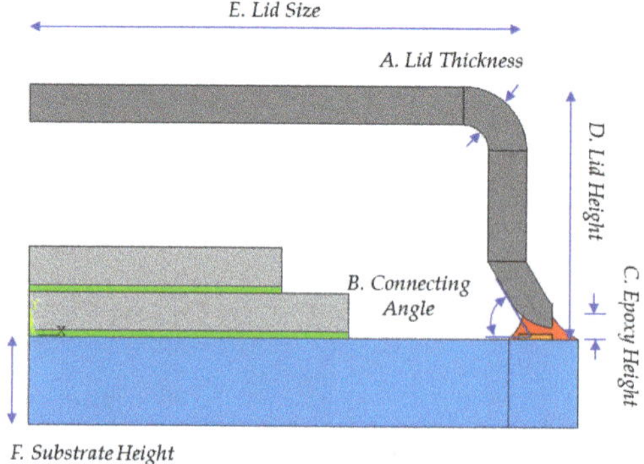

Figure 10. The schematic diagram of design factors.

Table 2. The design factors for analyzing the contact stress.

	Factors	Range
The Contact Area Effect	A. Lid thickness	0.1~0.2 (mm)
	B. Connecting angle	50~70 (°)
	C. Epoxy height	0.7~0.9 (mm)
The Lid Geometry Effect	D. Lid height	0.8~1.1 (mm)
	E. Lid size	2.5~6.1 (mm)
	F. Substrate height	0.2~0.5 (mm)
The Material Property Effect	G. Lid modulus	120~160 (GPa)
	H. Epoxy modulus	3~9 (GPa)

The one factor designs for the contact area, the lid geometry, and the material property effects are shown in Figures 11–13, respectively. According to the assessments related to the contact area effect, greater lid thickness, and epoxy height, and a lower connecting angle reduce the contact stress by increasing the contact interface area. The lowest contact stress (2.45 MPa) is obtained with the connecting angle of 50°. According to the lid geometry effect, the lid size is sensitive to contact stress because the force induced by internal pressure is based on the lid size. Specifically, the contact stress increases from 1.36 to 7.15 MPa when the lid size increases from 2.5 to 6.1 mm. Finally, the results related to the material property effect indicate that the Young's modulus of the silver epoxy and the lid do not exert significant changes on contact stress. Thus, even though Young's modulus of the silver epoxy is not the critical factor, the moisture absorption, shear strength, and the material curing characteristics are important for the contact stress.

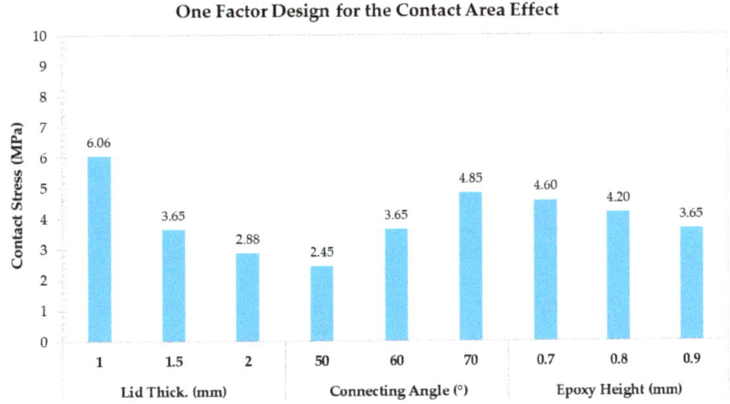

Figure 11. One factor design for discussing the contact area effect.

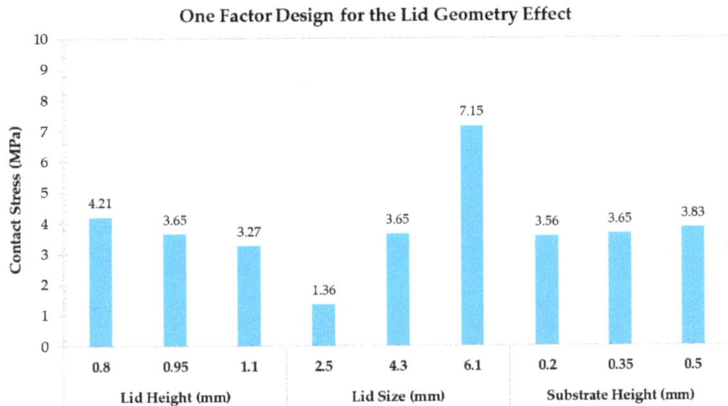

Figure 12. One factor design for discussing the lid geometry effect.

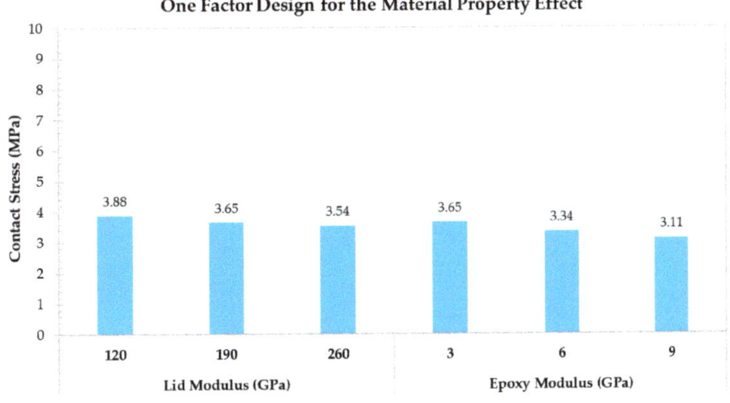

Figure 13. One factor design for discussing the material property effect.

6.2. Responsed Surface Method

The response surface method was also adopted to establish the relationship between the factors that are most influential on contact stress, as shown in Figures 14 and 15. As lid thickness, lid size, and the connecting angle are the critical factors for contact stress, their values were considered when interpreting the response surface results. By observing the relationship between lid thickness and lid size, it is evident that the slope of lid size is linear and is greater than the lid thickness. While the impact of lid thickness on contact stress is low, the curve flattens with increasing lid thickness. These results indicate that lid size is more significant than lid thickness. In addition, according to the response surface results based on the relationship between the connecting angle and the lid size, both factors exhibit linear distribution. Thus, for improving the contact stress, lid thickness should be increased, while its size and the connecting angle should be reduced.

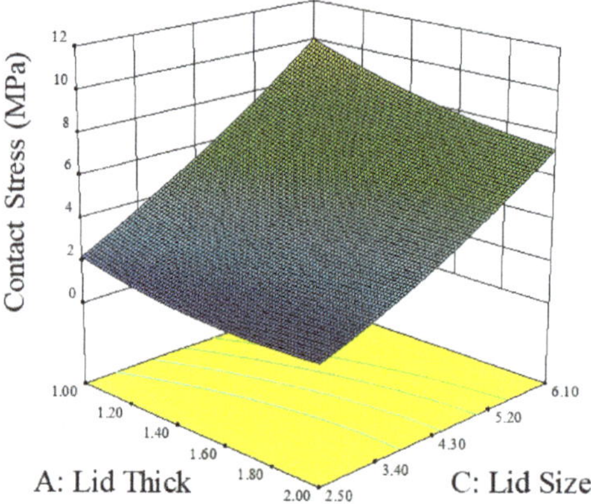

Figure 14. The response surface results for discussing the relationship between the lid thickness and the lid size.

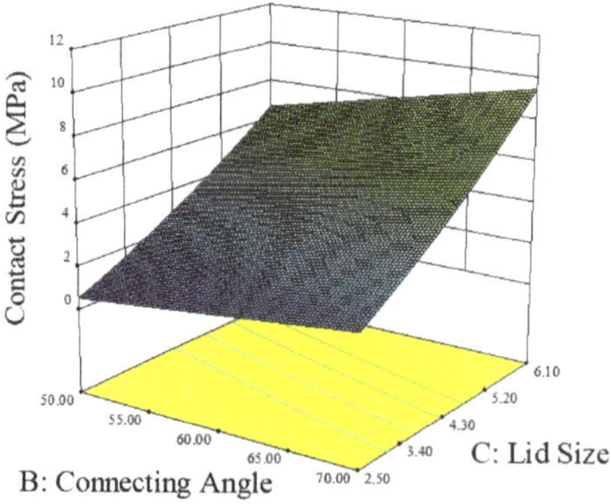

Figure 15. The response surface results for discussing the relationship between the connecting angle and the lid size.

7. Conclusions

In the research reported here, a hermetic MEMS package with a metal lid was designed and its reliability was tested. During the precondition test, lid-off failure occurred because moisture weakens the interfacial bonding strength and increases the internal pressure during the reflow process. The finite element method, which was adopted to simulate the contact stress of the metal lid/silver epoxy interface and the deformation of the metal lid, revealed that lid thickness, lid size, and the connecting angle are the critical factors for the contact stress. Although the Young's modulus of the silver epoxy is not the critical factor, the moisture absorption, shear strength, and material curing characteristics are important for the contact stress. In the experimental testing, the pre-heat step and fully curing the epoxy can enhance the maximum shear force by 60% and 48.5% under the precondition test. The findings further indicate that increasing the lid thickness, and decreasing the lid size and the connecting angle can decrease the contact stress, thus reducing the likelihood of lid-off failure under the precondition test.

Author Contributions: Conceptualization, M.-L.W.; methodology, M.-L.W. and J.-S.L.; software, J.-S.L.; validation, M.-L.W. and J.-S.L.; formal analysis, J.-S.L.; investigation, M.-L.W. and J.-S.L.; resources, M.-L.W.; data curation, J.-S.L.; writing—original draft preparation, J.-S.L.; writing—review and editing, M.-L.W.; visualization, J.-S.L.; supervision, M.-L.W.; project administration, M.-L.W.; funding acquisition, M.-L.W. All authors have read and agreed to the published version of the manuscript.

Funding: This research is partially supported by Advanced Semiconductor Engineering (ASE Group) collaborative research program, Taiwan, ROC, under grants N107174.

Institutional Review Board Statement: Not applicable.

Informed Consent Statement: Not applicable.

Data Availability Statement: Not applicable.

Acknowledgments: The authors gratefully acknowledge the financial support of the Advanced Semiconductor Engineering (ASE Group) collaborative research program (N107174). The authors would like to thank Director Louie Huang, Deputy Technical Manager Stuwart Fan, Supervisor Engineer Bobby Lee, and Engineer Chieh Hsu for their advice, technical support, and material support.

Conflicts of Interest: The authors declare no conflict of interest.

References

1. Hsu, H.C.; Lee, H.Y.; Hsu, Y.C.; Fu, S.L. Thermo-hygro-mechanical Design and Reliability Analysis for CMOS Image Sensor. *J. Therm. Stress.* **2008**, *31*, 917–934. [CrossRef]
2. Jeong, B.G.; Ham, S.J.; Moon, C.Y.; Kim, B.S. Reliability Verification of Hermetic Package with Nanoliter Cavity for RF-micro Device. *IEEE Trans. Adv. Packag.* **2009**, *33*, 64–71. [CrossRef]
3. Zhang, H.; Liu, Y.; Wang, L.; Fan, J.; Fan, X.; Sun, F.; Zhang, G. A New Hermetic Sealing Method for Ceramic Package Using Nanosilver Sintering Technology. *Microelectron. Reliab.* **2018**, *81*, 143–149. [CrossRef]
4. Al Farisi, M.S.; Hirano, H.; Tanaka, S. Low-temperature Hermetic Thermo-compression Bonding Using Electroplated Copper Sealing Frame Planarized by Fly-cutting for Wafer-level MEMS Packaging. *Sens. Actuators A Phys.* **2018**, *279*, 671–679. [CrossRef]
5. Huang, A.T.; Chou, C.K.; Chen, C. Hermetic Packaging Using Eutectic SnPb Solder and Cr/Ni/Cu Metallurgy Layer. *IEEE Trans. Adv. Packag.* **2006**, *29*, 760–765. [CrossRef]
6. Jang, C.; Goswami, A.; Han, B. Hermeticity Evaluation of Polymer-sealed MEMS Packages by Gas Diffusion Analysis. *J. Microelectromech. Syst.* **2009**, *18*, 577–587. [CrossRef]
7. Premachandran, C.S.; Chong, S.C.; Liw, S.; Nagarajan, R. Fabrication and Testing of a Wafer-level Vacuum Package for MEMS Device. *IEEE Trans. Adv. Packag.* **2009**, *32*, 486–490. [CrossRef]
8. Jiang, X.; Wang, C.; Liu, W. A Laser-assisted Bonding Method Using a Liquid Crystal Polymer Film for MEMS and Sensor Packaging. *IEEE Trans. Compon. Packag. Manuf. Technol.* **2015**, *5*, 583–591. [CrossRef]
9. Sandvand, Å.; Halvorsen, E.; Aasmundtveit, K.E.; Jakobsen, H. Influence of Glass-frit Material Distribution on the Performance of Precision Piezoresistive MEMS Pressure Sensors. *IEEE Trans. Compon. Packag. Manuf. Technol.* **2015**, *5*, 1559–1566. [CrossRef]

 *materials*

Article

Transient Electro-Thermal Coupled Modeling of Three-Phase Power MOSFET Inverter during Load Cycles

Hsien-Chie Cheng [1,*], Siang-Yu Lin [1] and Yan-Cheng Liu [1,2]

[1] Department of Aerospace and Systems Engineering, Feng Chia University, Taichung 407, Taiwan; ytalanben44556@gmail.com (S.-Y.L.); P0700100@o365.fcu.edu.tw (Y.-C.L.)
[2] Ph.D Program of Mechanical and Aeronautical Engineering, Feng Chia University, Taichung 407, Taiwan
* Correspondence: hccheng@fcu.edu.tw

Abstract: This study introduces an effective and efficient dynamic electro-thermal coupling analysis (ETCA) approach to explore the electro-thermal behavior of a three-phase power metal–oxide–semiconductor field-effect transistor (MOSFET) inverter for brushless direct current motor drive under natural and forced convection during a six-step operation. This coupling analysis integrates three-dimensional electromagnetic simulation for parasitic parameter extraction, simplified equivalent circuit simulation for power loss calculation, and a compact Foster thermal network model for junction temperature prediction, constructed through parametric transient computational fluid dynamics (CFD) thermal analysis. In the proposed ETCA approach, the interactions between the junction temperature and the power losses (conduction and switching losses) and between the parasitics and the switching transients and power losses are all accounted for. The proposed Foster thermal network model and ETCA approach are validated with the CFD thermal analysis and the standard ETCA approach, respectively. The analysis results demonstrate how the proposed models can be used as an effective and efficient means of analysis to characterize the system-level electro-thermal performance of a three-phase bridge inverter.

Keywords: electro-thermal coupling analysis; power MOSFET inverter; power loss; circuit simulation; computational fluid dynamics; Foster thermal network

Citation: Cheng, H.-C.; Lin, S.-Y.; Liu, Y.-C. Transient Electro-Thermal Coupled Modeling of Three-Phase Power MOSFET Inverter during Load Cycles. *Materials* **2021**, *14*, 5427. https://doi.org/10.3390/ma14185427

Academic Editor: Kuo-Ning Chiang

Received: 22 June 2021
Accepted: 14 September 2021
Published: 19 September 2021

Publisher's Note: MDPI stays neutral with regard to jurisdictional claims in published maps and institutional affiliations.

1. Introduction

Power electronics are widely used as powertrain components in the increasingly popular electric vehicles (EVs) and hybrid EVs, such as in electronic switches, converters, and inverters, to control and regulate electricity. In particular, three-phase voltage source inverters applied to control three-phase asynchronous induction motors are widely used in alternating current (AC) motor drives. Power semiconductors/modules inside inverters are the most crucial devices controlling the power conversion efficiency. In response to the urgent need for high-performance power conversion applications, the power semiconductor industry has recently seen rapid technological developments, such as insulated-gate bipolar transistors (IGBTs) [1,2], metal-oxide semiconductor field effect transistors (MOSFETs) [3,4], and even wide bandgap (WBG) silicon carbide (SiC) [5,6] and gallium nitride (GaN) power devices [7]. In contrast to IGBTs, MOSFETs comprise a number of advantageous features, such as a higher switching frequency and lower switching loss; accordingly, they have been used in a wide range of industrial applications, such as converters and inverters.

Power devices unavoidably result in great losses in power during operation, including conduction losses resulting from the on-state resistance and switching losses stemming from simultaneous current and voltage waveforms and the influence of input/output capacitances and inductances. The trend for high power and downsizing in power devices is likely to bring about high power densities [8] and thus great power losses. Furthermore, a high power loss together with extreme operating conditions may potentially give rise to a high device junction temperature [6,7], which can cause various thermal and mechanical challenges, such as thermal instability and even unreliability in terms of thermal fatigue. For example, as a result of increased phonon concentration and lattice scattering, a high device junction temperature may lower the carrier mobility and thus raise the temperature-sensitive on-state resistance, which, in turn, increases the conduction loss and further elevates the device junction temperature. This process may, in the worst case, trigger thermal runaway reactions, ultimately leading to device breakdown. As well as this, a high device junction temperature can deteriorate the electrical performance and even be detrimental to the thermal–mechanical reliability of power devices (see, e.g., [5,9,10]). Hence, the temperature is one of the most important issues for power device applications. In order to ensure the safe and normal operation of power devices, the device junction temperature should be operated below the nominal rated temperature [11].

Pulse width modulation (PWM) three-phase bridge inverters are used in AC motor drive systems to convert the direct current (DC) power of batteries to a three-phase AC output with variable frequency and voltage for speed control. In conjunction with the high power density trend in power electronics, the wide variation in the frequency and phase current during load cycles can drive the device junction temperature beyond the temperature limit of power electronics, i.e., the maximum junction temperature rating, which would cause damage to or the failure of the inverters. Thus, there is a critical need for a more thorough comprehension of the thermal behavior of the power devices of inverters during operation. Before looking into the thermal issues of the inverters, a more in-depth understanding of their switching characteristics and power losses during load cycles is required. Several studies have reported that, in addition to supply voltage and gate resistance in the current loop, parasitic parameters are highly susceptible to the ringing and overshoots in the switching transients and, thus, can impact the switching loss [4,12,13]. For example, Cheng et al. [4] explored the switching characteristics and power losses of a silicon (Si) power MOSFET packaged in SOT-227 and a three-phase MOSFET bridge inverter during a switching operation in an effective compact circuit simulation model. They found that parasitic parameters have a considerable influence on the switching loss because of their effect on the switching waveform and speed. In addition to the parasitic effect, temperature also plays an important role in the switching and conduction losses of power devices (see, e.g., [1,14]). The device junction temperature during load cycles greatly influences the switching transients and power losses of power devices, which are, in turn, highly dependent on their device junction temperatures and parasitics. Thus, an accurate understanding of electromagnetic and dynamic electro-thermal (ET) coupled behaviors over a long-term operation is crucial for the safe operation of power components and systems.

According to previous studies from the literature, there have been extensive efforts heavily focused on the component-level exploration of the electromagnetic (EM), switching (power loss), and thermal behaviors (see, e.g., [2,15–19], but very limited work has been done on the system level, such as on three-phase bridge inverters. Heat generation in a three-phase inverter fluctuates at two widely different frequencies: the load current modulation frequency at the level of tens to hundreds of Hz and the switching frequency at the level of 10–100 kHz, where the switching time is only about a few hundred nanoseconds. Precisely modeling the PWM and switching events so as to thoroughly capture the switching transients and power losses of the three-phase inverter during the six-step operation requires an extremely small time step and thus enormous computing time. The problem becomes even more severe for high-operation-frequency applications. Moreover, the device junction temperature generally needs between hundreds and thousands of seconds to reach a steady state, depending on the thermal time constant. Directly coupling the electrical circuit analysis with three-dimensional (3D) transient computational fluid dynamics (CFD) thermal analysis to calculate the ET coupled behavior presents a great computational challenge because of limitations on storage space and computational power. In the literature, the problem has been successfully eased using resistance–capacitance (RC) thermal networks [20–22], such as Foster and Cauer networks, instead of directly carrying out the CFD thermal analysis. RC thermal networks can be an effective and favorable means for junction temperature estimation due to their unparalleled computational efficiency and flexibility for both thermal and electrical models [22]. A direct coupling of the detailed circuit simulation model and an RC thermal network model forms the so-called standard ET coupling analysis (ETCA) approach [23–27]. The standard ETCA approach still cannot fully address the circuit simulation difficulty in the two widely different frequencies, which even makes it impossible to explore the ET coupled behavior of the three-phase PWM inverter in a long-term operation. Accordingly, a more effective approach that can ameliorate the circuit simulation difficulty is critical needed. Reichl et al. [23] attempted to improve the computational efficiency of the standard ETCA approach using a four-step iterative process and an average dissipated power over an electrical cycle. Later on, Reichl et al. [28] alternatively presented a full 3D multilayer and multichip thermal component model with asymmetrical power distributions for dynamic ET simulation, where the 3D heat conduction equation is solved using finite difference methods, and the thermal component model is parameterized in terms of structural and material properties to facilitate the development of a library of component models for any available power module. It has been found, however, that the circuit simulation difficulty still cannot be removed. Accordingly, this study proposes a more effective and efficient dynamic ETCA approach, in which a simplified equivalent circuit simulation model is developed and fully coupled with a Foster thermal network model to account for the effect of the instantaneous junction temperature on the instantaneous power losses (switching and conduction). The proposed, simplified equivalent circuit simulation model can address the computational difficulty associated with the two significantly different frequencies and, therefore, can greatly reduce the computational cost and make the multi-temporal and long-term ETCA of a power conversion system much more feasible. In addition, to address the effects of parasitics on the switching transients and power losses, the proposed ETCA approach can be integrated with a 3D EM model. The proposed ETCA approach is demonstrated through the estimation of the ET coupled behavior of a voltage source three-phase bridge MOSFET inverter (see Figure 1) for brushless DC (BLDC) motor drive under natural and forced convection during a six-step operation. The established Foster thermal network model and the proposed ETCA approach are validated using CFD thermal analysis and the standard ETCA approach, respectively.

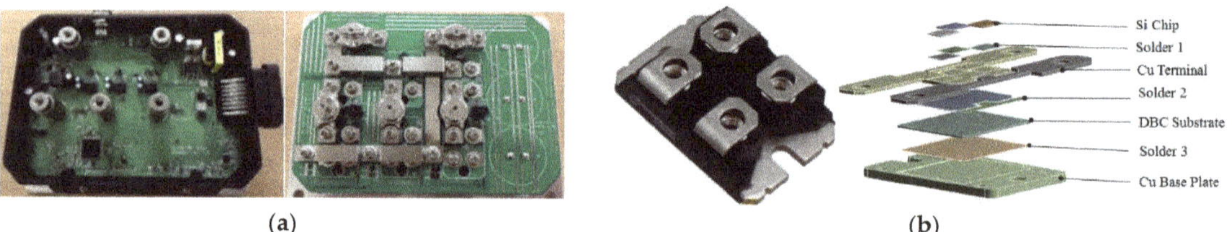

Figure 1. (**a**) Three-phase bridge inverter and (**b**) power MOSFET module and explosive view.

2. Three-Phase MOSFET Bridge Inverter

The voltage source three-phase bridge inverter, which transforms DC power from a DC source into AC power for an AC load, is shown in Figure 1a. It comprises three parallel legs for phases a, b, c, and each of them contains two semiconductor switches (100 V and 350 A SOT-227 power MOSFET modules, as illustrated in Figure 1b): one at the upper side and the other at the lower side. These two switches in each leg are complimentarily operated. In total, there are six switches (S1–S6) in the inverter to create a three-phase bridge circuit with six switching arms that turn the current on and off, as displayed in Figure 2a. In detail, three of these six switches (S1, S3, and S5) are connected to a high-voltage-side DC voltage (hereinafter referred to as "upper-side switches") and the others

(S2, S4, and S6) to a low-voltage one (hereinafter referred to as "lower-side switches"). These arms are linked to each other through a connection bridge. In each modulation cycle, there is an electrical cycle (360°) with six switching steps, each with a duration of 60°, creating a cyclic three-phase pattern, as depicted in Figure 2b. At any commutation sequence in the six-step commutation logic, only one upper switch and one lower switch are turned on to energize two motor phase windings. The upper-side switches' switching signals are kept discontinuously "on" (i.e., PWM "on") with a duty cycle whereas the lower-side switches' switching signals are always continuously "on" [29]. Thus, the upper switches are, alternatively, termed PWM power MOSFETs. The conduction sequence of one six-step commutation cycle is S1S4–S1S6–S3S6–S3S2–S5S2–S5S4, and the corresponding current states are ab, ac, bc, ba, ca, and cb.

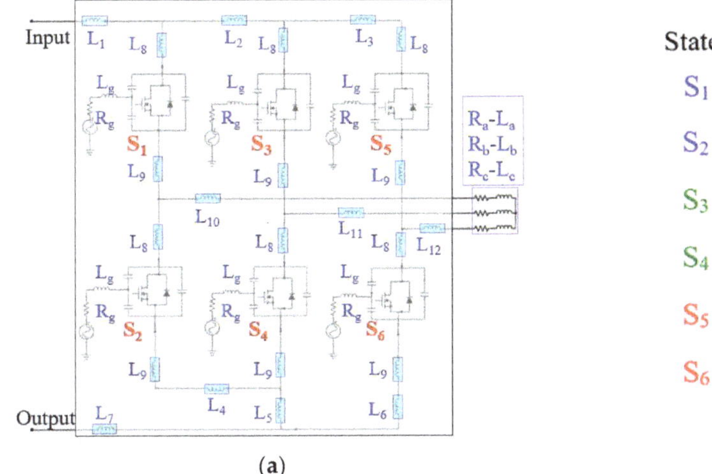
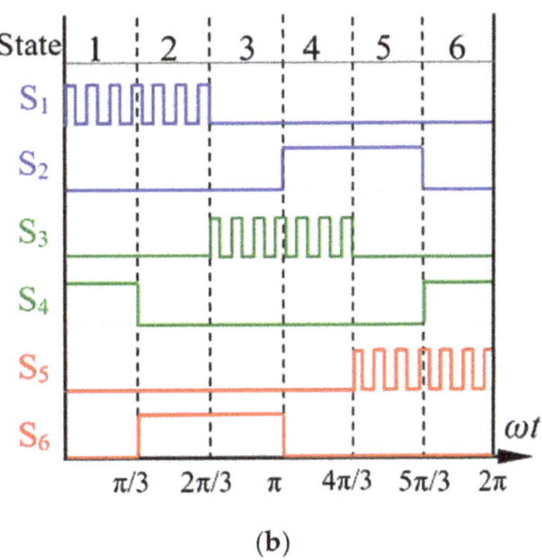

Figure 2. (**a**) Three-phase inverter circuit with parasitic inductances and (**b**) six-step SWPWM signal sequence.

In order to enhance the current rating [14,30], three Si power MOSFET chips connected in parallel are embedded in the power MOSFET module. When controlling the PWM power modules, the common rectangular-wave PWM (RWPWM) technique is employed to generate a square-wave pulse via a signal generator, and a microcontroller is used to supply the gate pulses to these semiconductor switches. The current supplied to the power MOSFET modules is PWM-regulated through the rapid switching on and off of these switches. The ratio of the pulse width to the total signal period is defined as the duty cycle (D). When D = 50%, it is a square wave PWM (SWPWM). An increased duty cycle raises the electrical power supply to the semiconductor devices. The temperature-dependent on-state resistance and the output, transfer, and body diode characteristics of the power MOSFET module provided in the manufacturer's datasheet and also in [4] are presented in Figure 3. Figure 3a,d reveal that the I–V characteristics of the power MOSFET and body diode show a strong temperature coefficient.

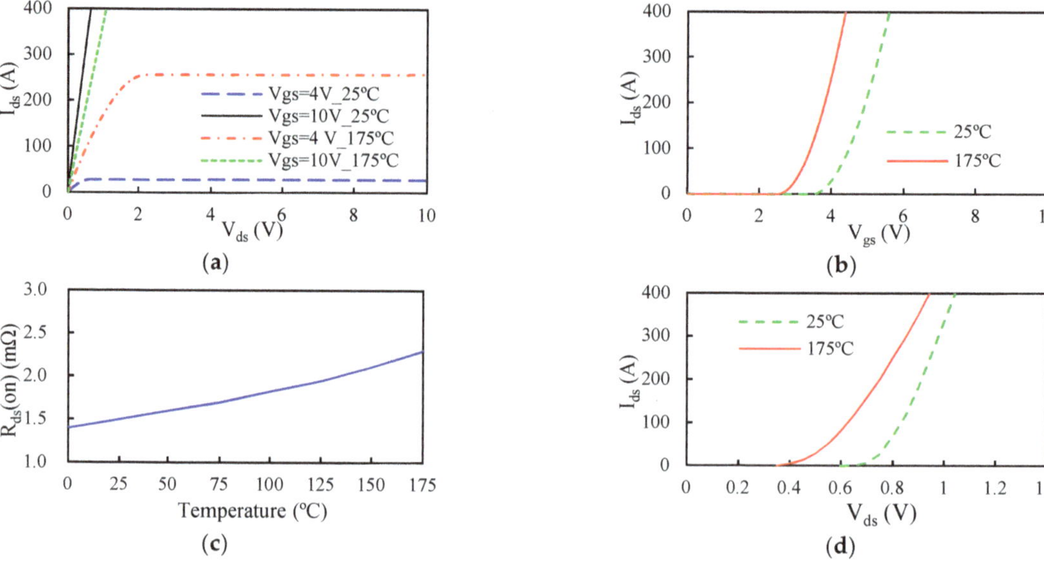

Figure 3. Characteristics of the power MOSFET and body diode presented in the manufacturer's datasheet and also in [4]: (**a**) power MOSFET output characteristic; (**b**) power MOSFET transfer characteristic; (**c**) power MOSFET on-state resistance; (**d**) diode characteristic.

To facilitate heat dissipation, these six SOT-227 power MOSFET modules are bonded onto a thick heat spreader made of aluminum (Al) metal. The power MOSFET module contains one gate, one drain, and two source terminals for electrical connection. In addition, it is primarily composed of three Si power MOSFET chips; an Al_2O_3-based direct bonded copper (DBC) substrate; Al bond wires; bond pads made of Al metal; a Cu base plate; Cu terminal leads; three Sn-3.0Ag-0.5Cu (SAC305) solder layers for the bonding between the Si power MOSFET chips and the Cu terminal leads, between the Cu terminal leads and the DBC substrate, and between the DBC substrate and the Cu base plate; a polyphenylene sulfide (PPS) housing; and a quick-drying rubber-based adhesive applied to fill the cavity between the housing and the DBC/Cu terminal leads. The power MOSFET chips, DBC substrate, terminal leads, pads, and base plate have thicknesses of 0.33, 0.45, 0.8, 0.01, and 2.0 (mm). The thicknesses of the three solder layers are 0.05, 0.1, and 0.1 (mm). In total, there are twelve Al wires with the same lengths and cross-sectional areas on the Al pads of these three power MOSFET chips.

3. Power Loss Prediction

The main types of power loss generated from power MOSFETs during operation include conduction, switching, and current leakage losses and diode conduction and reverse recovery losses. The leakage current loss is typically much lower than the conduction loss at low junction temperatures [14] and thus can be negligible if the junction temperature is appropriately controlled. The estimation of the conduction loss and switching loss of power MOSFETs, i.e., P_C and P_S, during operation can be briefly demonstrated in the following. When power MOSFETs are switched on by the gate voltage, drain-source current flows across the resistive components, causing Joule heating and resulting in heat conduction loss. For a particular switching period, the conduction loss can be calculated from the drain-source current I_{ds}, on-state resistance $R_{ds(on)}$, and duty cycle D as

$$P_C = \frac{D}{t_s} \int_0^{t_s} I_{ds}^2(t) R_{ds(on)}(T) dt \qquad (1)$$

Since the on-state resistance has a large and positive temperature correlation, as seen in Figure 3c, the conduction loss is a strong function of temperature. For modeling simplicity, an average power loss is generally utilized in computation through the application of a root-mean-square (RMS) average current (I_{RMS}) during a PWM operation. For an SWPWM control technique, I_{rms} is denoted as

$$I_{rms} = I_{ds}\sqrt{D} \tag{2}$$

With the RMS average current, the corresponding conduction loss can be expressed as

$$P_C = I_{rms}^2 R_{ds(on)}(T) \tag{3}$$

As a result of the simultaneous rise in current, from the leakage current to the on-state current I_{DS}, and fall in voltage, from the off-state voltage to the on-state voltage, power devices can induce considerable switching loss. Moreover, the PWM switching frequency has a positive and almost linear effect on the switching loss. A higher switching frequency causes a greater switching loss. As mentioned earlier, in addition to the device parameters, reverse recovery current, and gate drive current, the parasitic effect plays a significant role in the switching loss. Figure 4 shows typical voltage and current transients during turn-on and turn-off periods, where V_{gs} is the gate-source voltage; V_{TH} is the threshold voltage; V_{gp} is the gate-plateau voltage; V_{DD} is the supply voltage; I_{peak} is the current spike (overshoot); V_{ON} is the conduction voltage, which is equal to $I_{DS}R_{DS(on)}$; V_{GS} is the gate drive voltage; and V_{spike} is the voltage spike. The time increments $t_2 - t_1$ and $t_6 - t_5$ are defined as the rise time t_{ir} and fall time t_{if} of the on-state current I_{ds}, respectively, and the time increments $t_3 - t_2$ and $t_5 - t_4$ are defined as the fall time t_{vf} and rise time t_{vr} of the drain-source voltage V_{ds}. Accordingly, the turn-on switching period t_s^{on} is equal to $t_3 - t_1$, and the turn-off switching period t_s^{off} is equal to $t_6 - t_4$. These switching transients are largely determined by parasitic parameters, such as the gate-drain capacitance C_{gd}, the gate-source capacitance C_{gs}, the drain-source capacitance C_{ds}, the drain inductance L_d, the gate inductance L_g, and the source inductance L_s. These parasitic capacitances are closely related to the input capacitance C_{iss} ($C_{iss} = C_{gs} + C_{gd}$), output capacitance C_{oss} ($C_{oss} = C_{gd} + C_{ds}$), and reverse transfer capacitance C_{rss} ($C_{rss} = C_{gd}$). Basically, they somewhat vary with the drain-source voltage V_{ds}, as shown in Figure 5. Finally, the switching energy loss E_S during a switching cycle is given as

$$E_S = E_{on} + E_{off} = \int_0^{t_s^{on}} V_{ds}(t)I_{ds}(t)\,dt + \int_0^{t_s^{off}} V_{ds}I_{ds}(t)\,dt \tag{4}$$

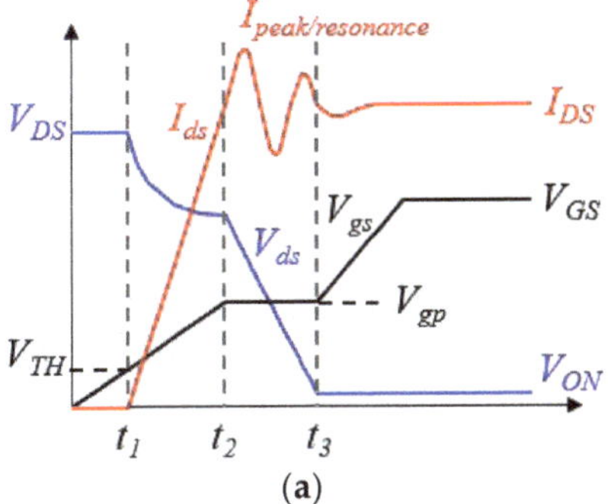

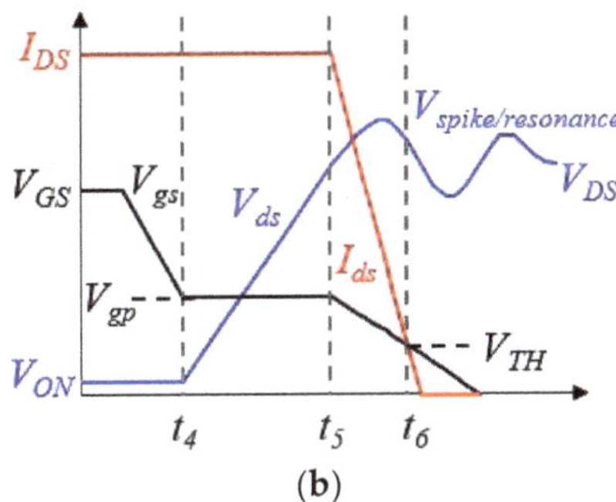

Figure 4. Power MOSFET switching transient: (**a**) turn-on waveform and (**b**) turn-off waveform.

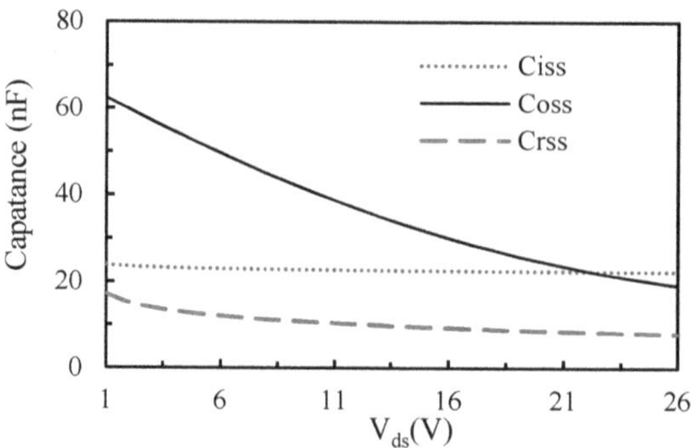

Figure 5. V_{ds}-dependent input, output, and reverse transfer capacitances.

The body diode can also contribute to the conduction loss and reverse recovery power loss. The former is produced when the upper switches (i.e., PWM power MOSFETs) are switched off and the current passes via the complementary lower switches (i.e., freewheeling diodes (FWDs)) [31]. The body diode conduction loss P_C^{BD} across the switching period t_S can be written as

$$P_C^{BD} = \frac{1}{t_S} \int_0^{t_S} \left(V_{BD}^0 I_{BD}(t) + R_{BD}(t) I_{BD}{}^2(t) \right) dt \tag{5}$$

where I_{BD} is the current passing through the body diodes, V_{BD} is the voltage of the body diodes, and V_{BD}^0 and R_{BD} are the on-state zero-current voltage and resistance of the body diodes, respectively, which can be read from the diagrams in the package datasheet. Furthermore, when the body diodes are switched off, the charge stored in the drain-source capacitor of the FWDs must be released. The reverse recovery current is absorbed by the PWM power MOSFTs when they are switched on again. In fact, the reverse recovery effect is included in the power loss calculation for the upper-side switches that are turned on.

4. EM Electro-Thermal Analysis
4.1. EM Modeling

Maxwell's equations, consisting of a set of coupled partial differential equations, are generally used to depict macroscopic electromagnetism phenomena. The equations indicate that EM waves moving along a field depend on time, space, the electric field, and the magnetic field [32]:

$$\nabla \cdot D = \bar{\rho} \tag{6}$$

$$\nabla \cdot B = 0 \tag{7}$$

$$\nabla \times E = -\frac{\partial B}{\partial t} \tag{8}$$

$$\nabla \times H = \bar{J} + \frac{\partial D}{\partial t} \tag{9}$$

where D denotes the electric displacement field or electric flux density, B the magnetic field density, E the electric field, $\bar{\rho}$ the free charge density (not including the bound charge), H the magnetic field intensity, and $\bar{J}$ the free current density (not including the bound current). Equations (1)–(4) are called Gauss's law, Gauss's law for magnetism, the Maxwell–Faraday equation, and the Ampère circuital law. The Ampère circuital law is also known as the Maxwell–Ampère law. The left-hand side of the Ampère circuital law possesses zero divergence due to the div–curl identity. Further expanding the divergence of the right-hand side, exchanging the derivatives, and applying Gauss's law yields:

$$0 = \nabla \cdot (\nabla \times H) = \nabla \cdot \bar{J} + \nabla \cdot \frac{\partial D}{\partial t} \tag{10}$$

This leads to

$$\nabla \cdot \bar{J} = -\frac{\partial \bar{\rho}}{\partial t} \tag{11}$$

The free charge density does not vary with time (i.e., $\partial \bar{\rho}/\partial t = 0$) for a stable current, and thus Equation (11) can be re-expressed as

$$\nabla \cdot \bar{J} = 0 \tag{12}$$

Note that $\bar{J} = \sigma E$ and $E = -\nabla V$ based on Ohm's law. If the conductivity σ of the conductor material is assumed to be constant and evenly distributed, the equation governing the steady-state electric field can be derived as

$$\nabla^2 V = 0 \tag{13}$$

4.2. CFD Modeling

The mass, momentum, and energy conservation laws are solved in the CFD analysis using finite volume method. The conservation equations, namely mass, momentum, and thermal energy, in the Cartesian coordinate system under the assumption of Newtonian, incompressible, and steady fluid can be described as

$$\nabla \cdot \mathbf{v} = 0 \tag{14}$$

$$\rho \frac{D\mathbf{v}}{Dt} = -\nabla p + \mu \nabla^2 \mathbf{v} + \rho g \tag{15}$$

$$\rho \frac{De}{Dt} = -p \nabla \cdot \mathbf{v} + \nabla \cdot (k \nabla T) + \Phi \tag{16}$$

In the above equations, $\mathbf{v}$ is the velocity; $D/Dt = \partial/\partial t + (\mathbf{v} \cdot \nabla)$, the so-called material derivative; p is the pressure; ρ is the density; μ is the viscosity; g is the gravity; T is the temperature; k is the thermal conductivity; e is the internal energy; and Φ is the dissipation function, defined as

$$\Phi = \nabla \cdot (\tau_{ij} \cdot \mathbf{v}) - (\nabla \cdot \tau_{ij}) \cdot \mathbf{v} = \tau_{ij} \frac{\partial v_i}{\partial x_j} \tag{17}$$

where τ_{ij} is the viscous stress component

$$\tau_{ij} = \mu \left(\frac{\partial v_i}{\partial x_j} + \frac{\partial v_j}{\partial x_i} - \frac{2}{3} \frac{\partial v_k}{\partial x_k} \delta_{ij} \right) \tag{18}$$

The body-force term in the Navier–Stokes equation, i.e., ρg, can be neglected for natural convection.

4.3. Foster Thermal Network Model

For a multiple-chip power system containing n power semiconductor devices, these devices will be subjected to temperature rise due to self-heating and cross-heating effects. More specifically, any chip in the module with a power dissipation P will undergo self-heating, causing a junction temperature rise T_j, whereas the other devices will experience cross-heating, likewise leading to junction temperature elevation. In this work, a compact RC thermal network model in the form of a Foster network is applied for quick thermal simulation and easy implementation. The Foster network comprises a number of RC elements, where R is the thermal resistance (K/W) and is C the thermal capacitance (J/K).

The Foster thermal network model does not have any physical meaning or represent the physical structure of power devices. In order to develop a Foster thermal network, it is necessary to obtain the transient thermal impedance curves for both the self- and cross-heating responses. In the transient thermal characterization, the thermal impedance $Z(t)$ at a time t is used to determine the temperature variations $\Delta T(t)$

$$Z(t) = \frac{\Delta T(t)}{P(t)} = \frac{T_j(t) - T_a}{P(t)} \tag{19}$$

Using a Foster RC model, the above time-dependent thermal impedance $Z(t)$ can be described as

$$Z(t) = \sum_{i=1}^{n} R_i \left(1 - \exp(-\frac{t}{\tau_i}) \right) \tag{20}$$

where $\tau_i (i = 1, \ldots, n)$ are the i-th time constants, equivalent to the product of $R_i C_i$ in the Foster network. For the three-phase inverter, consisting of six switching devices, the value of n is 6. The thermal impedance matrix of the three-phase inverter is shown below

$$\left\{ \begin{array}{c} T_1(t) \\ \vdots \\ T_n(t) \end{array} \right\} = \left[\begin{array}{ccc} Z_{11}(t) & \cdots & Z_{1n}(t) \\ \vdots & \ddots & \vdots \\ Z_{n1}(t) & \cdots & Z_{nn}(t) \end{array} \right] \left\{ \begin{array}{c} P_1(t) \\ \vdots \\ P_n(t) \end{array} \right\} + \left\{ \begin{array}{c} T_a \\ \vdots \\ T_a \end{array} \right\} \tag{21}$$

where T_a is the ambient temperature. In the thermal impedance matrix, the diagonal components, namely Z_{ii}, denote the self-heating impedance of the i-th switching device and the off-diagonal components, namely $Z_{ij}(i \neq j)$, stand for the cross-impedance between the i-th and j-th switching devices. The thermal impedance matrix can be established by applying a power step to the switching devices one by one and then measuring the corresponding temperature responses of each of them.

In this work, the CFD code ANSYS Icepak (ANSYS Icepak 2020R2, Canonsburg, PA, USA) was used for the transient heat transfer simulation. The ANSYS Icepak CFD 3D model of the three-phase inverter is presented in Figure 6. The initial power at time zero ($t = 0$) was set to the estimated total power loss of the inverter at room temperature T_a. Subsequently, curve fits of the simulated transient heating curves were performed to identify the parameters (i.e., R and C) and thus produce RC networks for all six of the power MOSFET switching devices in the inverter, with which the time-dependent thermal impedance matrix, as listed in Equation (21), was built. Using the characterized time-dependent thermal impedance matrix, the junction temperatures of these switching devices can be simply estimated with given power losses. In fact, this approach implies limitations. For example, the thermal model is established based on a linear system assumption, and the accuracy of the prediction actually relies on the degrees of nonlinearity, such as convection, radiation, and temperature-dependent material nonlinearity.

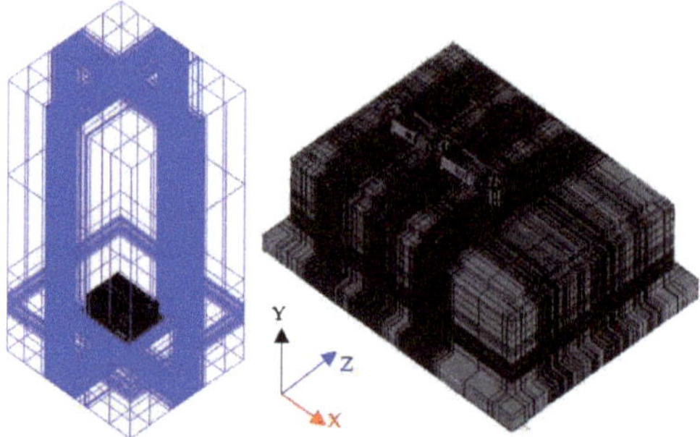

Figure 6. CFD thermal analysis 3D model of the three-phase inverter.

5. Electro-Thermal Coupling Analysis (ETCA)

The analysis flow of the proposed ETCA platform is shown in Figure 7 and comprises three analysis layers: EM modeling, electrical simulation, and thermal analysis based on an RC thermal network model. In order to account for the temperature effect on the switching transients and even power losses (conduction and switching), the latter two analysis layers, i.e., electrical simulation and thermal analysis, are fully coupled to co-simulate the ET coupled behavior of the three-phase power MOSFET inverter. In the switching loss estimation, the parasitic capacitances are also considered V_{ds}-dependent.

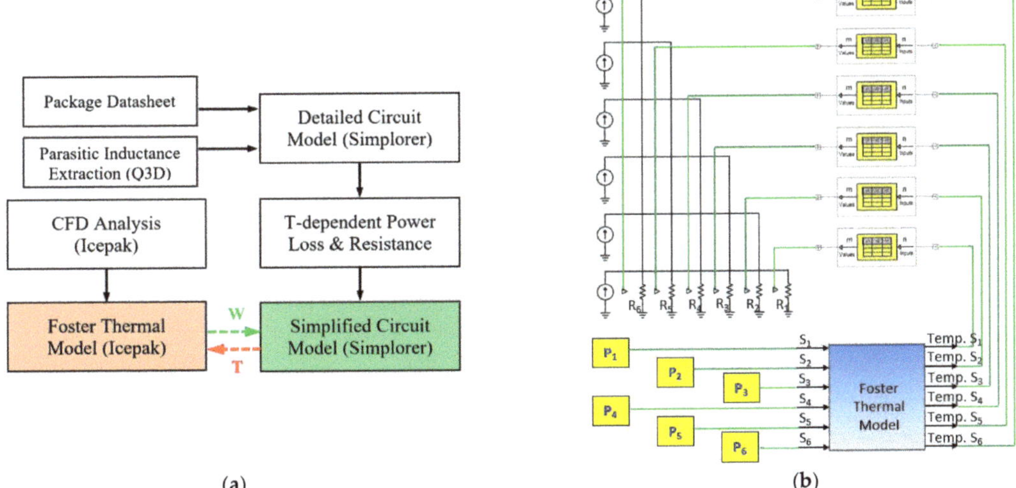

(a)

(b)

Figure 7. (**a**) Analysis flow of the proposed ETCA model and (**b**) simplified equivalent circuit model.

In the platform, the ETCA starts with the parasitic extraction (inductances) using ANSYS® Q3D Extractor, which is followed by the CFD thermal analysis and the fitting of the simulated heating curves in the time domain to establish the Foster thermal network model. ANSYS Icepak CFD software is responsible for solving the thermal problems in natural convection or forced convection and for deriving the transient thermal impedance curves. Instead of directly and iteratively performing the CFD analysis of natural or forced convection, the developed Foster thermal network model allows a rapid estimation of the junction temperature with different power conditions. Subsequently, with the characterized parasitic inductances together with the package model, including the output and transfer characteristics of the power MOSFET device, the diode characteristics, and the Vds-dependent parasitic capacitances, a detailed circuit simulation model of the three-phase inverter can be developed using ANSYS Simplorer to predict the switching transients and switching loss during the six-step operation. The detailed circuit simulation model of the three-phase inverter, together with the parasitic parameters (inductances) to be determined, is shown in Figure 2a.

The proposed ETCA approach can be applied to improve the computational efficiency of standard ETCA. In addition to the Foster thermal network model, it incorporates a simplified equivalent circuit model, as shown in Figure 7b, where the inverter switches (S1–S6) are simply modeled by resistors. The temperature-dependent equivalent electrical resistances of the resistors (R1–R6) are used to simulate the temperature dependence of the corresponding power losses (P1–P6) of the inverter switches during the six-step operation. The power loss of each of these inverter switches is composed of the conduction and switching losses of the power MOSFET modules and the conduction loss of the body diodes. Once the power loss–temperature relationships of these resistors are known, the power losses of each of these inverter switches at any temperature can be readily determined, which suggests that there is no longer a need to perform a tedious and complex detailed circuit simulation to predict the temperature-dependent power losses. The established power loss–temperature relationships of these resistors are implemented in the simplified equivalent circuit model. The interactions between the Foster thermal network model and the simplified equivalent circuit model, which exchanges the power and temperature data, are fulfilled through ANSYS Simplorer as the linking layer. It is important to note that for the common 120-degree square-wave commutation, each inverter switch conducts for 120 electrical degrees in each periodic cycle, indicating that the inverter switch is turned off in the rest of the periodic cycle. The calculated power losses of these power switches during the 120 electrical degrees are averaged across the periodic cycle. In this work, the temperature-dependent power losses of these power switches during one PWM six-step commutation cycle are derived using the abovementioned detailed circuit model under different temperature conditions, and with these the equivalent electrical resistance–temperature relationship can be determined based on Ohm's law.

6. Results and Discussion

6.1. Construction of Foster Thermal Network Model

Transient CFD thermal analysis of the three-phase inverter under natural convection was carried out using ANSYS Icepak. Then, constant power levels were sequentially set for each of the six switches, constituting six different power conditions. Accordingly, six parametric transient CFD analyses under natural convection associated with these six power conditions were performed using ANSYS Icepak and the corresponding transient junction temperature history profiles were collected. These temperature history profiles were further converted into transient thermal impedance curves. Two examples of the transient thermal impedance curves associated with $Z_{1i}(t)$ and $Z_{2i}(t)$ ($i = 1, \ldots 6$) are presented in Figure 8. Subsequently, these transient thermal impedance curves were used to extract the corresponding parameters in Equation (21), namely the time constants and resistances, by curve fitting in the time domain. The fit of the least squares regression analysis was outstanding, with a calculated multiple determination coefficient over 0.998, suggesting that the variation in the thermal impedance data was well-explained. Two examples of the curve-fitted values of these parameters associated with the transient thermal impedance curves $Z_{1i}(t)$ and $Z_{2i}(t)$ ($i = 1, \ldots 6$) shown in Figure 8 are presented in Table 1. According to Equation (21), these 36 time-dependent thermal impedance elements form the thermal impedance matrix, which was used to predict the junction temperatures of the power MOSFET chips under natural convection during load cycles.

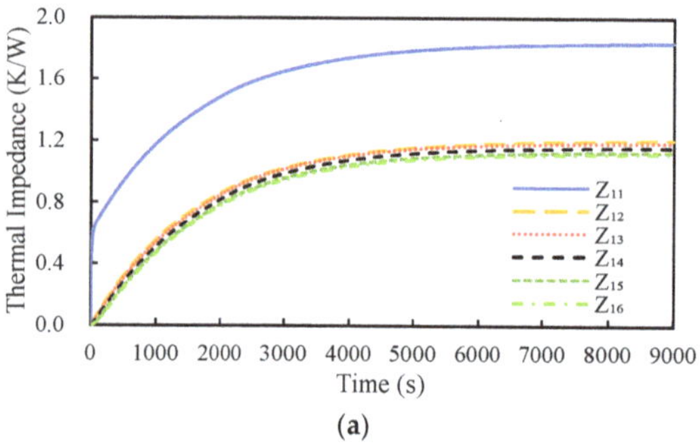

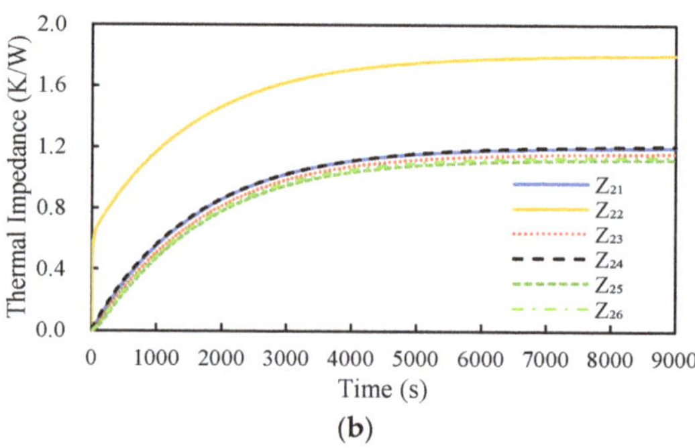

Figure 8. Two examples of the transient thermal impedance curves: (**a**) $Z_{1i}(t)$, $i = 1, \ldots 6$ and (**b**) $Z_{2i}(t)$, $i = 1, \ldots 6$.

Table 1. Curve-fitted resistances and time constants associated with thermal impedances Z_{1i} and Z_{2i} ($i = 1, \ldots 6$).

	Z_{11}	Z_{12}	Z_{13}	Z_{14}	Z_{15}	Z_{16}
R_i	1.82	1.211	1.199	1.173	1.147	1.137
τ_i	1151	1605	1629	1678	1730	1749
	Z_{21}	Z_{22}	Z_{23}	Z_{24}	Z_{25}	Z_{26}
R_i	1.207	1.791	1.175	1.213	1.139	1.153
τ_i	1600	1253	1677	1604	1745	1707

The feasibility of the developed Foster network thermal model based on the linear system assumption was demonstrated by comparing it with the CFD thermal analysis results associated with these six inverter switches (S_1–S_6) obtained using ANSYS Icepak at two different power settings, i.e., [13.2, 13.2, 13.2, 20.1, 20.1, 20.1] (W) with a total power (P^T) of 99.9 W and [11.4, 13.2, 15.9, 21.3, 24, 18] (W) with a total power of 103.8 W. The steady-state thermal analysis results are shown in Table 2. Note that the total power of the first power setting, i.e., P^T = 99.9 W, was the same as the initial preset power level used in the construction of the Foster thermal network model, while that of the second power setting (P^T = 103.8 W) was about 4% or 3.9 W larger than the initial preset power level. It can be clearly seen that for the first power setting, the developed Foster thermal network model produced a result that was very consistent with the CFD thermal analysis. By contrast, for the second power setting, there was a maximum deviation of 3% from the result of the CFD thermal analysis. If the discrepancy is over 5%, the Foster thermal network model may need to be updated or re-established for better accuracy, according to the power loss presented during the ETCA analysis. In other words, as long as the total power of applied power settings is similar to that used to create the Foster thermal network model, the derived result should be sufficiently accurate.

Table 2. Comparison of steady-state junction temperatures in the CFD analysis and Foster network (unit: °C).

P^T (W)	Method	S_1	S_2	S_3	S_4	S_5	S_6
99.9	Foster	149.8	154.7	150.1	155.3	149.1	154.1
	CFD	149.8	154.2	150.0	154.9	149.6	155.0
103.8	Foster	148.6	156.1	150.6	158.2	151.2	153.5
	CFD	153.2	159.6	154.6	161.7	155.9	158.4

6.2. ECTA Analysis of Three-Phase Inverter

The frequency-dependent parasitic parameters of the power MOSFET module and the three-phase inverter in a frequency sweep were explored using ANSYS® Q3D 3D quasi-static EM field solvers with various assigned conducting nets. In this parasitic analysis, three conducting nets were defined to describe the current paths of the power MOSFET module, i.e., drain, source, and gate (i.e., L_d, L_g and L_s), and ten conducting nets were assigned for the three-phase inverter in accordance with the switching sequence of the three-phase inverter, i.e., L_1–L_7 and L_{10}–L_{12}, as shown in Figure 2a. In the figure, L_8 and L_9 denote the drain and source inductances (L_d and L_s) of the power MOSFET module, respectively. It is worth mentioning that L_s represents the sum of the parasitic inductances of the source terminal leads and Al wires. Furthermore, the three-phase load is modeled as a resistor (R)–inductor (L) series impedance, i.e., R_a-L_a, R_b-L_b, and R_c-L_c, in Figure 2a. The parasitic inductances of the power MOSFET module extracted from the preceding inductive double-pulse test (DPT) circuit simulation at the working frequency of 20 kHz were 8.60, 5.47, and 7.53 nH and were associated with the gate, drain, and source terminals. As mentioned above, the source inductance is the sum of the parasitic inductances of the source terminal leads (i.e., 5.92 nH) and Al wires (i.e., 1.61 nH). The parasitic inductances associated with L_1–L_7 and L_{10}–L_{12} were calculated in the authors' previous work [4], and they are 23.34, 14.74, 25.52, 31.31, 6.93, 3.67, 54.89, 19.79, 19.52, and 19.78 (nH). These parasitic inductances, together with the package model (the output and transfer characteristics), the diode characteristics, and the V_{ds}-dependent parasitic capacitances, were applied in the detailed circuit simulation model, with which, together with the Foster thermal network model, the standard ETCA approach was constructed. The load condition of the inverter was a power supply voltage of 50 V, an SWPWM (D = 50%) switching frequency of 10 kHz, and an output frequency of 55 Hz. The inductance and resistance for these three-phase loads were 20 µH and 0.125 Ω, respectively. In addition, the switching frequency, gate resistance R_g, gate voltage V_g, gate inductance L_g, inductive load, and resistive load were set to 10 kHz, 1.6 Ω, 10 V, 8.6 nH, 20 µH, and 0.125 Ω. The ambient temperature was set to 25 °C.

The power losses of the switches in the first switching state of the six-step switching sequence were assessed first. The characterized power losses could then be applied to the other switching steps. The first switching state involved three inverter switches: S_1, S_2, and S_4. Basically, S_1 was a PWM power MOSFET in which the switching signal was discontinuously "on" (i.e., PWM "on") with a duty cycle of 50%, S_2 was an FWD switch in the commutation step, and S_4 was a commutation power MOSFET in which the switching signal was continuously "on". Accordingly, switching loss occurred only in S_1 (power MOSFET) and S_2 (diode), whereas conduction loss took place in all these three inverter switches. This switching state comprised two current loops during a single PWM cycle: PWM "on" and PWM "off". The parasitic inductances involved in the PWM "on" loop were L_1, L_8, L_9, L_{10}, L_{11}, L_8, L_9, L_5, and L_7 and those in the PWM "off" loop were L_{11}, L_8, L_9, L_4, L_9, L_8, and L_{10}. Next, circuit simulations of the power MOSFET inverter during the first switching state at eight different temperatures, i.e., 25, 50, 75, 100, 125, 150, 175, and 200 °C, were performed with the detailed circuit simulation model shown in Figure 2a. The calculated power losses of the inverter switches, S_1, S_2, and S_4, in the first switching state as a function of temperature are displayed in Figure 9a. In the figure, the legend of the light blue solid line with rectangle symbols, i.e., "Diode power loss", indicates the sum of the switching and conduction losses of the FWD switch. The results demonstrate that the switching and conduction losses of S_1, the diode power loss of S_2 (FWD), and the conduction loss of S_4 in the first switching state were around 10.8, 19.8, 51.6, and 17.7 W at 25 °C and increased or decreased to around 11.2, 25.7, 41.3, and 30.2 W at 200 °C. Specifically, in contrast to the diode power loss of S_2, the switching and conduction losses of the S_1 and S_4 switches tended to increase with increasing temperature. Noticeably, the diode conduction loss (S_2) showed a relatively strong and negative temperature coefficient, predominantly due to the diode characteristics shown in Figure 3d, where an increased temperature revealed a reduced drain-source voltage under the same drain-source current, thereby leading to a decreased conduction loss. Furthermore, it is interesting to note that temperature had a much smaller impact on the switching loss as compared to the conduction loss, that the switching loss of S_1 was much smaller than its conduction loss, and that the diode power loss outperformed the PWM (S_1) and commutation (S_4) power MOSFET modules.

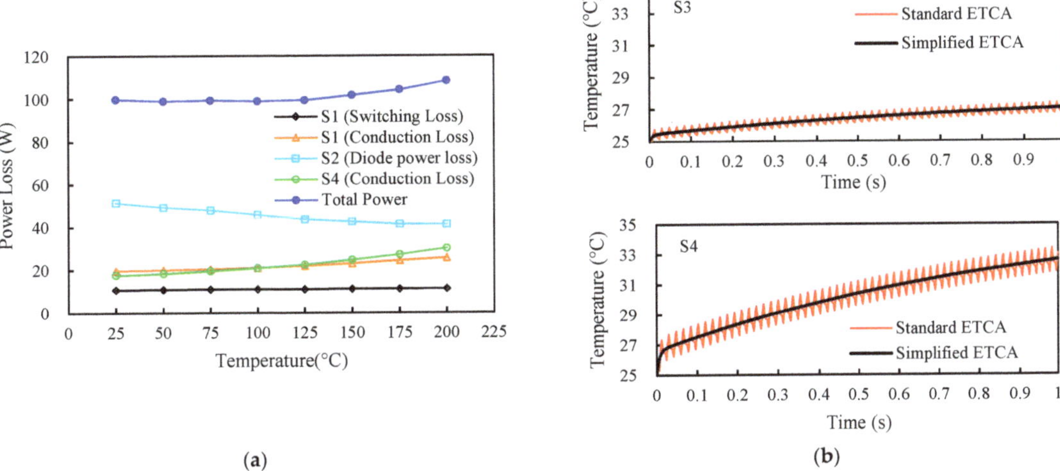

Figure 9. (a)Temperature-power loss dependence in the first switching state and (**b**) a comparison of the transient maximum junction temperatures of the switches S_3 and S_4 for the standard and proposed ETCAs during a one-second operation.

The total power loss of the inverter in the first switching state increased from about 99.9 W at 25 °C to about 108.4 W at 200 °C. The insignificant increase in the total power loss was mainly due to the negative temperature coefficient of the diode power loss. The total power loss at 25 °C was used as the initial power level for the development of the Foster thermal network model. Similarly, the power losses of these inverter switches at the other five switching states of the six-step switching sequence could also be derived in the temperature range of 25–200 °C. The calculated power losses during one PWM six-step commutation cycle at 25 and 200 °C are presented in Tables 3 and 4. It can be clearly seen that each inverter switch conducted for 120 electrical degrees in each periodic cycle for the common 120-degree square-wave commutation. For each inverter switch at each temperature, the power losses that occurred in the six switching states were averaged, and the results at 25 and 200 °C are also listed in the tables; with these, the equivalent electrical resistances (R_1–R_6) can be derived and the results at 25 and 200 °C are also demonstrated in the tables. The average power loss across one PWM six-step commutation cycle was used in the subsequent ETCA analysis.

Table 3. Power losses and equivalent resistances of these six inverter switches during one PWM six-step commutation cycle at 25 °C.

—	S_1	S_3	S_5	S_2	S_4	S_6	Total
Step 1	30.62	0.00	0.00	51.56	17.73	0.00	99.92
Step 2	30.62	0.00	0.00	51.56	0.00	17.73	99.92
Step 3	0.00	30.62	0.00	0.00	51.56	17.73	99.92
Step 4	0.00	30.62	0.00	17.73	51.56	0.00	99.92
Step 5	0.00	0.00	30.62	17.73	0.00	51.56	99.92
Step 6	0.00	0.00	30.62	0.00	17.73	51.56	99.92
Average (W)	10.21	10.21	10.21	23.10	23.10	23.10	—
R_i (Ω)	0.00102	0.00102	0.00102	0.00231	0.00231	0.00231	—

Table 4. Power losses and equivalent resistances of these six inverter switches during one PWM six-step commutation cycle at 200 °C.

—	S_1	S_3	S_5	S_2	S_4	S_6	Total
Step 1	36.90	0.00	0.00	41.31	30.17	0.00	108.38
Step 2	36.90	0.00	0.00	41.31	0.00	30.17	108.38
Step 3	0.00	36.90	0.00	0.00	41.31	30.17	108.38
Step 4	0.00	36.90	0.00	30.17	41.31	0.00	108.38
Step 5	0.00	0.00	36.90	30.17	0.00	41.31	108.38
Step 6	0.00	0.00	36.90	0.00	30.17	41.31	108.38
Average (W)	12.3	12.3	12.3	23.8	23.8	23.8	—
R_i (Ω)	0.00123	0.00123	0.00123	0.00238	0.00238	0.00238	—

Using the proposed ETCA approach, the transient maximum junction temperature profiles of the six inverter switches under natural convection over a time span of one second were calculated and compared with those of the standard ETCA approach. Two examples of the results associated with the inverter switches S_3 and S_4 are shown in Figure 9b. The reason for simply conducting the one-second test was that it is very difficult to perform the standard ETCA analysis for a longer period or to solve for the steady-state solution; hence, the more feasible ETCA approach was proposed. Evidently, there was a close agreement between them, suggesting the effectiveness of the proposed analysis approach. The calculated transient maximum junction temperature profiles of these inverter switches using the proposed ETCA are shown in Figure 10a for the time interval [0, 12000 s], and the corresponding temperature distributions in the power MOSFET chips of the inverter at the end of the simulation (t = 12,000 s) are illustrated in Figure 10b. Figure 10a reveals that the maximum junction temperatures of the power MOSFET chips would approach a steady state at around 4000 s. The maximum steady-state junction temperatures of the lower-side switches (namely S_2, S_4, and S_6) would be reached around 160 °C, while those of the upper-side switches (i.e., S_1, S_3, and S_5) would be reached at about 152 °C. These maximum junction temperatures exceed the maximum junction temperature rating of 150 °C and would not be permitted for device reliability and performance concerns. Active convection cooling, such as fans, or passive convection cooling, such as heat sinks and heat pipes, can be effective means to reduce the device junction temperature.

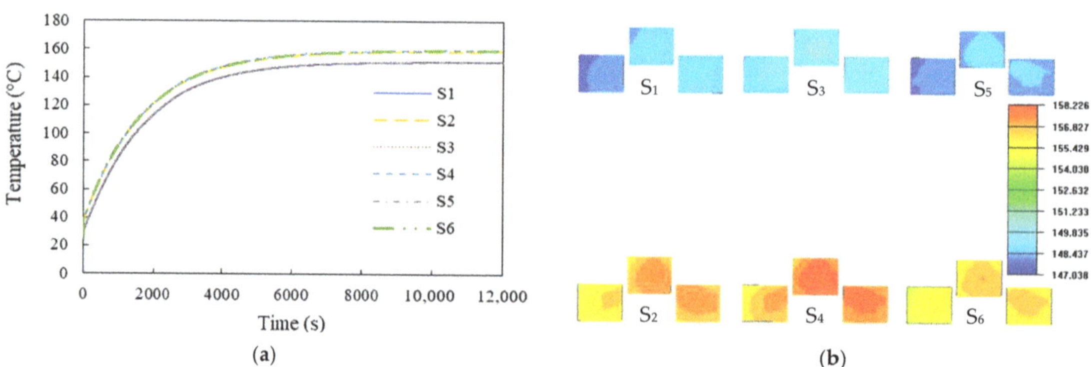

Figure 10. Thermal behavior of the six inverter switches: (**a**) transient maximum junction temperature profiles and (**b**) temperature distribution in the MOSFET chips.

The predicted maximum device junction temperatures of the three-phase inverter during the six-step operation unfavorably exceed the maximum junction temperature rating of 150 °C. The issue can be solved by active cooling with forced air. The CFD analysis of forced convection heat transfer was carried out with two wind speeds, 1.5

and 3.0 (m/s). The direction of the air flow was set to be horizontal, i.e., the x-axis in Figure 6. It can be noted that the Foster thermal network model derived above is no longer be applicable in this ETCA analysis due to its having different transient thermal impedance responses. Thus, a new Foster thermal network model was constructed. The total power loss at 25 °C, i.e., 99.9 W, was also applied as the initial power level to create the Foster thermal network model. The analysis results are displayed in Figure 11. For comparison, the natural convection result (i.e., wind speed = 0 m/s in the figure) is also demonstrated. The device junction temperature under natural convection is around 160 °C, and it is greatly reduced down to about 135 °C under forced convection with an air flow rate of 3 m/s. In addition, the increase in the air flow rate elevates the heat removal performance and thus lowers the device junction temperature.

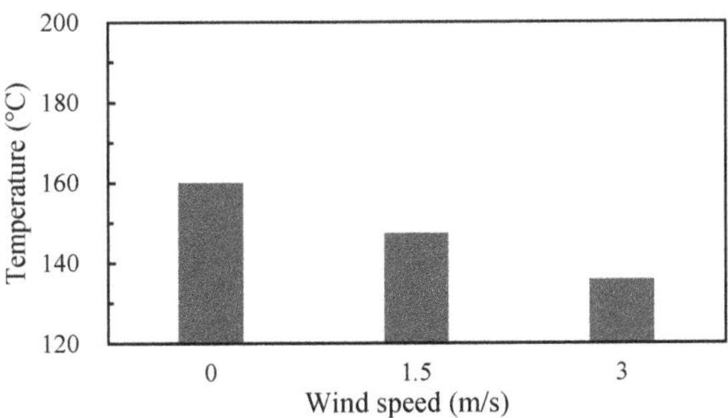

Figure 11. Maximum junction temperature of the inverter switches under forced convection with two different wind speeds.

7. Conclusions

This article presented an effective and efficient ETCA approach to characterize the ET coupled behavior of power systems under natural and forced convection during load cycles, which cannot be achieved using the conventional standard ETCA approach. The effect of temperature on the power losses and the influence of parasitics on the switching transients and power losses were all taken into account in the investigation. With this approach, the ET performance of a three-phase power MOSFET inverter for brushless DC motor drive under natural and forced convection during load cycles was explored. Additionally, both detailed and simplified circuit models were introduced, where the former was applied to develop the standard ETCA approach as well as the power loss–temperature relationship, while the latter was used to establish the proposed ETCA approach. Moreover, a Foster thermal network model for the three-phase inverter was created using the thermal impedance curves, which were derived through parametric transient CFD thermal analysis. The validity of the developed Foster thermal network model and the proposed ETCA approach was confirmed through the CFD thermal analysis and a standard ETCA approach.

The detailed circuit simulation demonstrated that the power losses (switching and conduction) of the PWM switches (e.g., S_1 in the first switching state) and the commutation switches (e.g., S_4 in the first switching state) had a positive temperature correlation while that of the PWM switches and the FWD switches (e.g., S_2 in the first switching state) had a negative temperature correlation. Moreover, in comparison with the PWM and commutation switches, the FWD not only had the largest power loss but also a relatively strong and negative temperature coefficient. This explains why the total power loss of the inverter would only slightly increase as temperature increases from 25 °C to 200 °C. It was also found that temperature played a much greater role in the conduction loss than the switching loss, and the switching loss of the PWM switches was considerably lower than its conduction loss.

The proposed ETCA analysis revealed that the maximum junction temperatures of the inverter switches would approach a steady state at around 4000 s, and the lower-side switches (namely S_2, S_4, and S_6) outperformed the upper-side switches (i.e., S_1, S_3, and S_5) in terms of the maximum steady-state junction temperature. Furthermore, these maximum junction temperatures of the inverter switches under natural convection with the specific load condition all exceeded the maximum junction temperature rating, and forced convection cooling with air was judged to be a very effective means to decrease the maximum junction temperatures.

Author Contributions: Conceptualization, H.-C.C.; methodology, H.-C.C.; software, S.-Y.L. and Y.-C.L.; validation, H.-C.C., S.-Y.L. and Y.-C.L.; formal analysis, H.-C.C., S.-Y.L. and Y.-C.L.; investigation, H.-C.C., S.-Y.L. and Y.-C.L.; resources, H.-C.C., S.-Y.L. and Y.-C.L.; data curation, H.-C.C., S.-Y.L. and Y.-C.L.; writing—original draft preparation, H.-C.C.; writing—review and editing, H.-C.C.; visualization, H.-C.C.; supervision, H.-C.C.; project administration, H.-C.C.; funding acquisition, H.-C.C. All authors have read and agreed to the published version of the manuscript.

Funding: This research is partially supported by Ministry of Science and Technology, Taiwan, ROC, under grants MOST-106-2221-E-035-052-MY3 and MOST-109-2221-E-035-004-MY3.

Institutional Review Board Statement: Not applicable.

Informed Consent Statement: No applicable.

Data Availability Statement: Data sharing not applicable.

Acknowledgments: Financial support from Ministry of Science and Technology, Taiwan is gratefully acknowledged.

Conflicts of Interest: The authors declare no conflict of interest.

References

1. Qi, J.; Yang, X.; Li, X.; Tian, K.; Mao, Z.; Yang, S.; Song, W. Temperature dependence of dynamic performance characterization of 1.2-kV SiC power MOSFETS compared with Si IGBTs for wide temperature applications. *IEEE Trans. Power Electron.* **2019**, *34*, 9105–9117. [CrossRef]
2. Liao, L.L.; Chiang, K.N. Material shear strength assessment of Au/20Sn interconnection for high temperature applications. *J. Mech.* **2019**, *35*, 81–91. [CrossRef]
3. Zhang, M.; Li, B.; Wei, J. New Power MOSFET with Beyond-1D-Limit RSP-BV Trade-Off and Superior Reverse Recovery Characteristics. *Materials* **2020**, *13*, 2581. [CrossRef] [PubMed]
4. Cheng, H.-C.; Shen, Y.-H.; Chen, W.-H. Parasitic extraction and power loss estimation of power devices. *J. Mech.* **2021**, *37*, 134–148. [CrossRef]
5. Hu, B.; Gonzalez, J.O.; Ran, L.; Ren, H.; Zeng, Z.; Lai, W.; Gao, B.; Alatise, O.; Lu, H.; Bailey, C.; et al. Failure and reliability analysis of a SiC power module based on stress comparison to a Si device. *IEEE Trans. Device Mater. Rel.* **2017**, *17*, 727–737. [CrossRef]
6. Wu, J.; Ren, N.; Guo, Q.; Sheng, K. A Comparative Study of Silicon Carbide Merged PiN Schottky Diodes with Electrical-Thermal Coupled Considerations. *Materials* **2020**, *13*, 2669. [CrossRef] [PubMed]
7. Oh, S.K.; Lundh, J.S.; Shervin, S.; Chatterjee, B.; Lee, D.K.; Choi, S.; Kwak, J.S.; Ryou, J. Thermal management and characterization of high-power wide-bandgap semiconductor electronic and photonic devices in automotive applications. *ASME J. Electron. Packag.* **2019**, *141*, 020801. [CrossRef]
8. Kanata, T.; Nishiwaki, K.; Hamada, K. Development trends of power semiconductors for hybrid vehicles. In Proceedings of the International Power Electronics Conference (IPEC), Sapporo, Japan, 21–24 June 2010.
9. US Department of Defense. *Military Standardization Handbook, MIL-HDBK-217C, Reliability Prediction of Electronic Equipment*; US Department of Defense: Arlington, VA, USA, 1980.
10. Yerasimou, Y.; Pickert, V.; Ji, B.; Song, X. Liquid metal magnetohydrodynamic pump for junction temperature control of power modules. *IEEE Trans. Compon. Packag. Manuf. Technol.* **2018**, *33*, 10583–10593. [CrossRef]
11. Blaabjerg, F.; Wang, H.; Vernica, I.; Liu, B.; Davari, P. Reliability of Power Electronic Systems for EV/HEV Applications. *Proc. IEEE* **2020**, *109*, 1–17. [CrossRef]
12. Merienne, F.; Roudet, J.; Schanen, J.L. Switching disturbance due to source inductance for a power MOSFET: Analysis and solutions. In Proceedings of the 1996 27th Annual IEEE Power Electronics Specialists Conference, Maggiore, Italy, 1 January 1996.
13. Rodríguez, M.; Rodríguez, A.; Miaja, P.F.; Lamar, D.G.; Zúniga, J.S. An insight into the switching process of power MOSFETs: An improved analytical losses model. *IEEE Trans. Power Electron.* **2010**, *25*, 1626–1640. [CrossRef]

14. Wang, H. Investigation of power semiconductor devices for high frequency high density power converters. Ph.D. Thesis, Virginia Polytechnic Institute and State University, Blacksburg, VA, USA, 2007.
15. Kibushi, K.; Hatakeyama, T.; Nakagawa, S.; Ishizuka, M. Analysis of heat generation from a power Si MOSFET. *Trans. Jpn. Inst. Electron. Packag.* **2013**, *6*, 51–56. [CrossRef]
16. Kibushi, R.; Hatakeyama, T.; Nakagawa, S.; Ishizuka, M. Calculation of temperature distribution of power Si MOSFET with electro-thermal analysis: The effect of boundary condition. *Trans. Jpn. Inst. Electron. Packag.* **2014**, *7*, 52–57. [CrossRef]
17. Cheng, H.-C.; Wu, C.-H.; Lin, S.-Y. Thermal and electrical characterization of power MOSFET module using coupled field analysis. *J. Mech.* **2019**, *35*, 641–655. [CrossRef]
18. Ibrahim, T.; Allard, B.; Morel, H.; MRad, S. VHDL-AMS model of IGBT for electro-thermal simulation. In Proceedings of the 2007 European Conference on Power Electronics and Applications, Aalborg, Denmark, 2–5 September 2007.
19. Wang, B. Research on electro-thermal model simulation of IGBT switching transient. *IOP Conf. Ser. Earth Environ. Sci.* **2021**, *702*, 012037. [CrossRef]
20. Touzelbaev, M.N.; Miler, J.; Yang, Y.; Refai-Ahmed, G.; Goodson, K.E. High-efficiency transient temperature calculations for applications in dynamic thermal management of electronic devices. *ASME J. Electron. Packag.* **2013**, *135*, 031001. [CrossRef]
21. Du, M.; Guo, Q.; Wang, H.; Ouyang, Z.; Wei, K. An Improved Cauer Model of IGBT Module: Inclusive Void Fraction in Solder Layer. *IEEE Trans. Compon. Packag. Manuf. Technol.* **2020**, *10*, 1401–1410. [CrossRef]
22. Alavi, O.; Abdollah, M.; Viki, A.H. Assessment of thermal network models for estimating IGBT junction temperature of a buck converter. In Proceedings of the 8th Power Electronics, Drive Systems & Technologies Conference (PEDSTC), Mashhad, Iran, 14–16 February 2017.
23. Reichl, J.; Lai, J.-S.; Hefner, A.; McNutt, T.; Berning, D. Inverter dynamic electro-thermal modeling and simulation with experimental verification. In Proceedings of the 2005 IEEE 36th Power Electronics Specialists Conference, Dresden, Germany, 12 June 2005; pp. 2208–2215.
24. Bouzida, A.; Abdelli, R.; Ouadah, M. Calculation of IGBT power losses and junction temperature in inverter drive. In Proceedings of the 8th International Conference on Modelling, Identification and Control (ICMIC-2016), Algiers, Algeria, 15–17 November 2016; pp. 768–773.
25. Shahjalali, M.; Lu, H.; Bailey, C. Electro-thermal modelling of multichip power modules for high power converter application. In Proceedings of the 2017 18th International Conference on Electronic Packaging Technology (ICEPT), Harbin, China, 16–19 August 2017.
26. Li, X.; Li, D.; Qi, F.; Packwood, M.; Luo, H.; Wang, Y.; Dai, X.; Luo, H.; Liu, G. EM-electrothermal analysis of semiconductor power modules. *IEEE Trans. Compon. Packag. Manuf. Technol.* **2019**, *9*, 1495–1503. [CrossRef]
27. Li, X.; Li, D.; Qi, F.; Packwood, M.; Luo, H.; Liu, G.; Wang, Y.; Dai, X. Advanced electro-thermal analysis of IGBT modules in a power converter system. In Proceedings of the 2019 20th International Conference on Thermal, Mechanical and Multi-Physics Simulation and Experiments in Microelectronics and Microsystems (EuroSimE), Hanover, Germany, 24–27 March 2019.
28. Reichl, J.; Ortiz-Rodríguez, J.M.; Hefner, A.; Lai, J.-S. 3-D thermal component model for electrothermal analysis of multichip power modules with experimental validation. *IEEE Trans. Power Electron.* **2015**, *30*, 3300–3308. [CrossRef]
29. Chen, H.-C.; Tsai, T.-Y.; Huang, C.-K. Comparisons of six-step square-wave PWMs in ultra-low-power SOC integration. In Proceedings of the International Symposium on Industrial Electronics (ISIE), IEEE, Seoul, Korea, 5–8 July 2009.
30. Chou, P.-C.; Cheng, S.; Chen, S.-H. Evaluation of thermal performance of all-GaN power module in parallel operation. *Appl. Therm. Eng.* **2014**, *70*, 593–599. [CrossRef]
31. Graovac, D.; Pürschel, M.; Kiep, A. MOSFET power losses calculation using the data-sheet parameters. *Infineon Appl. Note* **2006**, *1*, 1–23.
32. Inan, U.S. *Engineering Electromagnetics*; Pearson Education: Bangalore, India, 1998.

 *materials*

Article

Stress Impact of the Annealing Procedure of Cu-Filled TSV Packaging on the Performance of Nano-Scaled MOSFETs Evaluated by an Analytical Solution and FEA-Based Submodeling Technique

Pei-Chen Huang and Chang-Chun Lee *

Department of Power Mechanical Engineering, National Tsing Hua University, No. 101, Section 2, Kuang-Fu Road, Hsinchu 30013, Taiwan; mars420225@gmail.com
* Correspondence: cclee@pme.nthu.edu.tw

Abstract: Stress-induced performance change in electron packaging architecture is a major concern when the keep-out zone (KOZ) and corresponding integration density of interconnect systems and transistor devices are considered. In this study, a finite element analysis (FEA)-based submodeling approach is demonstrated to analyze the stress-affected zone of through-silicon via (TSV) and its influences on a planar metal oxide semiconductor field transistor (MOSFET) device. The feasibility of the widely adopted analytical solution for TSV stress-affected zone estimation, Lamé radial stress solution, is investigated and compared with the FEA-based submodeling approach. Analytic results reveal that the Lamé stress solution overestimates the TSV-induced stress in the concerned device by over 50%, and the difference in the estimated results of device performance between Lamé stress solution and FEA simulation can reach 22%. Moreover, a silicon–germanium-based lattice mismatch stressor is designed in a silicon p-type MOSFET, and its effects are analyzed and compared with those of TSV residual stress. The S/D stressor dominates the stress status of the device channel. The demonstrated FEA-based submodeling approach is effective in analyzing the stress impact from packaging and device-level components and estimating the KOZ issue in advanced electronic packaging.

Keywords: MOSFET; TSV; annealing process; finite element analysis; carrier mobility estimation

Citation: Huang, P.-C.; Lee, C.-C. Stress Impact of the Annealing Procedure of Cu-Filled TSV Packaging on the Performance of Nano-Scaled MOSFETs Evaluated by an Analytical Solution and FEA-Based Submodeling Technique. *Materials* **2021**, *14*, 5226. https://doi.org/10.3390/ma14185226

Academic Editor: Albonetti Cristiano

Received: 29 June 2021
Accepted: 8 September 2021
Published: 11 September 2021

Publisher's Note: MDPI stays neutral with regard to jurisdictional claims in published maps and institutional affiliations.

1. Introduction

Moore's law has been adopted for half a century, and it is still regarded as the target of transistor device performance. Silicon (Si) is the mainstream material for current semiconductor technology because of its low cost, mature fabrication process and acceptable performance. In the past decade, several advanced materials have been studied and used to replace Si as the new mainstream material in the semiconductor industry. Among the promising materials, germanium (Ge) and groups III–V are the most feasible due to their superior initial carrier transmission capability [1,2]. Strain engineering has been proposed to further enhance device performance under the same technology node through the lattice-mismatch mechanism. The four-point-bending technique is commonly utilized to extract the piezoresistance behaviors of device materials and estimate the stress-induced performance variation quantitatively [3–5]. Notably, the measured piezoresistance on bulk wafer and wafer with actual devices can differ considerably depending on the device type [3,5]. The stress sensitivity of different semiconductor materials has also been studied [4]. In electronic packaging architecture, the interconnect system plays an important role in signal transmission and delay time; notably, the overall delay time of an electronic packaging is determined by device and interconnect scaling [6]. Hence, the stability of the interconnect system is also an issue of electronic packaging. Through-silicon via (TSV) is the main interconnect architecture in 3D integrated circuit packaging, and the current mainstream TSV is fabricated with electroplated copper (Cu) [6–11]. Protrusion and thermal stress are the major mechanical reliability issues in TSV. The protrusion and thermal stress of TSV generally depend on the fabrication and annealing procedure and can generate cohesive and interfacial cracking on TSV [12–16]. Raman spectroscopy is widely used to estimate experimentally the residual stress of Cu TSV and the stress impact on the surrounding wafer [17–19]. In the fabrication procedure of Cu TSV, the annealing process is a critical step to manage the material characteristics, residual stress and Cu pumping. Cu annealing promotes interdiffusion, grain growth and re-crystallization to accomplish the abovementioned goals in thermomechanical reliability management. The TSV procedure is of three types, namely, TSV-first, TSV-middle and TSV-last, and the fabrication step of the via depends on the front-end (FE) and back-end (BE) processes. In the TSV-first procedure, the TSV is etched and filled, followed by FE and BE processes and wafer thinning. In comparison, TSV etching and filling are performed between the FE and BE in the TSV-middle procedure, and the TSV-last procedure means the FE/BE processes and wafer thinning are performed before TSV fabrication. The annealing effects on Cu's characteristics, including microstructure, elastic modulus and hardness, have been explored [20]. The critical temperature of zero stress impact transferred from Cu TSV to the surrounding Si wafer has been studied through simulation and experimental measurement [21–23]. For the diameter-dependent stress status of narrow Cu TSV, whose diameter is below 8 μm, the measured mean hydrostatic stress ranges from 150MPa to 200 MPa [24]. In a previous study, residual tensile stress values of 234 and 167 MPa before and after 200 °C of annealing, respectively, were separately extracted through synchrotron X-ray microdiffraction, and the mechanism of residual stress relaxation was generated by the lattice reorganization behavior [25]. Another study revealed that residual stress can be increased to around 600–700 MPa after 420 °C of annealing and cooling down to room temperature [26]. A transient selective annealing technology was presented in another work, and its influence on the thermomechanical reliability of Cu TSV was analyzed [27]. The layout design dependence of the thermomechanical behavior of TSV has also been investigated [28–30]. The concept of keep out zone (KOZ) has been proposed to obstruct the stress influence of TSV on the surrounding wafer and improve the transistor performance, and many structural and material designs have been studied since then. A novel structural design of TSV called annular-trench-isolated TSV was designed in previous research to reduce the volume of filled Cu and decrease the corresponding coefficient of thermal expansion (CTE, α) mismatch between Cu TSV (α =16.7 ppm/K) and Si (α =2.3 ppm/K) wafer [31,32]. In the TSV architecture, the barrier is adopted to prevent Cu diffusion into the surrounding wafer, and several barriers can function as stress buffer layers at the same time [33–39]. Different barrier materials with various deposition pressures and rates have been investigated [38]. The advantages of using metal-based barriers in Cu TSV protrusion and thermal stress have been studied [39]. The results have shown that a barrier with high modulus and similar CTE as that of Cu can significantly reduce the protrusion of TSV, but high stresses transfer from TSV to the surrounding Si. Several analytical formulas have been utilized to investigate the stress impact of the TSV core and adjacent region, and their results have been compared with experimental and simulation results [40–43]. However, previous studies on KOZ estimation generally considered the actual transistor as a simple bulk Si. This means the layout design of the nano-scaled device was not considered, and the stress transfer efficiency from the TSV core to the device might have been overestimated or underestimated. A finite element analysis (FEA)-based submodeling technique is developed in this study, and its results are compared with analytical results. Moreover, device strain engineering is considered in the constructed FEA model for investigating the comprehensive performance change of the device under lattice strain and TSV residual stress.

2. Fundamental Theories of 2D Analytical Stress Solution, Lattice Stress Estimation Approach and Piezoresistance Behavior for Stress-Induced Performance Investigation

2.1. 2D Analytical Solution for Stress Estimation of TSV and the Surrounding Substrate

The Lamé analytical stress model is widely adopted to estimate the influence of TSV-induced stress and the corresponding KOZ. The model assumes an infinite TSV surrounded by an infinite interposer when investigating the stress magnitude in the interposer under the impact of temperature change in the entire TSV interposer. A schematic of the Lamé radial stress solution is illustrated in Figure 1. Under the plane strain assumption, the entire structure is integrated with the core and surrounding material on the basis of the superposition principle. Assuming that the core is in a triaxial and uniform stress field, this stress field can be further separated into two stress components, namely, longitudinal stress (σ_L) along the out-of-plane axis and transverse stress (σ_T) for any two perpendicular axes. Through a derivation based on elasticity theory, the Lamé radial stress solution for the TSV stress-affected zone can be expressed as follows [40]:

$$\sigma_r = \frac{-E_{Cu}(\alpha_{Cu} - \alpha_{Si})\Delta T}{1 - 2\nu_{Cu} + \frac{1+\nu_{Si}}{1+\nu_{Cu}}\frac{E_{Cu}}{E_{Si}}}\left(\frac{D_{TSV}}{2r}\right)^2, \tag{1}$$

$$\sigma_r = \sigma_{xx}, \sigma_\theta = \sigma_{yy}, \sigma_r = -\sigma_\theta, \tag{2}$$

where σ_r and σ_θ denote radial and circumferential stresses, respectively. Figure 2 shows a detailed diagram of how TSV-induced stress influences the concerned device location. Labels D_{TSV} and r, respectively, denote the diameter of TSV and the distance between the TSV origin and the concerned device location for KOZ estimation. E, α and ν pertain to Young's modulus, CTE and Poisson's ratio, respectively. ΔT is the temperature variation resulting from the fabrication and treatment process, and the ΔT considered in this study is generated from the annealing procedure.

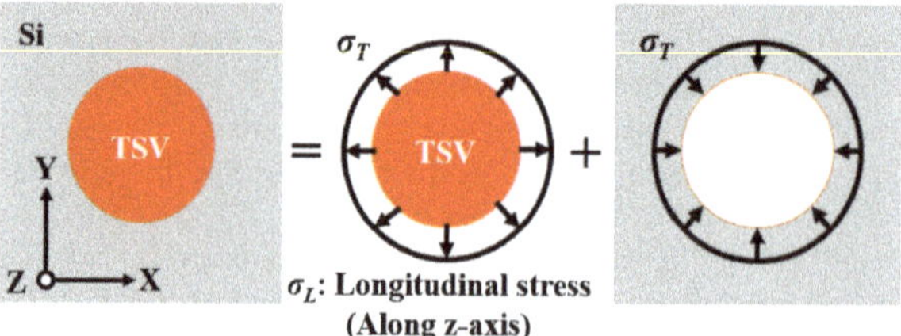

Figure 1. Schematic of Lamé radial stress solution based on the superposition principle. The stress field is integrated by the intrinsic stressed TSV core and reacted stress from the surrounding Si substrate.

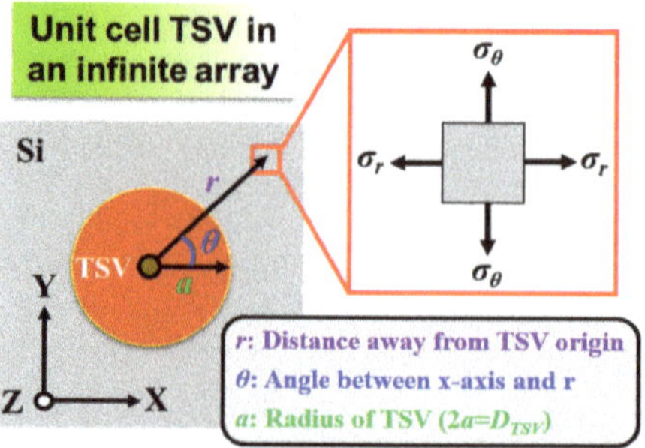

Figure 2. Schematic of TSV stress-affected zone calculation and corresponding structural parameters. The enlarged figure on the right side refers to the stress element in any location on the Si top surface.

These theoretical stresses are calculated based on analytical formulas integrated with structural and material parameters. On the basis of the foregoing equations, a semi-empirical formula with a similar form is presented for TSV stress-affected zone estimation. The semi-empirical formula is written as follows [19]:

$$\sigma_{Si,xx} = \sigma_{TSV} \cdot (\cos^2 \theta - \sin^2 \theta) \cdot \left(\frac{D_{TSV}}{2r}\right)^2,$$ (3)

$$\sigma_{Si,yy} = \sigma_{TSV} \cdot (\sin^2 \theta - \cos^2 \theta) \cdot \left(\frac{D_{TSV}}{2r}\right)^2.$$ (4)

The advantage of this form is that the experimentally measured residual stress of TSV can be interpolated directly. The sign θ is the angle between the x-axis and r.

Many assumptions are made in the aforementioned model and cause a significant difference from the real TSV interposer architecture. First, the analytical model only considers the filler material of TSV and surrounding interposer and does not include the components adjacent to the core of TSV. Second, the analytical model does not consider the relative orientation between the concerned device channel and TSV core. These simplifications influence the efficiency of stress transfer from the TSV core to the surrounding components from a mechanical perspective and make the accuracy of the estimated performance variation in excepted transistor location controversial. Accordingly, this research utilizes the FEA-based submodeling simulation approach to explore the stress impact of the TSV core on the surrounding interposer and compares this approach with the classic Lamé radial stress solution.

2.2. Theotical Calculation of Lattice Mismatch Strain on the Absis of Vegard's Law

Lattice mismatch strain simulation is performed in the present study to estimate the efficiency of lattice strain in device-level design and further compare it with the TSV stress-induced KOZ effect from the packaging-level structure. Lattice strain generation concentrates one element into another one to introduce volume expansion/shrinkage and corresponding stress; this phenomenon is approximated as the thermal stress mechanism. Accordingly, the virtual thermal strain approach has been proposed and validated, and it can be utilized to simulate lattice mismatch strain [44]. Its accuracy and feasibility have been validated through a comparison with literature data and analytically derived stress/strain formulas [45,46]. The lattice constant of concerned materials is calculated to estimate the subsequent lattice mismatch strain. Consequently, the formula for lattice parameter estimation of the concentrated component is computed in accordance with Vegard's law as follows:

$$a_{A_{1-x}B_x} = a_A \times (1 - x) + a_B \times (x),$$ (5)

where a_A and a_B refer to the lattice constant of the materials in pure form and $a_{A_{1-x}B_x}$ is the lattice constant of the combined components mixed by pure materials A and B. The sign x denotes the mole fraction of concentrated material B. Thus, the lattice mismatch strain amount, defined as parameter f, can be estimated as follows:

$$f = \frac{a_A - a_{A_{1-x}B_x}}{a_{A_{1-x}B_x}}.$$ (6)

From the abovementioned equation, the lattice mismatch strain generated from the designed lattice mismatch stressor is estimated and can be used to investigate the influence on device performance through the utilization of piezoresistance behavior.

2.3. Piezoresistance Characteristics of Semiconductor Material for Estimating Stress-Induced Carrier Mobility Variation

Piezoresistivity is a material characteristic in terms of stress and electric resistance. From the electric performance view, the increment in performance is proportional to the decrease in electric resistance. Accordingly, the approximation effect of stresses on device performance can be estimated through the integration of stresses in the device channel region and piezoresistance parameters. The detailed formula is expressed as follows:

$$\frac{\Delta\mu}{\mu_0} = -\frac{\Delta\rho}{\rho_0} = -(\sigma_{xx}\pi_{xx} + \sigma_{yy}\pi_{yy}), \tag{7}$$

where $\Delta\mu$ and μ_0, respectively, denote the stressed and initial mobilities of the device channel. Parameters π and σ are the piezoresistivity coefficient along the concerned direction of the device channel. With reference to the piezoresistance of the Si transistor, the piezoresistance values in longitudinal (x-direction) and transverse (y-direction) directions are significantly larger than that in the vertical (z-direction) direction. Two major stress components (σ_{xx} and σ_{yy}) of the device channel are extracted to investigate the stress-induced carrier mobility variation in Si pMOSFET in this study. The piezoresistance coefficients of Si pMOSFET for mobility change calculation are obtained from Literature [5].

3. FEA Modeling of Global TSV Interposer Packaging Architecture and Local Transistor De-Vice Model Based on the Submodeling Technique

Generally, the submodeling technique is a modeling approach to overcome the difficulty in the modeling and meshing of an FEA model with a significant size difference between concerned components. In this study, the approach is utilized to consider the stress influence and transfer efficiency of a micro-scaled TSV interposer and a nano-scaled transistor device in the same model. The micro-scaled TSV interposer is defined as the global model in the present submodeling procedure and illustrated in Figure 3. A single TSV unit in the entire symmetric array-type TSV interposer is extracted and constructed as an FEA model. The TSV unit model is composed of Cu-filled TSV, adjacent titanium nitride barrier layer and surrounding silicon dioxide (SiO_2) dielectric layer as shallow trench isolation (STI). The designed thickness of STI and the barrier layer are 0.16 μm and 40 nm, respectively. Notably, the features of STI prevent electronic signal leakage, and the barrier layer prevents the diffusion of the TSV core material from contaminating the surrounding Si interposer. In this study, the distance between the origin of the TSV core and the concerned device location is fixed at 20 μm (labeled as symbol r in Figure 3); this value is also referred to as the half of TSV pitch, and the TSV depth is fixed at 50 μm. Several TSV diameters, namely, 5, 10, 15 and 30 μm, are designed to estimate the stress influences on the transistor device by using the analytical Lamé radial stress solution and FEA-based submodeling simulation approach.

In the global TSV interposer model, the characteristics of the device region (labeled as the local model in Figure 3) is considered similar to the material of the device channel, Si, in this study. According to the procedure of the submodeling technique, a detailed transistor device local model also needs to be constructed, and the corresponding transistor model is shown in Figure 4. A half-symmetry FEA model of the device is constructed, but only a quarter model is illustrated in Figure 4 to introduce the structural parameters and structure components clearly. The components of the gate stacking structure of the device include the gate, liner, spacer and gate oxide, and the corresponding materials of the components considered in this study are poly-Si, SiO_2, silicon nitride and SiO_2, respectively. The thickness of the spacer, liner and gate oxide are 20, 2 and 1.5 nm, respectively. The heights of the gate and spacer are 70 nm, and the length of the gate is in accordance with the technology node of the 28 nm transistor device. The concerned device is fabricated on the (001)[110] lattice orientation of Si wafer.

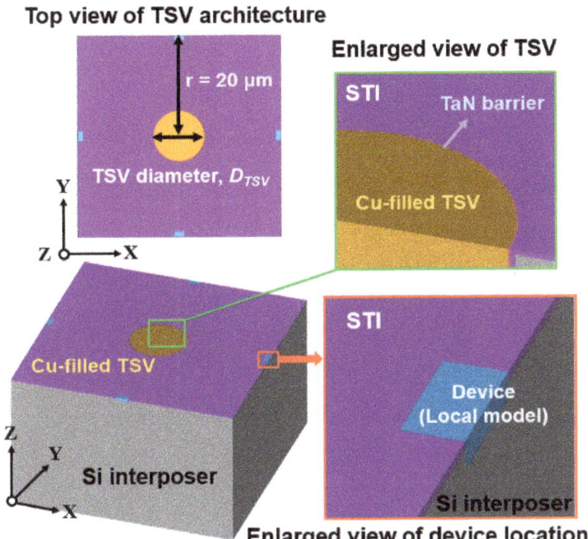

Figure 3. Schematic of a single TSV unit FEA model in an array-type TSV interposer and the corresponding structural parameters and materials. The TSV pitch is designed as 40 µm in this study.

3D view of Si pMOSFET local model

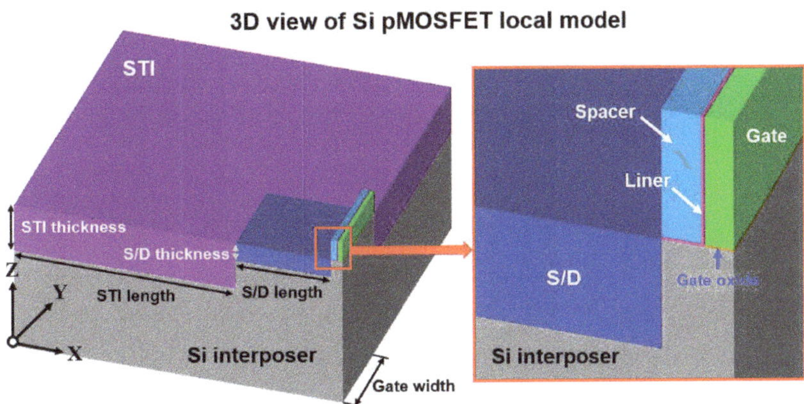

Figure 4. Schematic of a local Si pMOSFET FEA model and the corresponding structural parameters and components.

A general design for lattice strain generation incorporates the favorable element into the base material. In the Si pMOSFET architecture, Ge is adopted in Si and forms the silicon–germanium ($Si_{1-x}Ge_x$) concentrated alloy in the source/drain (S/D) region, which is a widely used and effective approach to introduce the preferred stress status of Si pMOSFET. The concentration of Ge in the $Si_{1-x}Ge_x$ stressor in this study is fixed at 25%, which is the most common design for Si pMOSFET. The S/D's length and thickness are fixed at 300 and 60 nm, respectively. As the functional isolation, STI length and thickness are fixed at 700 and 160 nm, respectively. Hence, the constructed local model can be regarded as a single unit in a device array layout. The gate width of the device channel is the major design parameter in analyzing the gate width dependence on stress influence from TSV residual stress and the relationship of lattice mismatch stress with performance variation. After constructing the global TSV interposer model and local transistor device model, the procedure of submodeling for linking the mechanical response between the TSV interposer and transistor device is illustrated in Figure 5 and described in detail as follows. All surfaces, expect for the top surface of the TSV interposer model, is regarded as symmetric planes for boundary condition consideration.

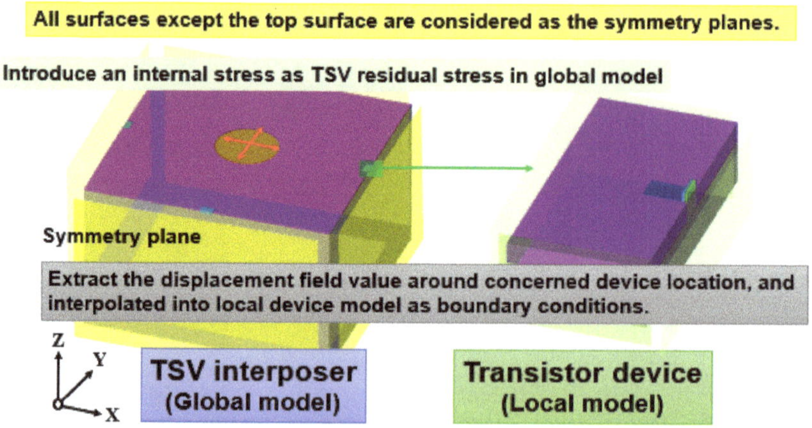

Figure 5. Schematic of how to introduce the TSV stress influence into the transistor device region based on the submodeling technique.

The annealing process-induced stress field of the TSV interposer should be generated in the FEA model by incorporating the considered residual stress (RS) magnitudes, 167 and 700 MPa, which, respectively, refer to the RS of Cu-filled TSV after 200 °C [25] and 420 °C [26] of annealing. Then, the displacement field around the device region of the global TSV interposer model is extracted and further interpolated into the local device model. Afterward, the TSV's RS impact is introduced to the concerned Si pMOSFET and can be integrated with the S/D lattice strain stressor comprehensively. The mechanical characteristics of all materials and the parameters for annealing stress calculation for analytical solution and FEA simulation are summarized in Table 1.

Table 1. Material characteristics utilized in this study for analytical stress calculation and FEA stress simulation.

Model	Components	E, GPa	ν	α, ppm/K
Global TSV interposer	TSV (Cu)	117	0.30	16.7
	Barrier (TaN)	186	0.342	6.48
	Interposer (Si)	169	0.26	2.3
	STI (SiO$_2$)	71.7	0.16	0.51
Local transistor device	Gate (Poly-Si)	160	0.22	2.3
	Liner (SiO$_2$)	71.7	0.16	0.51
	Spacer (SiN)	123.3	0.30	3.05
	S/D (Si$_{0.75}$Ge$_{0.25}$)	161	0.265	3.2
	Substrate (Si)	169	0.26	2.3
	STI (SiO$_2$)	71.7	0.16	0.51

4. Results and Discussions

4.1. Comparison of Analytical and FEA Submodeling Results on the Longitudinal and Trasnverse Stress of the Si pMOSFET Device Channel

The feasibility of the analytical/semi-empirical stress estimation formula and its difference from the FEA submodeling results are explored. From the viewpoint of the piezoresistance of Si pMOSFET, the stress sensitivity in the vertical direction of the device channel is at least 13 times lower than those in the longitudinal and transverse directions [10]. Moreover, because the 2D Lamé stress model cannot generate the analytic results of vertical directional stress, the two other major stress components (longitudinal and transverse stresses) are estimated and discussed using the aforementioned approaches. The TSV diameter-dependent channel stress is calculated and illustrated in Figure 6. Considering the 420 °C annealing process' impact on Cu-filled TSV with 30 μm diameter, the stress-free temperature is defined as the designed annealing temperature and subsequently cooled down to room temperature of 200 °C. The parameters $\Delta T = -400\ {}^{\circ\circ}$C and RS = 700 MPa are, respectively, interpolated into Equations (1) and (3). Thus, the estimated longitudinal stress introduced into the device channel provided by Equations (1) and (3) is 353.95 and 393.75 MPa, respectively. The two analytic results show similar magnitudes under the same temperature loading condition. The calculated stress magnitudes reveal that the stress estimation feasibility of Equations (1) and (3) is highly comparable, and a 10% difference exists between the pure analytic result [from Equation (1)] and semi-empirical formula result [from Equation (3)]. Moreover, the calculated results from Equation (1) are lower than the results from Equation (3) for all designed TSV diameters, but the variation for the narrow TSV whose diameter is less than 15 μm is small.

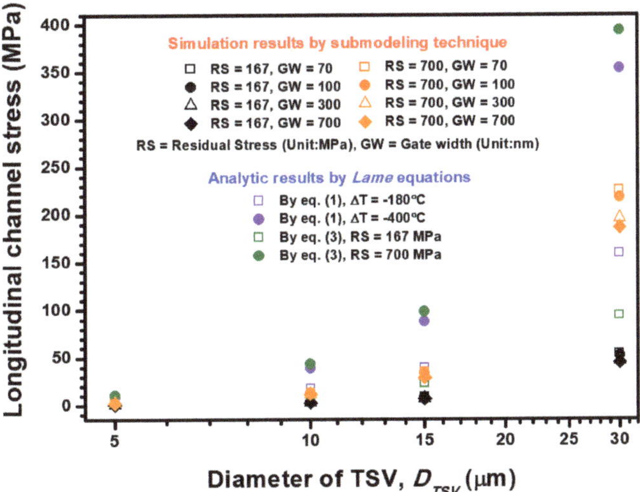

Figure 6. TSV-diameter, TSV-residual stress magnitude and channel gate width dependence longitudinal channel stress estimation (stress along the channel length direction) by using the analytical formula, semi-empirical formula and FEA submodeling simulation.

When the 200 °C annealing process-induced stress impact ($\Delta T = -180$ °C and RS = 167 MPa) generated by the 30 μm Cu-TSV on the Si device is considered, the estimated longitudinal stress in the device channel based on Equations (1) and (3) is calculated as 159.28 and 93.94 MPa, respectively. An opposite trend is observed compared with the situation that considers the 420 °C annealing stress impact. This phenomenon can be attributed to the pure analytic formula, which relies on the assumption that Cu-TSV is ideally stress-free under the considered annealing temperature. However, extant literature indicates that RS is not effectively relaxed by only the 200 °C annealing procedure, and -196 MPa (the minus mark refers to the compressed stress status) is measured at 200 °C [25]. Meanwhile, almost zero stress at 420 °C was experimentally obtained in another study [26]. Accordingly, the pure analytic and semi-empirical formulas show reasonable consistency when a high annealing temperature of over 400 °C is considered. However, when a relatively low-temperature (below 400 °C) annealing process is designed, the semi-empirical formula is more suitable for estimating the impact of RS on transistor device performance compared with the pure analytic formula.

The FEA submodeling results are also presented in Figure 6 to compare the estimated stress magnitudes with the analytic stress results. In consideration of packaging-level and device-level layout design, TSV diameter and channel gate width, the FEA submodeling stress results are much lower than the stress magnitudes calculated from Equations (1) and (3). When the 30 μm TSV diameter and 70 nm gate width are designed, 225.77 MPa of longitudinal stress is introduced into the Si pMOSFET channel. This result means that the analytical and semi-empirical formulas overestimated 56% of the longitudinal channel stress compared with the FEA submodeling simulation result. This mechanism can be explained by the stress buffer behavior of the barrier and STI structure between the TSV core and Si pMOSFET. Moreover, the layout design of Si pMOSFET plays a role in TSV stress transfer efficiency. When the channel gate width increases from 70 nm to 700 nm, the introduced longitudinal channel stresses decrease from 225.77 MPa to 186.83 MPa. These results indicate that the increased gate width leads to an increment in the structural stiffness of Si pMOSFET and further obstructs the RS influence from the TSV introduced into the device channel. Thus, the effects of TSV diameter and device gate width on longitudinal channel stress are systemically discussed.

The dependence of transverse channel stress on TSV diameter and channel gate width is illustrated in Figure 7. The same stress magnitude of transverse-direction channel stress but different stress status from tensile to compressive are calculated by the analytical and semi-empirical formulas. However, the channel gate width shows a positive influence on the increment in compressive transverse channel stress. The transverse channel stress is enhanced from -90.63 MPa to -163.52 MPa when the channel gate width increases from 70 nm to 700 nm. These results reveal that the enlarged gate width is beneficial to the stress obstruction in the longitudinal direction, but it aggravates the compression in the transverse direction because of its slim geometry with a large aspect ratio. The distance-to-radius ratio is also an important parameter to investigate annealing-induced thermal stress from TSV. In this study, the TSV pitch is fixed at 40 μm, which means the foregoing ratio is managed by the variation of TSV diameter. The distance-to-radius ratios of the four designed TSV diameters are calculated as 7, 3, 1.66 and 0.33, which correspond to 5, 10, 15 and 30 μm diameters, respectively. These ratios are attributed to the estimation of the critical criteria, which prevent the harsh thermal stress impact of TSV on the concerned device. For longitudinal and transverse channel stresses, the stress impact is suddenly increased when a distance-to-radius ratio of below 1 is considered. Accordingly, the aforementioned ratio is crucial for thermal stress management. On the basis of the piezoresistance of Si pMOSFET, the tensile and compressive stresses have a negative effect on stress-induced performance, which means the optimized gate width should be designed carefully to minimize the KOZ region under the annealing RS impact of Cu-filled TSV. Thus, KOZ estimation of Si pMOSFET is presented and discussed in the following section.

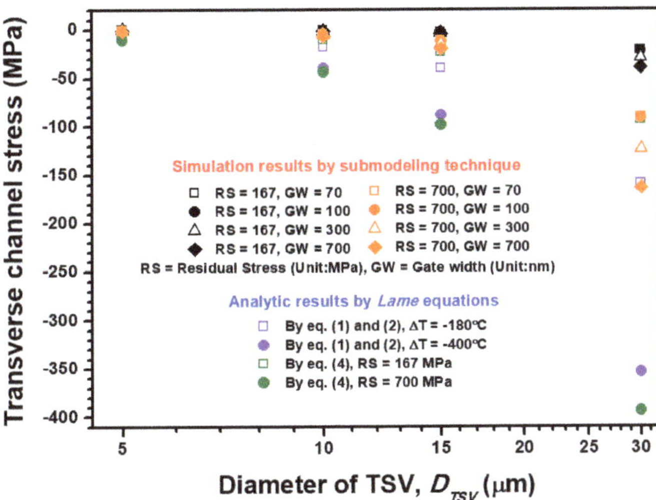

Figure 7. TSV-diameter, TSV-residual stress magnitude and channel gate width dependence transverse channel stress (stress along the channel width direction) estimated by the analytical formula, semi-empirical formula and FEA submodeling simulation.

4.2. Stress-Induced Hole Carrier Mobility Gain Change and KOZ Estimation Based on Analytical and FEA Submodeling Results

When the distance between the TSV origin and Si pMOSFET is fixed at 20 μm, the criterion for KOZ determination is considered to be a 10% change in carrier mobility gain [45]. As shown in Figure 8, the hole carrier mobility gain change is apparently unfavorable for the 30 μm TSV diameter design with 700 MPa RS. Carrier mobility gains of −36.46% and −40.56% are estimated by the analytical solution and semi-empirical formula, respectively. In comparison, the FEA submodeling results reveal nearly −18% hole carrier mobility gain change under the same TSV diameter and RS magnitude. This nearly −18% mobility gain change is not linearly proportional to the designed gate width because the longitudinal and transverse stress-induced mobility gain changes compete with each other. For the TSV diameter below 15 μm, the estimated mobility gain changes meet the design criteria of KOZ determination (10% mobility change) under 700 MPa RS impact (corresponding to the 420 °C annealing procedure). Notably, the nearly 22 μm TSV diameter design is acceptable according to the FEA submodeling results. The KOZ region can be further determined by the distance between the designed *r* (20 μm) and the edge of TSV. Therefore, the KOZ regions are, respectively, estimated as 12.5 μm (for D_{TSV} = 15 μm) and 9 μm (for D_{TSV} = 22 μm) by the analytical solution/semi-empirical formula and FEA submodeling approach. Moreover, the estimated results provided by the semi-empirical formula are similar to the FEA submodeling results when 167 MPa RS magnitude and 15 μm TSV diameter are considered; however, a −1.8% mobility change is still overestimated. Moreover, the mobility gain variation is almost independent of the designed D_{TSV} when the 167 MPa RS is considered. This phenomenon can be attributed to the piezoresistive behavior of Si pMOSFET, and the compressive and tensile stresses are attributed to the stress-induced performance. However, the longitudinal and transverse channel stresses are positively and negatively proportional to the increment of D_{TSV}, respectively, which means the positive and negative influences on device performance generated by longitudinal and transverse stresses are countervailed. Accordingly, the final carrier mobility gain is almost independent of channel gate width variation. These results indicate that the estimated hole carrier mobility change provided by the analytical solution and semi-empirical formula is significantly overestimated compared with the value from the FEA submodeling approach, which considers the actual structural characteristics of the layout design on packaging and device levels. These results are beneficial to accurately estimating the KOZ region and further increasing the integration density from packaging-level design in the TSV interposer architecture.

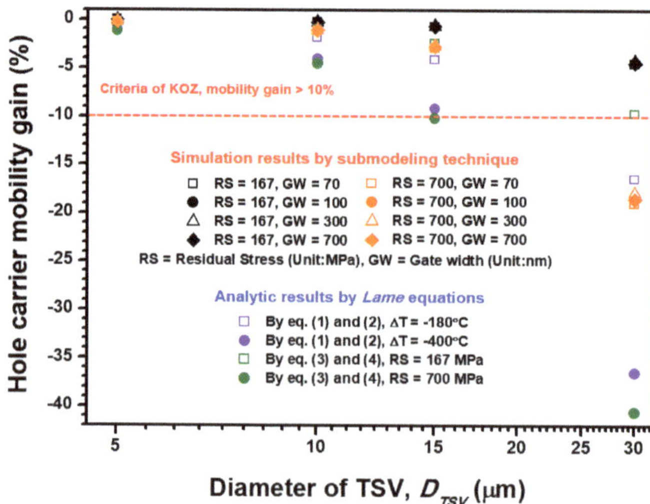

Figure 8. Comprehensive stress-induced impact on Si pMOSFET performance under different combinations of TSV/device layout design and TSV residual stress magnitude. The hole carrier mobility gain is estimated by the simulated stress components integrated with the piezoresistive behavior of Si.

4.3. FEA Submodeling Results on the Longitudinal and Transverse Stresses of the Si pMOSFET Device Channel under the Integrated Effect of TSV RS and Strain Engineering S/D Lattice Stressor

The influence of TSV RS on device performance is systemically discussed in this section. On the basis of an unstrained Si pMOSFET, the TSV RS-induced performance degradation is estimated to be 0.05 % to 18.93 % depending on the designed TSV diameter and channel gate width. This phenomenon is not favorable because the performance degradation of pMOSFET is difficult to determine using nMOSFET with a high initial carrier transport capability. For this reason, the S/D lattice-strained $Si_{0.75}Ge_{0.25}$ stressor is used in this study to analyze the comprehensive effect on stress-induced performance variation under the integrated stresses of TSV RS and S/D lattice mismatch. As shown in Figure 9, the S/D lattice stress dominates the longitudinal stress magnitudes in the Si pMOSFET channel. This dominance is attributed to the high stress transfer efficiency between S/D and the device channel in accordance with the direct contact between the foregoing components. According to the analytic results presented in Figure 7, the longitudinal channel stress impact introduced by 5 μm diameter TSV is almost zero. Hence, the longitudinal stress magnitude for the S/D strained Si pMOSFET with 5 μm TSV diameter (shown in Figure 9) can be regarded as the lattice mismatch stress generated by the S/D stressor. The lattice mismatch strain amount is proportional to the increment in gate width because an increased gate width enlarges the width of the S/D region and generates more lattice mismatch strain subsequently. For a narrow TSV design with a 5 μm diameter, the longitudinal channel stress is completely determined by the S/D lattice stress when the two different TSV RS magnitude (167 and 700 MPa) are considered. However, given that the design TSV diameter is enlarged from 5 μm to 30 μm, the longitudinal channel stress varies from −870.43 MPa to −647.66 MPa for a Si pMOSFET with a 70 nm gate width. These results are due to the enlarged TSV diameter shrinking the relative distance between TSV and the device location, thereby enhancing the tensile RS impact of TSV introduced into the device channel and weakening the compressive stress induced by the S/D stressor. Moreover, the narrow gate width cannot obstruct the tensile RS impact of TSV transferred to the concerned Si pMOSFET channel. Accordingly, the longitudinal channel stresses of lattice-strained Si pMOSFET are mainly dominated by the S/D stressor, but the RS impact of TSV also plays an important role when enlarged TSV diameters and RS magnitudes are utilized.

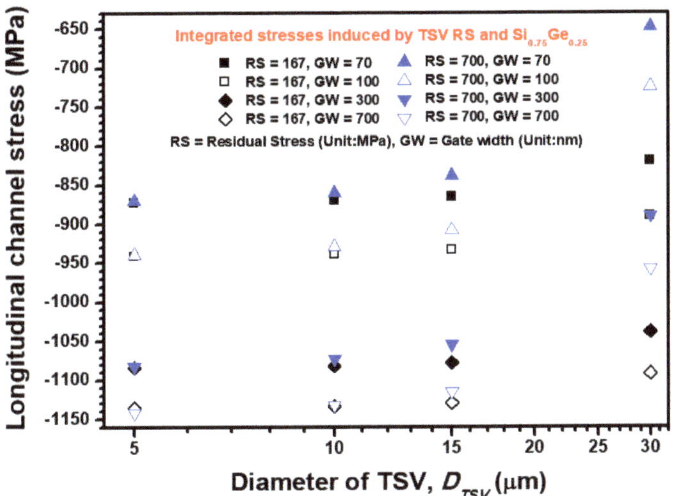

Figure 9. Dependence of the layout and annealing process design on the longitudinal channel stress of Si pMOSFET estimated by the analytical formula, semi-empirical formula and FEA submodeling simulation approach.

The transverse channel stress introduced by the TSV RS impact and S/D strained Si$_{0.75}$Ge$_{0.25}$ stressor is further illustrated in Figure 10. Similar to the stress trend shown in Figure 9, the transverse channel stress remains stable because the RS impact generated by the Cu-filled TSV is limited to almost zero when the TSV diameter of below 15 μm is utilized. For a narrow gate width of 70 nm, a significant tensile transverse channel stress is observed. This stress status can be attributed to the Poisson's ratio mechanism based on generalized Hooke's law. When the gate width is enlarged from 70 nm to 700 nm, the aspect ratio of the device channel increases and limits the Poisson's ratio mechanism. The enlarged gate width also extends the width of the channel and further degrades the uniformity and concentration of S/D-induced lattice mismatch stress. Notably, the compressive and tensile stress status along the longitudinal and transverse directions is favorable for stress-induced performance enhancement in accordance with the piezoresistance characteristics of Si pMOSFET. The optimized mobility gains of the considered Si pMOSFET are not linearly proportional to the increase or decrease in gate width. Thus, the gate width-dependent carrier mobility gains under the integrated stress generated by RS of TSV and the S/D lattice stressor are estimated and discussed in the following section.

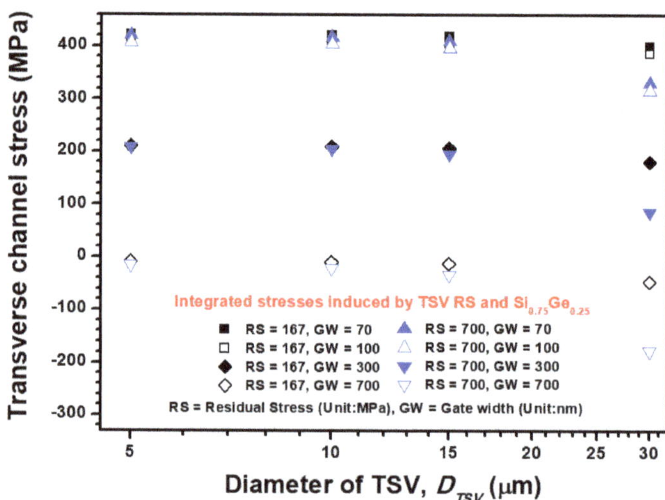

Figure 10. Dependence of the layout and annealing process design on the transverse channel stress of Si pMOSFET estimated by the analytical formula, semi-empirical formula and FEA submodeling approach.

4.4. Stress-Induced Hole Carrier Mobility Gain Change in the S/D Lattice-Strained Si pMOSFET Under the RS Impact Generated by Cu-Filled TSV

On the basis of the longitudinal and transverse channel stresses presented in Figures 9 and 10, the stress-induced carrier mobility gain generated by the RS of TSV and lattice mismatch stress of the S/D $Si_{0.75}Ge_{0.25}$ stressor is illustrated in Figure 11. The RS of TSV significantly degrades the hole carrier mobility of Si pMOSFET, and its influence is proportional to the increment in the designed TSV diameter and RS magnitude. In accordance with the gate width-dependent stresses discussed in the previous section, the optimized gate width is determined to be 300 nm. In consideration of the 420 °C annealing procedure for TSV (corresponding to 700 MPa RS) and 300 nm gate width for Si pMOSFET, the carrier mobility gain varies from 83.54% to 65.89% when the TSV diameter is enlarged from 5 μm to 30 μm. A −17.65% difference in carrier mobility gain results from the RS impact of TSV, and a similar change of nearly −18% to −19% is observed for all designed gate widths. Notably, the adopted piezoresistance parameter is regarded as constant, but in actual experimental measurements on piezoresistance extraction from transistors, it is simultaneously influenced by the doping density of the device channel, gate effective field, applied drain voltage and measurement uncertainty. On the basis of the piezoresistance of the Si pMOSFET in Literature [5], a maximum of 20% estimation uncertainty is explored. These results reveal that the performance of lattice-strained Si pMOSFET is dominated by the designed S/D stressor adjacent to the concerned device channel. However, a narrow TSV diameter and pitch design are the main factors for further increasing the integration density in electronic packaging. The TSV-induced stress impact on the performance of an advanced device will be harsh if the relative distance between TSV and the concerned device is narrowed to a few micrometers. The major contribution of the present study is that it demonstrates an FEA submodeling-based approach to estimate the stress-induced performance impact under the integration of packaging-level and transistor-level stresses. Moreover, the feasibility of the widely adopted analytical solution for TSV stress-affected zone estimation, the Lamé radial stress solution, and its derived semi-empirical formula are utilized and discussed in comparison with the present FEA submodeling approach. The analytic results presented in this study reveal that the abovementioned analytical solutions significantly overestimate the stress transfer efficiency from TSV to the device. The FEA submodeling technique demonstrated in this study provides an effective approach to analyze the stress-induced performance impact for high-integration-density design in electronic packaging and overcomes the difficulty of FEA model construction with a significant size difference between considered components from micro- to nano-level dimensions.

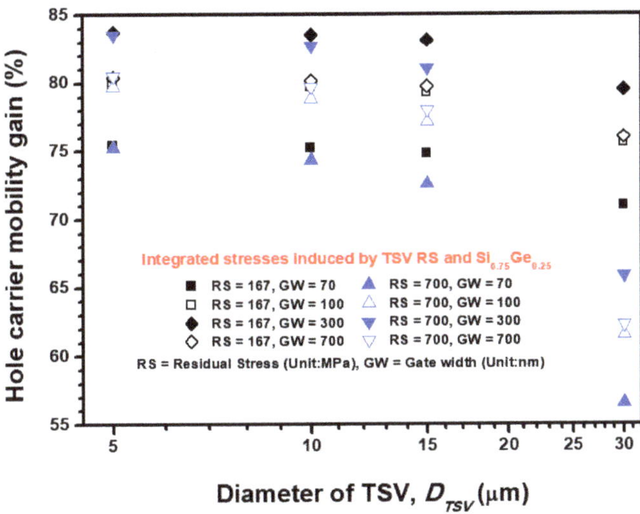

Figure 11. Comprehensive hole carrier mobility gain variation in the S/D lattice-strained Si pMOSFET.

5. Conclusions

An FEA-based submodeling approach was demonstrated to estimate the stress impact from a packaging-level interconnect to a nano-scaled transistor device. The widely utilized Lamé radial stress solution and its derived semi-empirical formula were adopted to analyze the stress-affected zone generated by Cu-filled TSV design, and their results were compared with the results of the FEA-based submodeling approach. The analytic results revealed that the Lamé analytical solution overestimated the stresses transferred from TSV to the concerned device by over 50%. This result means that the stress transfer efficiency between TSV and the transistor device was overestimated because the stress buffer mechanism from the barrier, STI and the layout of the device are neglected in the abovementioned analytical solution. Under the same layout and RS impact of the designed TSV interposer packaging, −36.46%, −40.56% and −18% carrier mobility gains were estimated by the analytical solution, semi-empirical formula and FEA-based submodeling approach, respectively. The highly accurate estimation of the TSV stress-affected zone and its impact on device performance by the presented submodeling approach is due to the analysis of the KOZ region and increased integration density of the device with good usage of the wafer area. Moreover, the comprehensive stress impact generated by the RS of TSV and device-level strain engineering was investigated. The S/D lattice-strained Si$_{0.75}$Ge$_{0.25}$ stressor was utilized to generate a favorable stress status in Si pMOSFET, and its effect was compared with the RS impact from the Cu-filled TSV. The results showed the S/D lattice stressor dominated the status of stress components in the device channel, but the stress impact generated by TSV could be harsh if the integration density of devices and TSV interconnects is further increased. These issues can be further investigated using advanced 3D device architectures and the present FEA-based submodeling approach.

Author Contributions: Conceptualization, P.-C.H. and C.-C.L.; methodology, P.-C.H. and C.-C.L.; investigation, P.-C.H.; writing—original draft preparation, P.-C.H.; writing—review and editing, C.-C.L.; supervision, C.-C.L. Both authors have read and agreed to the published version of the manuscript.

Funding: This research was funded by the Ministry of Science and Technology (MOST), Taiwan, R.O.C., which provided financial support under contract numbers MOST 108-2221-E-007-081-MY3 and MOST 109-2218-E-007-007.

Institutional Review Board Statement: Not applicable.

Informed Consent Statement: Not applicable.

Data Availability Statement: Data sharing is not applicable to this article.

Conflicts of Interest: The authors declare no conflict of interest.

References

1. Lubow, A.; Ismail-Beigi, S.; Ma, T.P. Comparison of Drive Currents in Metal-Oxide-Semiconductor Field-Effect Transistors Made of Si, Ge, GaAs, InGaAs, and InAs Channels. *Appl. Phys. Lett.* **2012**, *96*, 122105. [CrossRef]
2. Alamo, J.A. Del. Nanometre-scale Electronics with III–V Compound Semiconductors. *Nature* **2011**, *479*, 317–323. [CrossRef]
3. Suthram, S.; Ziegert, J.C.; Nishida, T.; Thompson, S.E. Piezoresistance Coefficients of (100) Silicon nMOSFETs Measured at Low and High (∼1.5 GPa) Channel Stress. *IEEE Electron Device Lett.* **2007**, *28*, 58–61. [CrossRef]
4. Nainani, A.; Yum, J.; Barnett, J.; Hill, R.; Goel, N.; Huang, J.; Majhi, P.; Jammy, R.; Sarawat, K.C. Study of Piezoresistance Under Unixial Stress for Technologically Relevant III-V Semiconductors Using Wafer Bending Experiments. *Appl. Phys. Lett.* **2010**, *96*, 242110. [CrossRef]
5. Chu, M.; Nishida, T.; Lv, X.; Mohta, N.; Thompson, S.E. Comparison between High-Field Piezoresistance Coefficients of Si Metal-Oxide-Semiconductor Field-Effect Transistors and Bulk Si Under Uniaxial and Biaxial Stress. *J. Appl. Phys.* **2008**, *103*, 113704. [CrossRef]
6. Gambino, J.P.; Adderly, S.A.; Knickerbocker, J.U. An Overview of Through-Silicon-Via Technology and Manufacturing Challenges. *Microelectron. Eng.* **2015**, *135*, 73–106. [CrossRef]
7. Shi, P.; Enloe, J.; van den Boom, R.; Sapp, B. Direct Copper Electrodeposition on a Chemical Vapor-Deposited Ruthenium Seed Layer for Through-Silicon Vias. In Proceedings of the IEEE International Interconnect Technology Conference, San Jose, CA, USA, 4–6 June 2012.
8. Lau, J.H. Evolution and Outlook of TSV and 3D IC/Si Integration. In Proceedings of the Electronics Packaging Technology Conference, Singapore, 8–10 December 2010.
9. Lau, J.H. Evolution, challenge, and outlook of TSV, 3D IC integration and 3D Silicon Integration. In Proceedings of the International Symposium on Advanced Packaging Materials, Xiamen, China, 25–28 October 2011.
10. Li, L.; Ton, P.; Nagar, M.; Chia, P. Reliability Challenges in 2.5D and 3D IC Integration. In Proceedings of the IEEE Electronic Components and Technology Conference, Orlando, FL, USA, 30 May–2 June 2017.
11. Leduc, P.; Crécy, F.D.; Fayolle, M.; Charlet, B.; Enot, T.; Zussy, M.; Jones, B.; Barbé, J.C.; Kernevez, N.; Sillon, N.; et al. Challenges for 3D IC Integration: Bonding Quality and Thermal Management. In Proceedings of the IEEE International Interconnect Technology Conference, Burlingame, CA, USA, 4–6 June 2007.
12. Heryanto, A.; Putra, W.N.; Trigg, A.; Gao, S.; Kwon, W.S.; Che, F.X.; Ang, X.F.; Wei, J.; Made, R.I.; Gan, C.L.; et al. Effect of Copper TSV Annealing on Via Protrusion for TSV Wafer Fabrication. *J. Electron. Mater.* **2012**, *41*, 2533–2542. [CrossRef]
13. Che, F.X.; Putra, W.N.; Heryanto, A.; Trigg, A.; Zhang, X.; Gan, C.L. Study on Cu Protrusion of Through-Silicon Via. *IEEE Trans. Compon. Pack. Manuf. Technol.* **2013**, *3*, 732–739. [CrossRef]
14. Ranganathan, N.; Prasad, K.; Balasubramanian, N.; Pey, K.L. A Study of Thermo-Mechanical Stress and Its Impact on Through-Silicon Vias. *J. Micromech. Microeng.* **2008**, *18*, 75018. [CrossRef]
15. Amagai, M.; Suzuki, Y. TSV Stress Testing and Modeling. In Proceedings of the Electronic Components and Technology Conference, Las Vegas, NV, USA, 1–4 June 2010.
16. Xu, X.; Karmarkar, A. 3D TCAD Modeling for Stress Management in Through Silicon Via (TSV) Stacks. *AIP Conf. Proc.* **2011**, *1378*, 53–66.
17. Zhu, Y.; Zhang, J.; Li, H.Y.; Tan, C.S.; Xia, G. Study of Near-Surface Stresses in Silicon Around Through-Silicon Vias at Elevated Temperatures by Raman Spectroscopy and Simulations. *IEEE Trans. Device Mater. Reliab.* **2015**, *15*, 142–148. [CrossRef]
18. Saettler, P.; Hecker, M.; Boettcher, M.; Rudolph, C.; Wolter, K.J. μ-Raman Spectroscopy and FE-Modeling for TSV-Stress-Characterization. *Microelectron. Eng.* **2015**, *137*, 105–110. [CrossRef]
19. Okoro, C.; Levine, L.E.; Xu, R.; Hummler, K.; Obeng, Y.S. Nondestructive Measurement of the Residual Stresses in Copper Through-Silicon Vias Using Synchrotron-Based Microbeam X-ray Diffraction. *IEEE Trans. Electron Devices.* **2014**, *61*, 2473–2479.
20. Song, Y.; Abbaspour, R.; Bakir, M.S.; Sitaraman, S.K. Thermal Annealing Effects on Copper Microstructure in Through—Silicon-Vias. In Proceedings of the IEEE Intersociety Conference on Thermal and Thermomechanical Phenomena in Electronic Systems, Las Vegas, NV, USA, 31 May–3 June 2016.
21. Mercha, A.; Van der Plas, G.; Moroz, V.; Wolf, I.D.; Asimakopoulos, P.; Minas, N.; Domae, S.; Perry, D.; Choi, M.; Redolfi, A.; et al. Comprehensive Analysis of the Impact of Single and Arrays of Through Silicon Vias Induced Stress on High-k/Metal Gate CMOS Performance. In Proceedings of the International Electron Devices Meeting, San Francisco, CA, USA, 6–8 December 2010.

22. Okoro, C.; Yang, Y.; Vandevelde, B.; Swinnen, B.; Vandepitte, D.; Verlinden, B.; Wolf, I.D. Extraction of the Appropriate Material Property for Realistic Modeling of Through-Silicon-Vias using μ-Raman Spectroscopy. In Proceedings of the International Interconnect Technology Conference, Burlingame, CA, USA, 1–4 June 2008.
23. Jung, M.; Pan, D.Z.; Lim, S.K. Through-Silicon-Via Material Property Variation Impact on Full-Chip Reliability and Timing. In Proceedings of the IEEE International Interconnect Technology Conference, San Jose, CA, USA, 20–23 May 2014.
24. Okoro, C.; Levine, L.E.; Xu, R.; Obeng, Y.S. Experimentally, How Does Cu TSV Diameter Influence its Stress State? In Proceedings of the IEEE Electronic Components and Technology Conference, San Diego, CA, USA, 26–29 May 2015.
25. Budiman, A.; Shin, H.A.S.; Kim, B.J.; Hwang, S.H.; Son, H.Y.; Suh, M.S.; Chung, Q.H.; Byun, K.Y.; Tamura, N.; Kunz, M. Measurement of Stresses in Cu and Si around Through-Silicon Via by Synchrotron X-ray Microdiffraction for 3-Dimensional Integrated Circuits. *Microelectron. Reliab.* **2012**, *52*, 530–533. [CrossRef]
26. Wolf, I.D.; Croes, K.; Pedreira, O.V.; Labie, R.; Redolfi, A.; Van De Peer, M.; Vanstreels, K.; Okoro, C.; Vandevelde, B.; Beynea, E. Cu Pumping in TSVs: Effect of Pre-CMP Thermal Budget. *Microelectron. Reliab.* **2011**, *51*, 1856–1859. [CrossRef]
27. Lee, C.C.; Huang, C.C. Induced Thermo-Mechanical Reliability of Copper-Flled TSV Interposer by Transient Selective Annealing Technology. *Microelectron. Reliab.* **2015**, *55*, 2213–2219. [CrossRef]
28. Pan, Y.; Li, F.; He, H.; Li, J.; Zhu, W. Effects of Dimension Parameters and Defect on TSV Thermal Behavior for 3D IC packaging. *Microelectron. Reliab.* **2017**, *70*, 97–102. [CrossRef]
29. Jung, M.; Mitra, J.; Pan, D.Z.; Lim, S.K. TSV Stress-Aware Full-Chip Mechanical Reliability Analysis and Optimization for 3D IC. *Commun. ACM* **2014**, *57*, 107–115. [CrossRef]
30. Che, F.X.; Xie, L.; Chen, Z.H.; Wickramanayaka, S. Study on Warpage and Stress of TSV Wafer with Ultra-Fine Pitch Vias for High Density Chip Stacking. In Proceedings of the IEEE Electronics Packaging Technology Conference, Singapore, 6–9 December 2017.
31. Feng, W.; Bui, T.T.; Watanabe, N.; Shimamoto, H.; Aoyagi, M.; Kikuchi, K. Fabrication and stress analysis of annular-trench-isolated TSV. *Microelectron. Reliab.* **2016**, *63*, 142–147. [CrossRef]
32. Feng, W.; Watanabe, N.; Shimamoto, H.; Kikuchi, K.; Aoyagi, M. Methods to Reduce Thermal Stress for TSV Scaling ~TSV with Novel Structure: Annular-Trench-Isolated TSV. In Proceedings of the IEEE Electronic Components and Technology Conference, San Diego, CA, USA, 26–29 May 2015.
33. Li, Z.; Tian, Y.; Teng, C.; Cao, H. Recent Advances in Barrier Layer of Cu Interconnects. *Materials* **2020**, *13*, 5049. [CrossRef]
34. Lee, K.W.; Wang, H.; Bea, J.C.; Murugesan, M.; Sutou, Y.; Fukushima, T.; Tanaka, T.; Koike, J.; Koyanagi, M. Barrier Properties of CVD Mn Oxide Layer to Cu Diffusion for 3-D TSV. *IEEE Electron Device Lett.* **2014**, *35*, 114–116. [CrossRef]
35. Lee, K.W.; Nagai, C.; Nakamura, A.; Bea, J.C.; Murugesan, M.; Fukushima, T.; Tanaka, T.; Koyanagi, M. Effects of Electro-less Ni Layer as Barrier/Seed Layers for High Reliable and Low Cost Cu TSV. In Proceedings of the International 3D Systems Integration Conference, Kinsdale, Ireland, 1–3 December 2014.
36. Li, F.; He, H.; Tian, Q.; Xiao, C.; Li, J.; Zhu, W. A Measurement Method on Nanoscale Thickness of the Ti Barrier Layer of TSV Structure for 3-D IC. *IEEE Trans. Compon. Pack. Manuf. Technol.* **2016**, *6*, 954–958. [CrossRef]
37. Djomeni, L.; Mourier, T.; Minoret, S.; Fadloun, S.; Piallat, F.; Burgess, S.; Price, A.; Zhou, Y.; Jones, C.; Mathiot, D.; et al. Study of Low Temperature MOCVD Deposition of TiN Barrier Layer for Copper Diffusion in High Aspect Ratio Through Silicon Vias. *Microelectron. Eng.* **2014**, *120*, 127–132. [CrossRef]
38. Battegay, F.; Fourel, M. Barrier Material Selection for TSV Last, Flipchip & 3D—UBM & RDL Integrations. In Proceedings of the IEEE Electronic Components and Technology Conference, San Diego, CA, USA, 26–129 May 2015.
39. Zare, Y.; Sasajima, Y.; Onuki, J. Evaluation of Cu-TSV Barrier Materials as a Solution to Copper Protrusion. *J. Electron. Mater.* **2020**, *49*, 2076–2085. [CrossRef]
40. Tsai, M.Y.; Huang, P.S.; Huang, C.Y.; Jao, H.; Huang, B.; Wu, B.; Lin, Y.Y.; Liao, W.; Huang, J.; Huang, L.; et al. Investigation on Cu TSV-Induced KOZ in Silicon Chips: Simulations and Experiments. *IEEE Trans. Electron Devices.* **2013**, *60*, 2331–2337. [CrossRef]
41. Wolf, M.J.; Dretschkow, T.; Wunderle, B.; Jürgensen, N.; Engelmann, G.; Ehrmann, O.; Uhlig, A.; Michel, B.; Reichl, H. High Aspect Ratio TSV Copper Filling with Different Seed Layers. In Proceedings of the IEEE Electronic Components and Technology Conference, Lake Buena Vista, FL, USA, 27–30 May 2008.
42. Lu, K.H.; Zhang, X.; Ryu, S.K.; Im, J.; Huang, R.; Ho, P.S. Thermo-Mechanical Reliability of 3-D ICs containing Through Silicon Vias. In Proceedings of the IEEE Electronic Components and Technology Conference, San Diego, CA, USA, 26–29 May 2009.
43. Ryu, S.K.; Lu, K.H.; Zhang, X.; Im, J.H.; Ho, P.S.; Huang, R. Impact of Near-Surface Thermal Stresses on Interfacial Reliability of Through-Silicon Vias for 3-D Interconnects. *IEEE Trans. Device Mater. Reliab.* **2011**, *11*, 35–43. [CrossRef]
44. Chiang, K.N.; Chang, C.H.; Peng, C.T. Local-Strain Effects in Si/SiGe/Si Islands on Oxide. *Appl. Phys. Lett.* **2005**, *87*, 191901. [CrossRef]
45. Lee, C.C.; Huang, P.C.; Lin, Y.C. Analytical Model Developed for Precise Stress Estimation of Device Channel Within Advanced Planar MOSFET Architectures. *IEEE Trans. Electron Devices* **2020**, *67*, 1498–1505. [CrossRef]
46. Lee, C.C.; Huang, P.C. Layout Study of Strained Ge-Based PMOSFETs Integrated with S/D GeSn Alloy and CESL by Using Process-Oriented Stress Simulations. *IEEE Trans. Electron Devices* **2018**, *65*, 4975–4981. [CrossRef]

Article

Solder Joint Reliability Risk Estimation by AI-Assisted Simulation Framework with Genetic Algorithm to Optimize the Initial Parameters for AI Models

Cadmus Yuan [1,*], **Xuejun Fan** [2] **and Gouqi Zhang** [3]

[1] Department of Mechanical and Computer-Aided Engineering, Feng Chia University, Taichung 407082, Taiwan
[2] Department of Mechanical Engineering, Lamar University, Beaumont, TX 77710, USA; xuejun.fan@lamar.edu
[3] Department of Electronic Components, Technology, and Materials, Delft University of Technology, 2628 CD Delft, The Netherlands; g.q.zhang@tudelft.nl
* Correspondence: cayuan@fcu.edu.tw; Tel.: +886-939-873-055

Abstract: Solder joint fatigue is one of the critical failure modes in ball-grid array packaging. Because the reliability test is time-consuming and geometrical/material nonlinearities are required for the physics-driven model, the AI-assisted simulation framework is developed to establish the risk estimation capability against the design and process parameters. Due to the time-dependent and nonlinear characteristics of the solder joint fatigue failure, this research follows the AI-assisted simulation framework and builds the non-sequential artificial neural network (ANN) and sequential recurrent neural network (RNN) architectures. Both are investigated to understand their capability of abstracting the time-dependent solder joint fatigue knowledge from the dataset. Moreover, this research applies the genetic algorithm (GA) optimization to decrease the influence of the initial guessings, including the weightings and bias of the neural network architectures. In this research, two GA optimizers are developed, including the "back-to-original" and "progressing" ones. Moreover, we apply the principal component analysis (PCA) to the GA optimization results to obtain the PCA gene. The prediction error of all neural network models is within 0.15% under GA optimized PCA gene. There is no clear statistical evidence that RNN is better than ANN in the wafer level chip-scaled packaging (WLCSP) solder joint reliability risk estimation when the GA optimizer is applied to minimize the impact of the initial AI model. Hence, a stable optimization with a broad design domain can be realized by an ANN model with a faster training speed than RNN, even though solder fatigue is a time-dependent mechanical behavior.

Keywords: solder joint fatigue risk estimation; wafer level chip-scaled packaging; artificial neural network; recurrent neural network; generic algorithm; principle component analysis; time/temperature-dependent nonlinearity

Citation: Yuan, C.; Fan, X.; Zhang, G. Solder Joint Reliability Risk Estimation by AI-Assisted Simulation Framework with Genetic Algorithm to Optimize the Initial Parameters for AI Models. *Materials* **2021**, *14*, 4835. https://doi.org/10.3390/ma14174835

Academic Editor: Kuo-Ning Chiang

Received: 15 July 2021
Accepted: 23 August 2021
Published: 26 August 2021

1. Introduction

Solder joint reliability is one of the most critical issues for most ball-grid array packaging types. The time dependency of this failure mechanism requires considerable experiment time to obtain statistically reliable results. On the other hand, the nonlinear material/geometry properties are required for the finite element (FE) modeling to retrieve trustable results, which can be validated by the experimental results. Hence, both the reliability experiment and numerical modeling require unique expertise to conduct the relevant tasks, which creates a technical barrier for the design for reliability.

The neural network (NN)-based AI algorithms were applied to assist the design and simulation of the solder joint risk assessment. Chou and Chiang [1] and Hsiao and Chiang [2] developed an AI-assisted design and simulation framework. It includes the virtual prototyping of solder fatigue failure mode with the geometrical/material nonlinearity and the proper validation by the experimental results. The training database is generated from the parametric FE model. Next, AI modeling is trained by the selected data points from the database and validated by the rest. Careful validation works should be conducted to prove the representation capability of the AI model to the FE dataset.

The NN approaches of AI modeling received more and more attention due to the capability of abstracting the knowledge from the database without the pre-defined framework nor prior knowledge/experience. ANN is a basic architecture of NN with weak representation capability for sequential events. However, the solder joint fatigue mechanism is highly time-dependent. Accordingly, the sequential NN techniques, including the recurrent neural network (RNN), GRU, and LSTM, are successfully applied to the time-dependent failure mechanism for electronic packaging. However, these sequential NN methods consider the recurrent parameters and iterations that induce learning difficulties and require considerable computation resources. Yuan and Lee [3] applied the sequential NN to model the time-dependent nature of the solder joint fatigue, and the average error norms below 1.213×10^{-4} were achieved. Yuan et al. [4] developed a gated neural network technique to learn the performance shifting of the solid-state lighting (SSL) lamp over time. Meszmer et al. [5] applied many NN techniques to study which is the best for the electronic packaging, and the sequential NN performed best, including the gate recurrent unit (GRU) and long short-term memory (LSTM) architectures because of their capability to capture the characteristics of the sequential dataset. Selvanayagam et al. [6] applied the AI-assisted modeling concept for the improvement of the packaging warpage. Tabaza et al. [7] applied the non-sequential NN to simulate the time-dependent hysteretic response of a viscoelastic material. The possibility of using the non-sequential NN, such as NN, upon a time-dependent engineering problem remains a challenge.

Considering the learning procedure, NN is a parametric AI modeling method. In addition to the network structure, NN utilizes the parameter, including the weightings and bias, to learn the knowledge from the database. However, NN requires the initial guessing of parameters to start the learning process, and the improper selection of the initial parameter results in a slow convergence speed and bumpy learning procedure. The genetic algorithm (GA) is always applied. A genetic algorithm (GA) is an optimization method proposed by John Holland to find the approximate solutions. This algorithm is a specific form of an evolutionary algorithm in which evolutionary biology techniques such as inheritance and mutation are used. In genetic algorithms, to obtain the optimal solution, the appropriate responses of a generation are combined based on the principle of the survival of the fittest environment [8]. White and Ligomenides proposed a topology and weighting optimization algorithm for neural networks [9]. Juang [10] hybrids the GA and particle swarm optimization (PSO). The new generation of the GA can be generated not only by the crossover and mutation but also by PSO. Ding et al. [11], Ahmadizar et al. [12] and Arabasadi et al. [13] apply the GA for weighting optimization of NN. However, few literature had covered GA for the sequential network nor utilized the continuously evolving nature of the NN backpropagation.

Based on the AI-assisted simulation framework, this research investigates the possibility of using the ANN instead of the RNN in physically sequential issues, such as the solder joint fatigue mechanisms. Secondly, this research develops a genetic algorithm method to obtain the optimal parameters for NN learning.

This paper is organized as follows: the fundamental scientific issues and the literature review are described in the Introduction. The following section, Theory, provides the basic theoretic approaches that have been applied in this research. The execution of the AI-assisted simulation framework is explained in the sections The AI-assisted simulation framework and FE dataset preparation and The design of the AI modeling. The AI model training with GA optimized initial parameters section summarizes the learning experience of the AI modeling. The conclusion of this paper is given in the last section.

2. Theory

In GA, the fitness criterion is first defined to quantify the members of the current generation with more compatibility are more likely to generate the next population [8,9]. The fitness criterion (*F*) is set as follows:

$$F = 1/r \tag{1}$$

where the *r* is the error norm of learning. The members with higher *F* values are more likely to generate the next population by the crossover and mutation operators.

The member is also called the chromosome, which is made up of many genes. The gene is constructed by many base-pairs (bps) [12]. Given a pair of parent chromosomes with *m* genes, the crossover operator will generate 2^m different offsprings by the recombination of the genes that are from the parent [13]. To cover the genetic representation ultimately, the whole 2^m offsprings are forming to the next generation without any possibility. As illustrated in Figure 1, there is one pair of parents with three genes. Eight ($= 2^3$) offsprings are generated by the recombination of the parents' gene, which is the definition of the crossover operator.

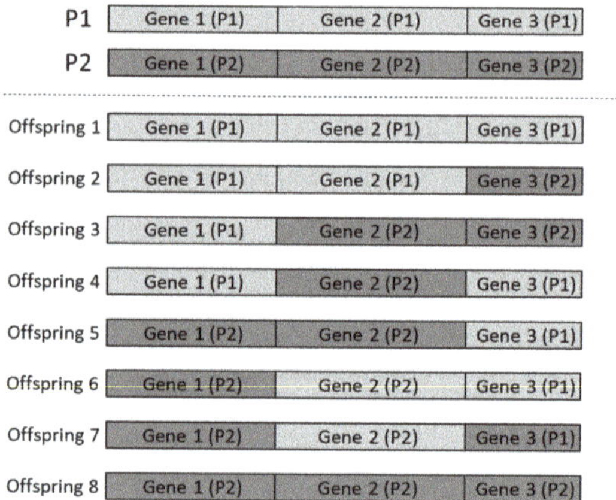

Figure 1. The illustration of the crossover operator.

Therefore, if the top *n* chromosomes are selected, there are $\frac{n(n-1)}{2}$ parent combinations and there are maximum $\frac{n(n-1)}{2} \times 2^m$ offsprings possibilities. However, if the parent chromosomes consisted of many similar genes, it might induce many duplicated offsprings. These offsprings with the same genes will be removed to save the computation resource.

The mutation operator [13] is used to make changes in the genes of a member of the current generation to produce a new member. The mutation occurs at the bps level and is controlled by the mutation rate [12]. When the mutation is invoked at certain bps, the representation bps will be replaced by the opposite parent bps. For example, as illustrated in Figures 1 and 2, the 9th bps of the first gene of offspring 7 will mutate, and the original bps will be replaced by the 9th bps of the first gene of parent 1 (P1).

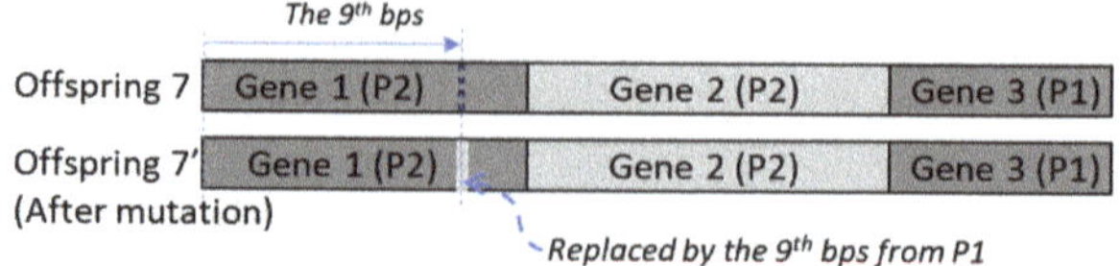

Figure 2. The illustration of the crossover and mutation operator.

After few completed genetic algorithm iterations, these best chromosomes of each iteration are not the same due to the nature of the neural network. The principal component analysis (PCA) algorithm is applied to build a super chromosome based on these best chromosomes. If the complete chromosome, including the weightings and bias, are considered as vectors, and the covariance matrix K of all the chromosomes are formed based on the squared exponential kernel function:

$$K(x, x') = \sigma \cdot \exp\left(-\frac{\| x - x' \|^2}{2l^2}\right) \tag{2}$$

where the x and $x\prime$ are the chromosome vectors. The parameter σ and the characteristic length l are both set as 1. An eigenvector analysis is applied to the K matrix. The super chromosome, called the PCA gene, is obtained as the inner product of the best chromosomes and the eigenvector of the first eigenvalue.

3. The AI-Assisted Simulation Framework and FE Datasets Preparation

In this section, a practical engineering case will be applied to analyze the capability of GA and PCA to generate the initial parameters of the ANN and RNN. However, due to the multiphysical and multiscaled characteristics of the engineering questions, this research obeys a reliable AI-assisted simulation framework [1–3], illustrated in Figure 3, to improve the predicting accuracy of the AI model.

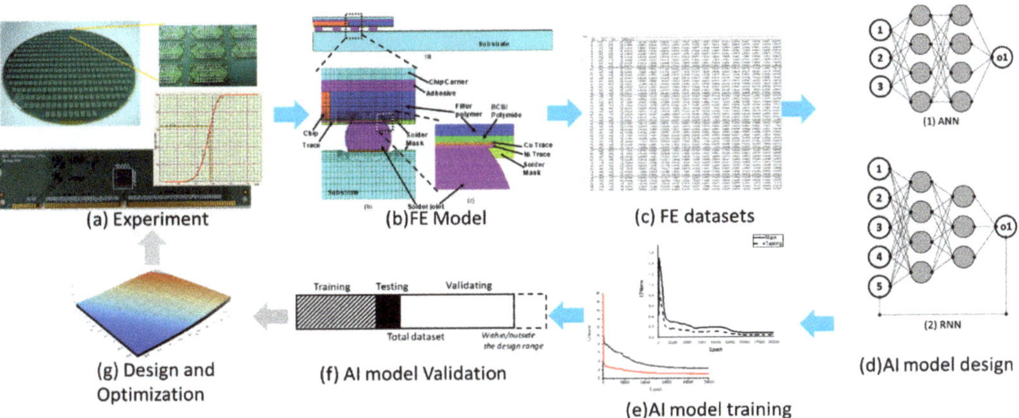

Figure 3. The AI-assisted simulation framework [3,14] (**a**) Experiment; (**b**) FE model; (**c**) FE datasets; (**d**) AI model design; (**e**) AI model training (**f**) AI model validation; (**g**) Design and optimization.

Due to the limited resources, only limited actual samples with very few design parameter combinations are made for the experiments. The FE modeling method is applied to expand the design domain. Based on the experimental results (Figure 3a), a FE model (Figure 3b) can be established and validated. The validated FE model then can be parameterized, and the FE datasets (Figure 3c) can be obtained. However, the specified design parameter combination might induce bad aspect ratios of the elements, which cause the instability of the FE analysis results. The NN model is expected to broaden the design domain. The NN will be carefully designed (Figure 3d) based on the characteristics of the FE dataset and supervised trained (Figure 3e). Moreover, the AI model training's accuracy requirement(s) should be carefully defined based on the FE dataset. The datapairs that have not been included in the training procedure will be applied to validate the NN model. The validated NN model can be used for design and optimization (Figure 3f). When the new experimental result is available (Figure 3g), the whole procedure can be relaunched.

Figure 4 shows the G-WLCSP structure [14], where the IC is placed on a glass substrate with metal traces and solder bumps Figure 4a for redistribution purposes. Figure 4a′,b show the device's top view and cross-section view, respectively. Moreover, Figure 4b depicts the key structural components in G-WLCSP, including the glass substrate, the adhesive, the IC, polyimide (PI) for the stress buffer layer, solder mask, and the solder. After the wafer has been diced, individual packaging can be obtained.

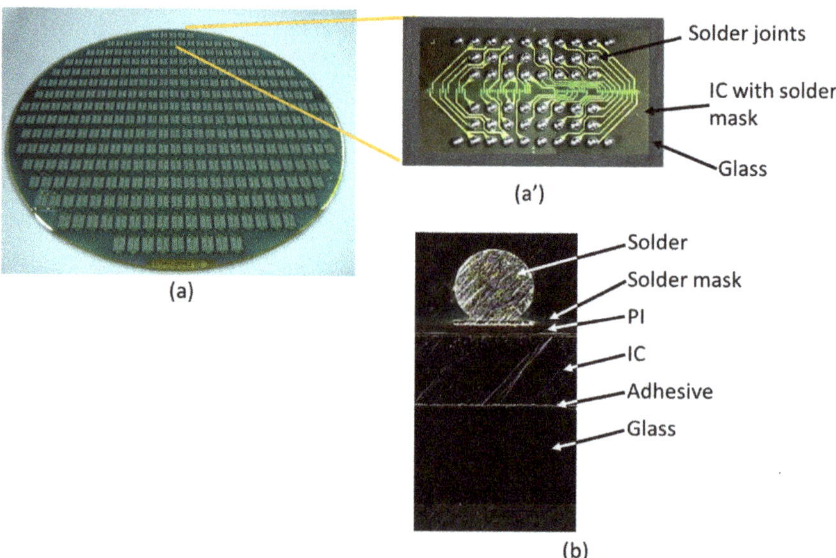

Figure 4. Glass distributed wafer level packaging: (**a**) wafer view with the detail (**a′**) and (**b**) cross-section of a single device [15] (Copyright 2020 EuroSimE).

An actual G-WLCSP structure was made for reliability testing. The sample consisted of a chip with the size of $5.77 \times 10.38 \times 0.3$ mm^3 and a glass thickness of 0.5 mm. The sample was attached to a 1.2 mm-thick test board, as shown in Figure 5a. The cross-section view shows the bonding condition, and no defect has been detected (Figure 5b). A 0.45 mm-diameter 63Sn/37Pb solder ball was applied onto the 0.37 mm die-side pad. The stand-off height for solder joints was reduced to 0.35 mm after reflow.

Thermal cycle testing of this G-WLCSP is performed between $-40\,°$C and $125\,°$C with a ramp rate of 11 $°$C/min and a dwelling time of 15 min. Figure 5c shows the Weibull solder fatigue failure experimental result of 21 samples; the 63.2% fatigue cycle number is approximately 1444.

We develop a two-dimensional FE model with a plane strain assumption to estimate the G-WLCSP solder joint risk under the thermal cycle loading by the incremental plastic strain. The initial stress-free reference temperature equals 25 $°$C. In the finite element model, all materials except the solder joint and the PI are linear, as shown in Table 1. Moreover, the solder joints and PI are treated as temperature-dependent, elastic-plastic materials [15,16], as shown in Figures 6 and 7, respectively. As seen in Liu et al. [16], the solder joint failure risk can be estimated at a certain accuracy level without the time-dependent material properties. Due to the symmetrical condition, one-half of the full-scaled two-dimensional FE model is used, and the analysis result is obtained by the commercial finite element code ANSYS® (version 15, ANSYS, Inc., Canonsburg, PA, USA). The mesh density of the most critical solder is shown in Figure 8.

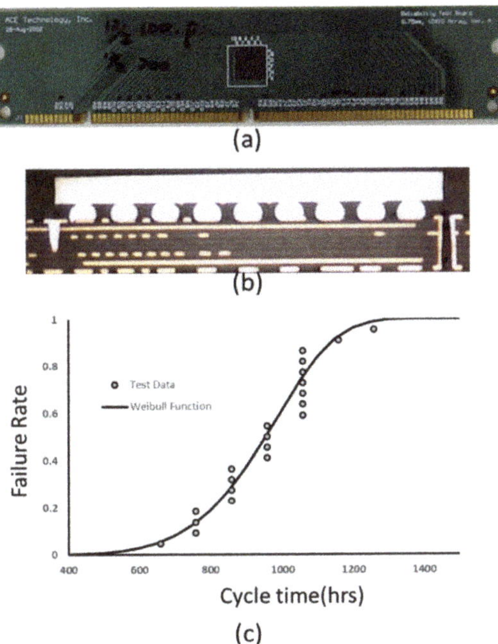

Figure 5. Thermal cycling results: (**a**) the test board, (**b**) the Weibull plot, (**c**) Weibull solder fatigue failure experimental result of 21 samples.

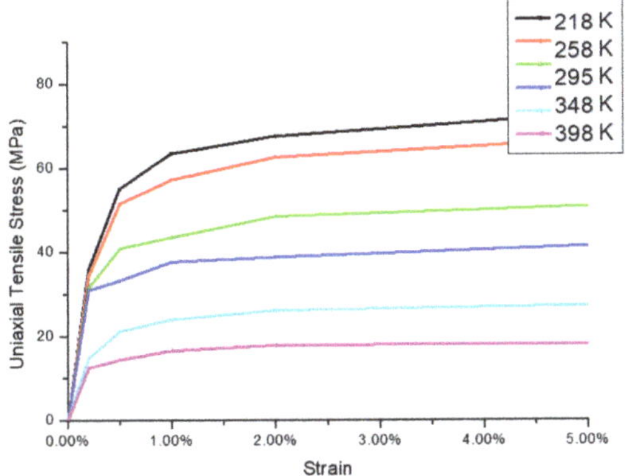

Figure 6. Nonlinear mechanical response of solder joint (63Sn/37Pb) with different temperatures [15] (Copyright 2020 EuroSimE).

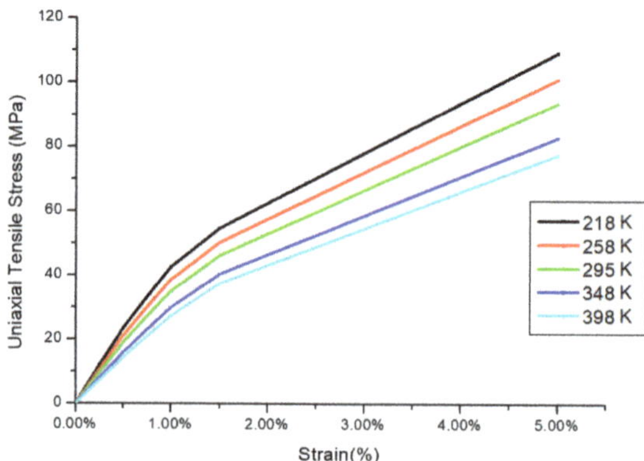

Figure 7. Nonlinear mechanical response of PI with different temperatures [15] (Copyright 2020 EuroSimE).

Table 1. Material properties [14].

	Young's Modules (MPa)	CTE (ppm)	Poisson's Ratio
Solder joint	Temperature dependent and nonlinear (Figure 6)	23.90	0.35
Copper	76,000	17.00	0.35
Solder mask	3400	30.00	0.35
FR4 PCB	18,200	16.00	0.19
PI (Stress Buffer Layer)	Temperature dependent and non-linear (Figure 7)	150.00	0.40
BCB	3000	50.00	0.34
Silicon	112,400	2.62	0.28
Adhesive	0.7	300.00	0.45
Glass	63,000	3.25	0.28
Epoxy	80	250.00	0.34

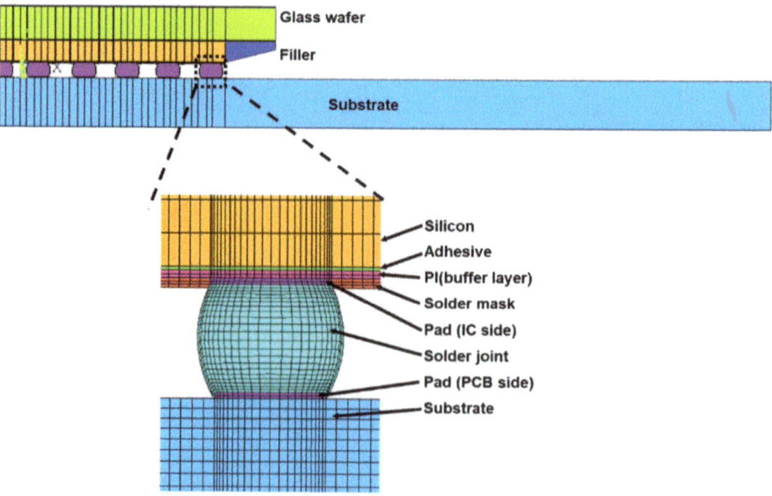

Figure 8. Finite element model for conventional WLCSP and proposed glass WLCSP.

By fine-tuning the mesh density and the solution parameters, the FE model can achieve good agreement with the experimental results by the empirical Coffin–Mason equation [17–19], shown in Table 2. Afterward, the validated FE with the solution parameters can be parametrized.

Table 2. Finite element model validation.

	Experimental Result (Cycles)	Simulation Prediction (Cycles)
Cycles	1007	1444

Key parameters with the levels and noise factors are listed in Table 3. These three key design parameters have been chosen in response to the packaging industry requirements and manufacturing capabilities. A finite element model based on experimental validation is first used to broaden the domain of parameters, and then the neural network model is applied. Each simulation comprises a complete five thermal cyclic loading. To build the FE dataset, 81 parametric finite element models, according to the parameters in Table 3, are executed with a controlled mesh density of the most critical solder joint.

Table 3. Parametric model settings.

Parameter Name	Parameter Alias	Level 1	Level 2	Level 3	Noise Factor Levels
Die thickness	1	0.25 mm	0.375 mm	0.5 mm	±0.015 mm
Glass thickness	2	0.3 mm	0.5 mm	0.8 mm	±0.03 mm
PI thickness	3	0.04 mm	0.025 mm	0.015 mm	±0.005 mm

Figure 9 shows the averaged incremental plasic strain of each loading cycle, where the plastic strain is only induced by plasticity deformation. After the third cycle, the averaged incremental plastic strain incremental becomes stable. From these 81 data points, the average strain increment is 3.01% ($\Delta\epsilon_{avg}$), with a standard deviation of 1.17% (σ_ϵ). The empirical Coffin–Mason equation [17–19] converts to the reliability cycles, as

$$N_f\left(\Delta\varepsilon_p\right) = 0.4405\cdot\left(\Delta\varepsilon_p\right)^{-1.96} \tag{3}$$

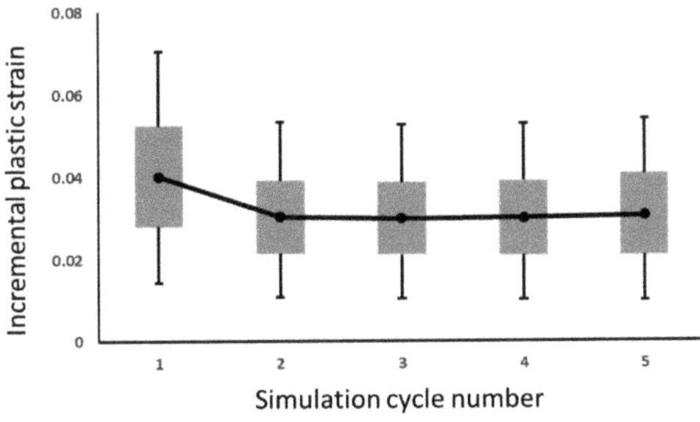

Figure 9. The boxplot of the plastic strain incremental of the 81 data points [15] (Copyright 2020 EuroSimE).

On the other hand, Table 2 indicates that the difference between the experimental and simulation result is 437 cycles. Based on Equation (3), we define a max-min problem:

$$f = \underbrace{\arg}_{\Delta\epsilon_{accu} \in \mathbb{R}} \left\{ \min\left[\max\left(\Delta N_f - \Delta\right)\right]\right\} \tag{4}$$

where ΔN_f is defined as $\left|N_f(\epsilon + \Delta\epsilon_{accu}) - N_f(\epsilon - \Delta\epsilon_{accu})\right|$, $\epsilon = \mathcal{N}\left(\Delta\epsilon_{avg}, \sigma_\epsilon\right)$, and $\mathcal{N}(\cdot,\cdot)$ represents the Gaussian distribution. The $|f|$ in Equation (4) is expected to be zero. By the regression computation, $\Delta\epsilon_{accu} \sim 0.18\%$ is obtained, and it is assigned as the accuracy requirement for machine learning.

4. The Design of the AI Model

Figure 10 schematically illustrates the characteristics of the datapairs. There are three design parameter inputs, including the die, glass, and PI thickness. Since the FE model is under five thermal cycle loadings, denoted as the cycle 1–5 illustrated in Figure 10. At each cyclic thermal loading, there is an equivalent plastic strain ($\Delta\epsilon_{pl}$) with respect to each temperature.

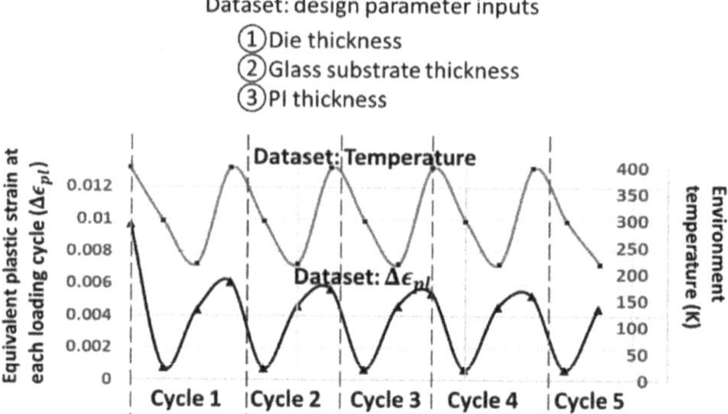

Figure 10. The FE dataset.

Hence, the ANN structure is designed as follows: the three geometric design parameters are considered as the inputs. Referring to the plastic strain incremental of the 81 data points shown in Figure 9, the average equivalent plastic strain increment of the last three loading cycles is selected as the output. The design concept of ANN is to capture the relationship directly from the design parameters to the solder joint fatigue cycle, which is represented by the equivalent plastic strain and converted by the Coffin-Manson equation (Equation (3)).

The ANN structure is designed as "3,4,4,1". There are three inputs and one output, and there are two hidden layers, including the weightings and bias, to capture the feature characteristics of the training datasets. The sigmoid is selected as the activation function because it is stable for the initial parameter studies in the next paragraph. To keep the simplicity during learning, the ANN optimizer is limited to forward computation and backpropagation. The data normalization is applied to the datasets.

Each prediction error is defined as $e = o - t$, where o is the prediction obtained from the ANN output, and t is the ground truth from the FE datasets. The cost function is defined as the Euclidian sum of each prediction error:

$$C = \sqrt{\frac{\sum_{i=1}^{n} e_i}{n}} \tag{5}$$

where e_i is the prediction error of ith prediction and there are total n predictions.

On the other hand, the plastic strain accumulated from the previous cycle will impact the system's mechanical response. Hence, RNN is applied to predict the equivalent plastic strain at each cycle. The equivalent plastic strain (per cycle) is the output. Moreover, there are five inputs, including the three geometrical design parameters, the temperature of the current cycle, and one recurrent parameter. Referring to the top-right schematic drawing in Figure 3, the recurrent parameter of RNN converts the previous output of equivalent plastic strain into the input of the next cycle.

The structure of RNN is set to "5,4,3,1". There are five outputs and one input, and two hidden layers with four and three neurons, respectively. The design concept of the hidden layer is to keep similar numbers of weightings and bias, to compare to the ANN results directly. The sigmoid is selected as the activation function. To maintain the simplicity during learning, the ANN optimizer is limited to forward computation and backpropagation through time. The data normalization is applied to the datasets. One extra post-processing is applied to the ANN. Only the average of the last three outputs is considered to compare to the ANN prediction accuracy directly.

5. AI Model Training with GA Optimized Initial Parameters

In neural network-based AI modeling, the initial parameters are required to launch the machine learning process. This initial parameter is usually chosen randomly in the literature. In this paper, a GA is proposed to achieve the best initial parameter for the AI modeling. The GA chromosome is defined as the combination of genes. Each gene is the combination of the weightings or bias between two layers [12]. For instance, if there are two layers with l_{m-1} and l_m neurons, the gene occupies $l_m \cdot l_{m-1}$ individuals as illustrated in Figure 11. Each individual in the gene is defined as a base-pair (bps). The chromosome is a combination of genes from various layers.

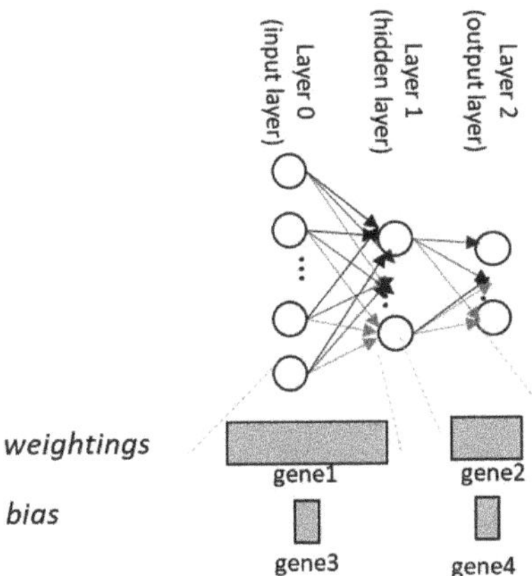

Figure 11. The definition of the chromosomes in GA.

The GA optimization procedure is illustrated in Figure 12. Each GA step starts with the old population of many chromosomes (Ch). The fitness ranking by Equation (1) limits that only m the best chrmosomes can enter the next population. By the crossover and mutation operators, a new population can be generated. The backpropagation of NN learning provides new fitness rankings for the next GA step.

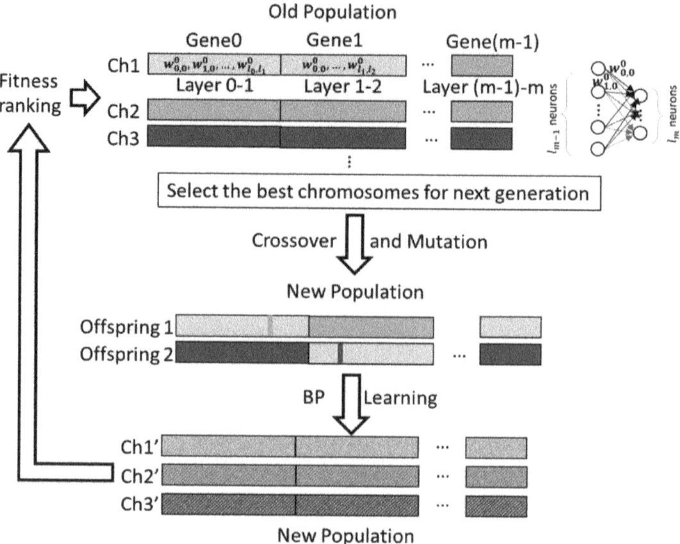

Figure 12. The GA flow chart.

In the AI-learning algorithm, the parameters, including the weightings and bias, are updated each iteration by the backpropagation process. Hence, there is an option to select which parameter sets generate the next GA generation. In this research study, two conditions are considered:

- "back-to-original" condition: only the parameters that initially input to the AI model are applied to generate the next generation.
- "progressing" condition: the parameters that after n backpropagation iterations are applied.

5.1. The "Back-to-Original" GA Optimizer

Due to the design of the neural network structure, 2000 initial parameter sets are firstly generated by the random generator, which follows a zero-mean Gaussian distribution of the standard deviation of $\sqrt{2/(l_m \cdot l_{m+1})}$ ($0 \le m \le n$), where n is the total layer number, and l_m is the neuron numbers at m^{th} layer. It also indicates that 2000 initial chromosomes have been generated for each case to initialize the GA optimization procedure. The learning rate of ANN is fixed to 0.3 with the sigmoid activation function, and backpropagation is selected as the learning optimizer.

When the RNN architecture is applied with the "back-to-original" GA optimizer, Figure 13 shows convergence curves among random select chromosomes from the 2000 members (random selection), the best chromosome from the first generation (generation 1 best), and the best one from the whole GA optimization (GA best). A clear contribution of the GA optimization to the convergence speed can be confirmed.

A typical performance of the GA generations under the "back-to-original" optimizer is illustrated in Figure 14, where the lightest grey curve is best of the 2000 initials, and it evolves continuously from the light grey to the darker ones by the GA. The mutation rate for generating the next generation is fixed by 0.001. After the seventh generation, the evolving of the best chromosome stops. The inset of Figure 14 shows the average error norm among the GA generation and generation size of the GA. The average error norms reduce through each generation. The size of GA generation reduces accordingly as the duplicated chromosomes in the generation are removed. Moreover, the standard derivation of average norm varies during the GA optimization, but it dramatically decreases when the GA evolving stops (GES). This is because the optimized chromosome dominates the next GA generation.

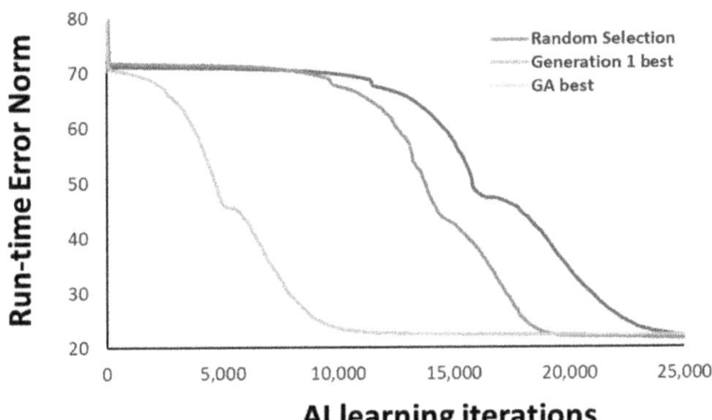

Figure 13. The contribution of GA optimization.

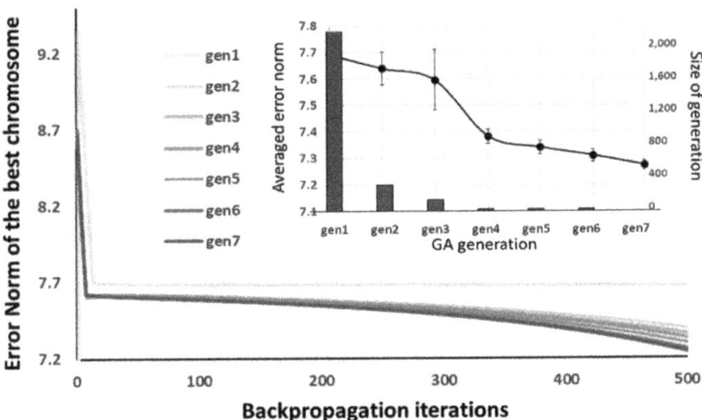

Figure 14. The performance of the GA generations under the "back-to-original" optimizer.

We conducted four GA optimization procedures and listed them in Table 4, where the "Run-time error norm" column indicates the error norm of the best chromosome against the training sets after the 500 cycle training and before the denormalization.

Table 4. The ANN training result under GA "back-to-original" optimizer.

	The Best Chromosomes for the Next Generation	GES	Run-Time Error Norm	Denormalized Difference
Case 1	4	6	7.47	0.001530
Case 2	4	6	7.48	0.001287
Case 3	6	7	7.42	0.001225
Case 4	6	7	7.39	0.001098
PCA gene	–	–	1.72	0.001264

Case 1 and 2 introduce the four best chromosomes to the next generation, and Case 3 and 4 introduce six. Due to that, the initial 2000 members were generated independently by the random number generator for each case in Table 4, a different optimization result is achieved under the same learning parameters. This phenomenon is clearly depicted in column "Run-time error norm" of Table 4.

After the GA optimization procedure, a PCA process is applied based on these four best chromosomes. The covariance matrix K is formed by Equation (2). An eigenvector analysis is applied to the K matrix. Then, The eigenvector of the first eigenvalue is selected and applied to those four best chromosomes. The PCA gene can be obtained by

$$[ch_1, \ ch_2, ch_3, ch_4] \cdot \vec{v_e} \tag{6}$$

where ch_1, ch_2, ch_3, and ch_4 are the best chromosomes from the GA optimization, and $\vec{v_e}$ is the eigenvector of the first eigenvalue. The convergence curve is shown in Figure 15, where a relatively fast convergence can be achieved. Moreover, from the inset of Figure 15, which extended the convergence curve to 100k iteration, one can identify that these best chromosomes approach similar error norm levels after considerable iteration numbers.

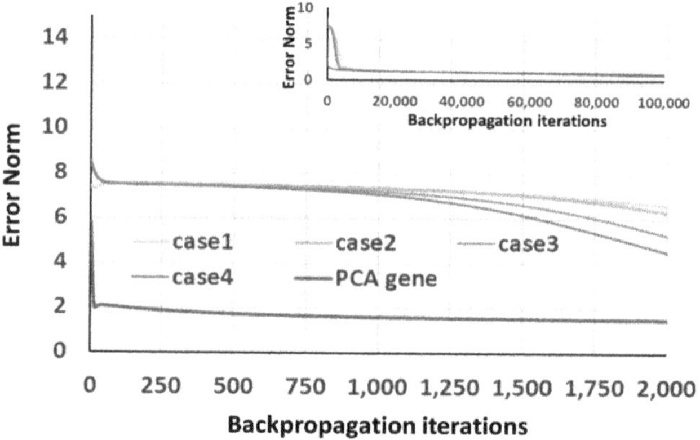

Figure 15. The convergence of 4 GA "back-to-original" optimized chromosomes and PCA gene under ANN architecture.

The "Denormalized difference" column in Table 4 indicates the difference between the best chromosome against the whole 81 datasets after 100,000 training cycles and after the denormalization process. Moreover, no matter the "Run-time error norm", the best chromosomes can consistently achieve the 0.18% requirement. Moreover, when more best chromosomes are put into the next generation, the denormalized difference improves.

Under the similar GA condition of "back-to-original", the RNN architecture has been implemented. Each GA generation is optimized by 1000 backpropagation through time iterations. The rest of the learning parameters remain the same as the ANN case. Table 5 lists the learning results of the four GA optimizations. Note that the run-time error norm is much higher than the ANN ones (listed in Table 4); this is because the definition of the run-time error norm in RNN consists of all recurrent cycles. Figure 10 shows that there are 15 recurrent cycles.

Table 5. The RNN training result under GA "back-to-original" optimizer.

	The Best Chromosomes for the Next Generation	GES	Run-Time Error Norm	Denormalized Difference
Case 1	4	6	69.92	0.002017
Case 2	4	2	70.72	0.001189
Case 3	6	4	69.61	0.001343
Case 4	6	3	70.05	0.001226
PCA gene	–		37.57	0.001090

The PCA analysis procedure is applied to these four GA optimization listed in Table 5. The performance of the PCA gene, obtained by Equation (6) is listed in the last row of Table 5 and Figure 16. Case 1 in Table 5 does not satisfy the accuracy requirement, but case 2 does. Both cases 1 and 2 use the same GA parameters but different initial GA generation. Hence, the GA optimization results are influenced by the initial GA generation.

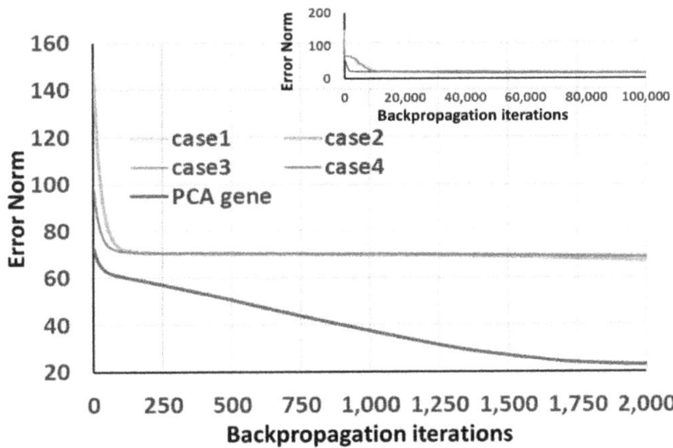

Figure 16. The convergence of 4 GA "back-to-original" optimized chromosomes and PCA gene under RNN architecture.

On the other hand, one can identify the fast convergence capability of the PCA gene, shown in Figure 16. To decrease the impact of the initial GA generation, it is recommended that one should execute multiple GP optimization processes and then extract these best chromosomes to the PCA gene. Moreover, although RNN provides more information than ANN, both give similar final prediction accuracy, as listed in the last row of Tables 4 and 5.

5.2. The "Progressing" GA Optimizer

The converging behavior of the "progressing" GA optimization demonstrates much difference from the "back-to-original" one. As illustrated in Figure 17, which is the performance of the GA generations under the "progressing" optimizer under an ANN architecture. It clearly depicts the evolving of the chromosomes moving along the backpropagation process. The dashed line of Figure 17 is the convergence curve of the randomly selected 2000 initial GA generation. Comparing the dashes curve to the solid ones, one can identify the continuous optimizing characteristics of the "progressing" GA. The inset of Figure 17 shows the average error norm and size of the generation with respect to the GA generation. Although a continuous convergence of the average error norm can be found, the size of the generation is larger than the "back-to-original" one. This is because the backpropagation algorithm provides various gene combinations with few duplications for the next generation, which might increase the optimization capability of GA with considerable computation resources.

The complete training results are listed in Table 6. All of the denormalized differences satisfy the accuracy requirement of 0.18%. Among all four cases, case 3 performs worse than the other three in terms of the denormalized difference, but it performs well in the run-time error norm. This is because the run-time error norm only reports the learning results of the chromosomes after 500 iterations, but the denormalized difference reflects the 100k learning. This phenomenon depicts the dilemma of selecting how many backpropagation iterations for each GA generation member.

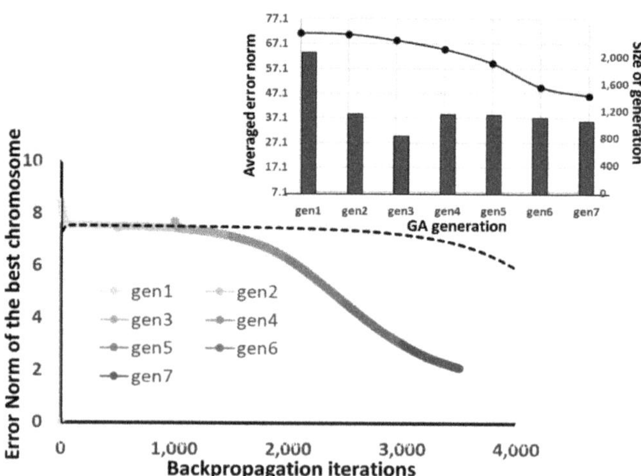

Figure 17. The performance of the GA generations under the "progressing" optimizer.

Table 6. The ANN training result under GA "progressing" optimizer.

	The Best Chromosomes for the Next Generation	GES	Run-Time Error Norm	Denormalized Difference
Case 1	4	6	6.87	0.001058
Case 2	4	4	6.58	0.001037
Case 3	6	6	4.93	0.001573
Case 4	6	8	1.70	0.001175
PCA gene	–	–	1.87	0.001098

Figure 18 shows the convergence of the four cases and the PCA gene obtained by the same method mentioned above. Although case 4 reveals a low run-time error at the beginning of the learning process, the PCA gene still performs well after approximately 750 learning cycles.

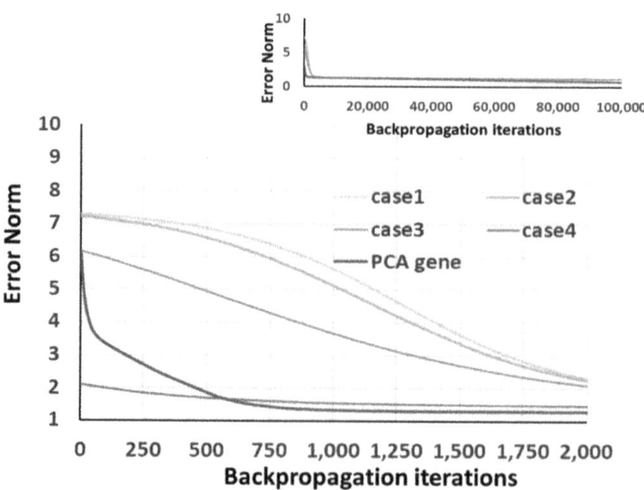

Figure 18. The convergence of 4 GA "progressing" optimized chromosomes and PCA gene under ANN architecture.

5.3. The Impact of the Initial GA Generation

The learning results of the RNN architecture with the "progressing" GA optimizer are shown in Table 7 and Figure 19. Although the overall characteristics, including the convergence curves, the denormalized difference, PCA gene, etc., behave similarly to the previous cases, the overall performance is not good enough, compared to the same RNN architecture with the "back-to-original" GA optimizer, shown in Table 5 and Figure 16.

Table 7. The RNN training result under GA "progressing" optimizer.

	The Best Chromosomes for the Next Generation	GES	Run-Time Error Norm	Denormalized Difference
Case 1	4	8	42.29	0.003281
Case 2	4	6	47.31	0.001663
Case 3	6	6	45.31	0.001592
Case 4	6	7	46.03	0.001254
PCA gene	–	–	26.78	0.001493

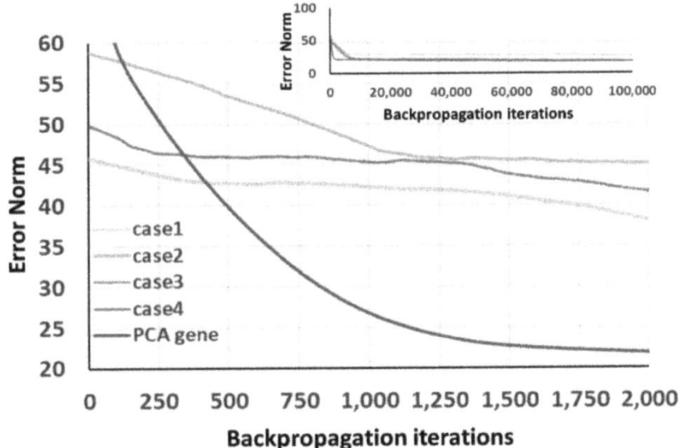

Figure 19. The convergence of 4 GA "progressing" optimized chromosomes and PCA gene under RNN architecture.

In order to investigate the impact of the initial GA generation, the ones belonging to Table 5 are applied to the "progressing" GA optimizer, and the learning results are listed in Table 8 and Figure 20. The PCA gene shows fast convergence capability, as illustrated in Figure 20. Comparing Tables 7 and 8, a clear difference can be identified under the "progressing" GA optimizer with the same GA optimization parameters in terms of run-time error norm and denormalized difference. The influence of the initial GA generation is proven to be significant. Therefore, the recommendation of using the independent initial GA generation for all cases is made.

Table 8. The RNN training result under GA "progressing" optimizer by the same initial 2000 GA generation in Table 5.

	The Best Chromosomes for the Next Generation	GES	Run-Time Error Norm	Denormalized Difference
Case 1	4	4	22.70	0.001406
Case 2	4	9	21.46	0.001022
Case 3	6	9	21.78	0.001263
Case 4	6	6	22.00	0.001204
PCA gene	–	–	20.07	0.001110

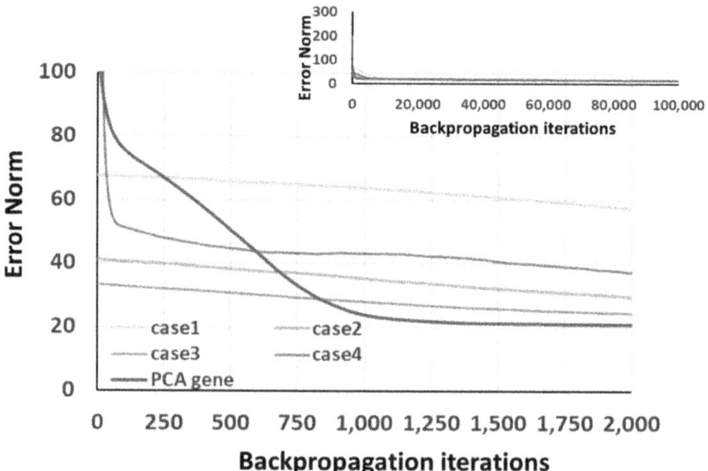

Figure 20. The convergence of 4 GA "progressing" optimized chromosomes and PCA gene under RNN architecture by the same initial 2000 GA generation in Table 5.

6. Conclusions

In this research, the WLCSP solder joint reliability risk is modeling by an AI model, following the AI-assisted simulation framework. ANN and RNN architectures are conducted to investigate their capability of abstracting the time-dependent solder joint fatigue knowledge from the dataset. The GA optimization is applied to decrease the influence of the initial guessings, including the weightings and bias of the neural network architectures. Due to the continuous learning characteristics of the backpropagation, the "back-to-original" and "progressing" GA optimizers are developed.

Both ANN and RNN architectures, with two hidden layers, are conducted with similar neural network structures. Two GA optimizers are applied to both ANN and RNN architectures with four and six best chromosomes to the next generation. Each GA optimization case starts with an independent 2000 initial GA generation, and each component of the chromosomes follows a zero-mean Gaussian distribution. Moreover, a PCA is applied to the GA optimization results to obtain the PCA gene. PCA gene shows high-speed convergence capability in all cases.

The investigation of the GA optimization shows that increasing the number of the best chromosomes to the next generation and choosing the "progressing" GA optimizer improve the GA optimization results. However, both increase a significant computation resource to conduct. The influence of the initial GA generation is proven to be significant. Therefore, using the independent initial GA generation for all cases and using the PCA gene obtained by several GA optimization processes are recommended.

Because ANN and RNN learnings are more robust due to the GA, these neural networks are suitable for generating response surfaces, as seen in Figure 3g. The predictability of the neural network model enables the exploration of the domain that is outside the training domain (the FEM domain) at a certain range due to the contribution of the nonlinear activation functions [3]. Moreover, due to the continuity of the neural network model, these models are feasible for the optimization procedure.

No matter ANN nor RNN architecture, after 100 k learning iterations, all the AI learning results satisfy the accuracy requirement of 0.18% when the PCA gene is applied as the initial parameter. Moreover, there is no clear evidence that RNN is statistically better than ANN in the WLCSP solder joint reliability risk estimation if the PCA gene is applied. Although RNN provides more information than ANN, RNN is influenced by the noise in the dataset during the learning, which limited the RNN to perform much better than

ANN. However, RNN learning requires more computation resources than ANN because of the backpropagation process under a similar neural network structure. Hence, a stable optimization with a broad design domain can be realized by an ANN model with PCA gene with a faster training speed than RNN, even though solder fatigue is a time-dependent mechanical behavior.

Author Contributions: Conceptualization, methodology, software, formal analysis, investigation, data curation, writing—original draft preparation, C.Y.; Problem identification, concept initialization, writing—review and editing, discussion during research, supervision, X.F. and G.Z. All authors have read and agreed to the published version of the manuscript.

Funding: This research is partially supported by the "Developing Smart Automation Mechatronics System and Internet of Things for Shoe Manufacturing" project of Feng Chia University, sponsored by the Ministry of Science and Technology, under the grand no. MOST109-2218-E-035-002, and the "Research of the industrial internet of things for the continuous manufacturing with heterogeneous machines, the production status prediction model and the smart decision-making algorithm" project of Feng Chia University, sponsored by the Ministry of Science and Technology, under the grand no. MOST 109-2221-E-035-008-MY2.

Institutional Review Board Statement: Not applicable.

Informed Consent Statement: Not applicable.

Data Availability Statement: The data presented in this research study are available in this article.

Conflicts of Interest: The authors declare no conflict of interest.

References

1. Chou, P.H.; Chiang, K.N.; Liang, S.Y. Reliability Assessment of Wafer Level Package using Artificial Neural Network Regression Model. *J. Mech.* **2019**, *35*, 829–837. [CrossRef]
2. Hsiao, H.Y.; Chiang, K.N. AI-assisted reliability life prediction model for wafer-level packaging using the random forest method. *J. Mech.* **2020**, *37*, 28–36. [CrossRef]
3. Yuan, C.; LEE, C.-C. Solder Joint Reliability Modeling by Sequential Artificial Neural Network for Glass Wafer Level Chip Scale Package. *IEEE Access* **2020**, *8*, 143494–143501. [CrossRef]
4. Yuan, C.C.A.; Fan, J.; Fan, X. Deep machine learning of the spectral power distribution of the LED system with multiple degradation mechanisms. *J. Mech.* **2021**, *37*, 172–183. [CrossRef]
5. Meszmer, P.; Majd, M.; Prisacaru, A.; Gromala, P.J.; Wunderle, B. Neural networks for enhanced stress prognostics for encapsulated electronic packages-A comparison. *Microelectron. Reliab.* **2021**, *123*, 114181. [CrossRef]
6. Selvanayagam, C.; Duong, P.L.T.; Raghavan, N. AI-Assisted Package Design for Improved Warpage Control of Ultra-Thin Packages. In Proceedings of the 2020 21st International Conference on Thermal, Mechanical and Multi-Physics Simulation and Experiments in Microelectronics and Microsystems (EuroSimE), Cracow, Poland, 5–8 July 2020; pp. 1–7.
7. Tabaza, T.A.; Tabaza, O.; Barrett, J.; Alsakarneh, A. Hysteresis modeling of impact dynamics using artificial neural network. *J. Mech.* **2021**, *37*, 333–338. [CrossRef]
8. Mitchell, T.M. *Machine Learning*, 1st ed.; McGraw-Hill: Boston, MA, USA, 1997.
9. White, D.; Ligomenides, P. GANNet: A genetic algorithm for optimizing topology and weights in neural network design. In Proceedings of the New Trends in Neural Computation, Sitges, Spin, 9–11 June 1993; pp. 322–327.
10. Juang, C.F. A hybrid of genetic algorithm and particle swarm optimization for recurrent network design. *IEEE Trans. Syst. Man Cybern. Part B (Cybernetics)* **2004**, *34*, 997–1006. [CrossRef] [PubMed]
11. Ding, S.; Su, C.; Yu, J. An optimizing BP neural network algorithm based on genetic algorithm. *Artif. Intell. Rev.* **2011**, *36*, 153–162. [CrossRef]
12. Ahmadizar, F.; Soltanian, K.; AkhlaghianTab, F.; Tsoulos, I. Artificial neural network development by means of a novel combination of grammatical evolution and genetic algorithm. *Eng. Appl. Artif. Intell.* **2015**, *39*, 1–13. [CrossRef]
13. Arabasadi, Z.; Alizadehsani, R.; Roshanzamir, M.; Moosaei, H.; Yarifard, A.A. Computer aided decision making for heart disease detection using hybrid neural network-Genetic algorithm. *Comput. Methods Programs Biomed.* **2017**, *141*, 19–26. [CrossRef] [PubMed]
14. Yuan, C.-A.; Han, C.N.; Yew, M.-C.; Chou, C.-Y.; Chiang, K.-N. Design, analysis, and development of novel three-dimensional stacking WLCSP. *IEEE Trans. Adv. Packag.* **2005**, *28*, 387–396. [CrossRef]
15. Yuan, C.; Lee, C. Solder joint reliability risk estimation by AI modeling. In Proceedings of the 2020 21st International Conference on Thermal, Mechanical and Multi-Physics Simulation and Experiments in Microelectronics and Microsystems (EuroSimE), Cracow, Poland, 5–8 July 2020.

16. Liu, C.-M.; Lee, C.-C.; Chiang, K.-N. Enhancing the Reliability of Wafer Level Packaging by Using Solder Joints Layout Design. *IEEE Trans. Compon. Packag. Technol.* **2006**, *29*, 877–885. [CrossRef]
17. Coffin, L.F., Jr. A Study of the Effects of Cyclic Thermal Stresses on a Ductile Metal. *Trans. ASME* **1954**, *76*, 931–950.
18. Manson, S.S. *Thermal Stress and Low Cycle Fatigue*; McGraw-Hill Book Co., Inc.: New York, NY, USA, 1966.
19. Solomon, H.D. Fatigue of 60/40 Solder. *IEEE Trans. Compon. Hybrids Manuf. Technol.* **1986**, *9*, 91–104. [CrossRef]

materials

Article

Improvement Prediction on the Dynamic Performance of Epoxy Composite Used in Packaging by Using Nano-Particle Reinforcements in Addition to 2-Hydroxyethyl Methacrylate Toughener

Chih-Ming Chen [1] , Huey-Ling Chang [2],* and Chun-Ying Lee [3]

[1] Department of Mechanical Engineering, National Chin-Yi University of Technology, Taichung 41170, Taiwan; cmchentc@gmail.com
[2] Department of Chemical and Materials Engineering, National Chin-Yi University of Technology, Taichung 41170, Taiwan
[3] Graduate Institute of Manufacturing Technology, National Taipei University of Technology, Taipei 10608, Taiwan; leech@mail.ntut.edu.tw
*   Correspondence: hlchangtc@gmail.com; Tel.: +886-9-81583585

Abstract: Epoxy with low viscosity and good fluidity before curing has been widely applied in the packaging of electronic and electrical devices. Nevertheless, its low flexibility and toughness renders the requirement of property improvement before it can be widely acceptable in dynamic loading applications. This study investigates the possible use of 2-hydroxyethyl methacrylate (HEMA) toughening agent and nano-powders, such as alumina, silicon dioxide, and carbon black, to form epoxy composites for dynamic property improvement. Considering the different combinations of the nano-powders and HEMA toughener, the Taguchi method with an L9 orthogonal array was adopted for composition optimization. The dynamic storage modulus and loss tangent of the prepared specimen were measured by employing a dynamic mechanical analyzer. With polynomial regression, the curve-fitted relationships of the glass transition temperature and storage modulus with respect to the design factors were obtained. It was found that although the raise in the weight fraction of nano-powders was beneficial in increasing the rigidity of the epoxy composite, an optimal amount of HEMA toughener existed for its best damping improvement.

Keywords: electronic packaging; dynamic storage modulus; loss tangent; optimization

Citation: Chen, C.-M.; Chang, H.-L.; Lee, C.-Y. Improvement Prediction on the Dynamic Performance of Epoxy Composite Used in Packaging by Using Nano-Particle Reinforcements in Addition to 2-Hydroxyethyl Methacrylate Toughener. *Materials* **2021**, *14*, 4193. https://doi.org/10.3390/ma14154193

Academic Editor: Mustafa Yavuz

Received: 26 June 2021
Accepted: 26 July 2021
Published: 27 July 2021

Publisher's Note: MDPI stays neutral with regard to jurisdictional claims in published maps and institutional affiliations.

1. Introduction

Epoxy is one of widely used synthesized resins for general purposes and industrial applications. Due to its low viscosity and good fluidity before curing and excellent strength and stiffness after curing, epoxy resin is commonly thought of as a good candidate for packaging use in electrical and electronic applications. However, with the increase in degree of crosslinking among its molecular chains after the reaction with hardener agent finished, the cured epoxy often becomes too brittle to be applied in some circumstances which require material toughness, such as the environments involving dynamic loading or vibrations.

By adding some "reinforcements" into the epoxy resin to form composites, the mechanical properties of the resin can be tuned. Liu and coworkers [1] investigated the effects on fracture toughness by adding nano-powders of silicon dioxide and rubber into the epoxy resin. The uni-axial tensile performance of the composite revealed both the increase in Young's modulus and fracture toughness. The incorporation of nano-rubber powder demonstrated a clear improvement on its fracture toughness. The other SiO_2 nano-powder has 3D network molecular structure and forms a floccular ball-like particle, which is commonly used as a reinforcement in composite [2]. The combination of SiO_2 nano-powder and glass fiber in a composite laminate has been reported to raise the fatigue life by 3–4 folds compared to its pristine epoxy resin counterpart [3]. The study by Chen et al. [4] also reported the increase in both stiffness and fracture toughness of epoxy composite by adding spherical SiO_2 nano-powders. However, the glass transition temperature was lowered if the content of SiO_2 nano-powder was greater than 5 wt.%.

In addition to simply adding inorganic powder for reinforcement, the interface usually needs treatment to improve the compatibility between phases. Liu and coworkers [5] used MXene (Ti_3AlC_2 and Ti_3C_2Tx) to treat the acidified short graphite fibers and reported the 100% and 67% increases in tensile and flexural strengths of the epoxy composite, respectively. The SiO_2 nano-powders also had good compatibility with the epoxy resin and its epoxy composite showed the improvements in tensile compressive strengths and fracture toughness [6,7]. Moreover, these well dispersed nano-powders between the reinforced fibers and the resin even presumed as the buffers for absorbing energy in the interfacial bonding [8].

The other approach for toughening the epoxy resin is by modifying its structure with interpenetrating polymer networks (IPNs). This structural modification down to the molecular level with uniform interlacing of toughening phase and the crosslinked epoxy phase could promote the material compatibility. Around the 1980s, Sperling and coworkers [9,10] used different monomers and polymers to prepare toughened epoxy with different types of IPNs, such as full-IPNs and semi-IPNs. They also studied the chemical compositions, microstructural morphology, phase behavior and the correlation with their mechanical property by employing various precision instruments. It was reported that the toughener and the epoxy resin were uniformly mixed at the molecular level or in microscopic phase separation. Since there was no chemical bonding between these two phases, the individual properties from these constituents were to be retained. However, since these two phases were intertwined closely, the cooperative effect emerged as an improvement in its toughness over its mechanically mixed counterpart.

The further modification on the microstructure of IPNs was proposed by Hsieh and Han [11] with the introduction of PU based on polybutylene adipate, PU(PBA) and PU based on polyoxypropylene, PU(PPG), into diglycidyl ether bisphenol A, DGEBA, to form grafted IPNs. The tensile test result revealed that the grafted IPNs could increase the α-and β-transition domain of the epoxy and PU and, subsequently, its tensile strength. Lin and Lee [12] based on the DGEBA matrix added ethylene dimethacrylate and 2-hydroxymethyl methacrylate (HEMA) to form full-IPNs and semi-IPNs. Their experimental results showed the material with full-IPNs had a higher Young's modulus, larger percent elongation and better fracture toughness to absorb more energy in dynamic loading. When the added reinforcement was the nano-powder, the epoxy composite not only had increases in tensile strength but also the toughness in impact.

Mimura and coworkers [13] reported the preparation of toughened epoxy with semi-IPNs by using synchronized polymerization of polyethersulfone (PES) and epoxy resin. A 60% increase in fracture toughness was obtained. Sometimes, the addition of toughener could create non-uniform phase separation and crosslinking density. In the work by Kwon and coworkers [14], the appearance of additional side peak in the tanδ of a viscoelastic measurement on the polytriazoleketone (PTK) and polytriazolesulfone (PTS) toughened epoxy resin demonstrated the probable concern.

In the electronic packaging applications, the use of different materials to tune the mechanical property was a common practice. However, the collateral effect arisen from the mismatch in the thermal expansion of its constituents could worsen or even fail the packaging function [15]. The other concern may lie on the viscosity of resin. Oh and coworkers [16] used two micro-fillers, Al_2O_3 and SiO_2, to tune the thermal conductivity of the epoxy resin. However, considering the Al_2O_3 filler had higher specific gravity than SiO_2, less volumetric fraction of the former could have similar weight fraction with the latter. The partial substitution of Al_2O_3 for the SiO_2 would keep the epoxy resin lower in viscosity or higher in fluidity, which was beneficial in packaging. The other report by

Khalil et al. [17] illustrated the incorporation of Al_2O_3 nano-powder within 2.0 wt.% in epoxy would improve the wetting behavior and tensile shear strength of the epoxy resin. Thus, the addition of nano-powders within 2.0 wt.% in the epoxy resin was adopted in this study.

As reviewed from the previous studies, the incorporation of inorganic nano-powders in epoxy resin seems beneficial to its packaging applications. However, the underlined mechanisms have not been well investigated. Moreover, the reliability performance of epoxy composites with different compositions are susceptible to the service temperature and loading conditions. Therefore, by using the dynamic mechanical analyzer, the main focus of this study is to investigate the viscoelastic behavior and energy dissipation from the internal friction within its microstructure of the epoxy composite over a specified temperature range. Three nano-powders of Al_2O_3, SiO_2, and carbon black were chosen to be the reinforcements of the epoxy resin in addition to a HEMA toughener. The design of experiments by using the Taguchi method was adopted and the analysis of variables (ANOVA) was performed to find the optimal compositions for the dynamic property improvement. The prediction of the dynamic properties of the prepared epoxy composite was established for the design purpose in packaging.

2. The Theory

2.1. The Reaction Mechanism of Epoxy

In this study, a simultaneous polymerization reaction was adopted to prepare the epoxy with IPNs. Three reactions were proceeded simultaneously in the crosslinking of epoxy, as presented in Figure 1. Firstly, the crosslinking started with the ring-opening reaction of the epoxide group in epoxy and the primary amine of hardener agent to form the hydroxyl group and the formation of secondary amine from the primary amine of the hardener as described in Figure 1a. The thus formed secondary amine could also take part in the ring-opening reaction with the epoxide group as denoted in Figure 1b. In the meantime, the hydroxyl group started a self-catalyzed ring-opening reaction as denoted in Figure 1c, which is also known as etherification, and was the least reactive among the three mentioned reactions. Subsequently, a crosslinkage in spatial dimensions was obtained with the described reactions.

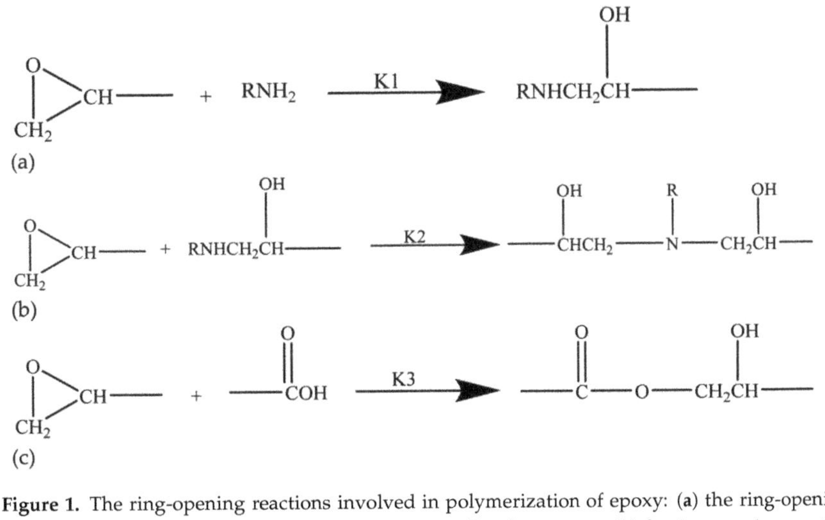

Figure 1. The ring-opening reactions involved in polymerization of epoxy: (**a**) the ring-opening reaction of the epoxide group and the primary amine of hardener agent; (**b**) the ring-opening reaction of the epoxide group with the secondary amine formed in (**a**); (**c**) the self-catalyzed ring-opening reaction from the hydroxyl group.

Secondly, benzoyl peroxide as a thermal initiator disintegrated into two free radicals upon being heated and are described in Figure 2. The generation of these free radicals induced the polymerization of 2-hydroxyethyl methacrylate to form the PHEMA polymer as denoted in Figure 3. These PHEMAs constituted the interpenetrating polymer network in the already spatially crosslinked structure of epoxy. Due to the thermoplastic property of the PHEMA, the thus prepared epoxy resin could improve the toughness over the pristine brittle nature.

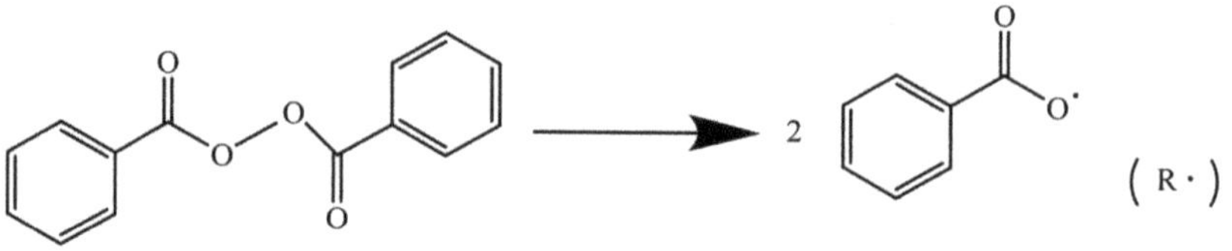

Figure 2. The disintegration reaction of benzoyl peroxide into free radical.

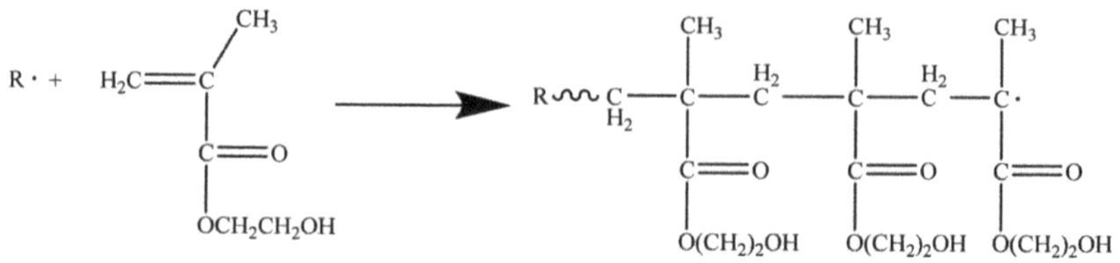

Figure 3. The polymerization reaction of 2-hydroxyethyl methacrylate to form the PHEMA polymer.

2.2. Dynamic Mechanical Property

A dynamic mechanical analyzer (TA DMA 2980) was used in this study to measure the dynamic properties of the prepared specimens in different time, temperature and loading frequency. In the measurement, the specimen was installed in a specified configuration, such as three-point bending and cantilevered fixture, and was subjected to a vibrational stress or strain. The responses of the specimen in deformation or loading were recorded both in magnitude and phase. Accordingly, the storage modulus E′, loss modulus E″, and coefficient of loss tangent tanδ of the specimen can be calculated as follows.

Storage modulus E′ is calculated as follows.

$$E' = \left(\frac{\sigma}{\varepsilon}\right)\cos\delta \tag{1}$$

The loss modulus E″ is calculated as follows.

$$E'' = \left(\frac{\sigma}{\varepsilon}\right)\sin\delta \tag{2}$$

The coefficient of loss tangent, tanδ, is calculated as follows.

$$\tan\delta = \frac{E''}{E'} \tag{3}$$

In the above equations, ε and σ represent the amplitudes of the applied strain and the measured stress, respectively. The larger the peak of tan, the more viscous the material behaves. In other words, the specimen with larger tanδ has more damping in response to dynamic loading. Moreover, the temperature where the tanδ attains its peak magnitude denotes the corresponding glass transition temperature of the specimen.

2.3. Design of Experiment Using Taguchi Method

Table 1 presents the $L_9(3^4)$ orthogonal table for this study. There were four design factors, i.e., the weight contents of A: nano-alumina powder; B: nano-silica powder; C: nano-carbon black powder; and D: HEMA toughener, respectively. Three levels for each factor were selected, which were 0, 1, and 2 wt.% and 0, 5, and 10 wt.% for the nano-powders and the toughener, respectively. Accordingly, the dynamic properties of the specimens were measured by using the DMA analyzer and the S/N ratios were calculated from the quality equation.

Table 1. $L_9(3^4)$ orthogonal table for the design of experiments.

No. of Specimen	Factor A	Factor B	Factor C	Factor D
1	1	1	1	1
2	1	2	2	2
3	1	3	3	3
4	2	1	2	3
5	2	2	3	1
6	2	3	1	2
7	3	1	3	2
8	3	2	1	3
9	3	3	2	1

The-higher-the-better (HB) characteristics were employed with decibels (db) as its unit, shown in the following.

$$S/N_{LTB} = -10 \times \log_{10}\left(\frac{1}{n}\sum_{i=1}^{n}\frac{1}{y_i^2}\right) \tag{4}$$

2.4. Coefficient of Variance

Three specimens for each measurement were employed in this study in order to evaluate the variation of the results. Therefore, the coefficient of variance (C.V.) was calculated accordingly:

$$C.V. = \frac{S}{\underline{X}} \tag{5}$$

where $\underline{X}$ and S are the mean and standard deviation for each set of measurements.

2.5. Multiple Regression Analysis

With the storage moduli as a substitute for the machining process factor, the regression fit of the object function could be preceded. For simplicity, the dependence of the object function on the factors was assumed to be quadratic and the coupling among the factors was neglected. These assumptions will be checked with the verification example presented later in this study. Thus, the regression function Y is written as follows:

$$Y = \sum_{i=A}^{G}\left(c_{i2}X_i^2 + c_{i1}X_i\right) + c_0 \tag{6}$$

where X_i denotes the value of the ith factor and c_{i2} and c_{i1} are the coefficients for the quadratic and linear terms, respectively. The constant term for the relationship is summed up in c_0. The object functions that will be evaluated later include the storage moduli.

2.6. Coefficient of Determination for Regression

The validity of the regression function needs to be examined. Herein, a coefficient of determination R^2 was adopted. The larger the R^2 is, the more accurate the regression function becomes. The coefficient of determination was defined as the ratio of regression variance to total variance. It can be expressed as follows:

$$R^2 = \frac{SS_T}{SS_R} \tag{7}$$

where SS_T is the total sum of squares and SS_R is the regression sum of squares. The magnitude of R^2 lies between 0 and 1, i.e., $0 < R^2 < 1$. The closer the R^2 approaches 1, the better the regression function represents the physical relationship.

3. Experimental

3.1. Materials

A diglycidyl ether of bisphenol A type epoxy resin (an epoxide equivalent weight of 180 g/equiv.) was purchased from Chang-Chun Plastics Co. Ltd. (Taipei, Taiwan). The epoxy resin was cured with the incorporation of an amine type epoxy curing agent (the amine hydrogen weight of 65 g/equiv. and amine value 430 g/equiv.). When the nano-powder used was a silicon dioxide (silica) powder, the surface of the fumed silica was chemically modified with a poly(dimethyl siloxane) coupling agent. The hydrophobic fumed silica had a specific surface area (BET) of 100 m^2/g, an average primary particle diameter of 14 nm, and a tapped density (according to DIN ISO 787/XI, August 1983) of 60 g/L. For the other nano-powders, the aluminum oxide (γ-Al$_2$O$_3$, alumina) had an average particle diameter of 50 nm and density 2.414 g/cm^3. The nano carbon black powder was with a specific surface area (BET) of 90 m^2/g, average particle diameter of 28 nm, and density of 1.719 g/cm^3. Moreover, the toughening agent was 2-hydroxyethyl methacrylate (HEMA) with a mass density of 1.106 g/mL, which was an acrylate with polar substations that was purchased from TCI Co. Ltd. Benzoyl peroxide (BPO) was used as a thermal initiator. It should be mentioned that all the purchased materials were used as-received without further purification.

3.2. Specimen Preparation

Firstly, the epoxy resin and HEMA toughener with specified weight fraction were put in a container of a deaerator and thoroughly mixed. Subsequently, the nano-powders with specified weight fractions were added and agitated mechanically. The well mixed solution, which was mixed with an ultrasonic and centrifugal mixer, was obtained as the pre-mixture for next process. In our previous study [18], the prepared material was found to have a well dispersion of nano-powders in the epoxy matrix. On the other hand, a specified amount of hardener and thermal initiator were mixed in the other container. This mixed hardener was then added into the epoxy pre-mixture and agitated in the centrifugal mixer. Subsequently, the prepared mixture was poured into a DMA specimen mold manufactured according to the ASTM D4065 standard. This casted mold was cured in a 120 °C oven for 1 h and the removed specimen was then post-cured at 140 °C for another 1 h. The prepared specimen has to be placed at room temperature for more than 24 h before it can be employed in the material testing.

Table 2 lists the compositions of the 9 types of specimens adopted from an L$_9$(3^4) orthogonal array of the Taguchi method. Four control factors were employed in the experiments: nano-alumina powder, nano-silica powder, nano-carbon black powder, and HEMA toughener. Three levels for each factor were selected, which were 0, 1, 2 wt.% and 0, 5, 10 wt.% for the nano-powders and the toughener, respectively.

Table 2. The compositions of the specimens used in an L9 orthogonal array.

Specimen	Epoxy (wt.%)	Hardener (wt.%)	Al_2O_3 (wt.%)	SiO_2 (wt.%)	Carbon Black (wt.%)	HEMA (wt.%)	BPO (wt.%)
T1	74.64	24.86	0	0	0	0	0
T2	69.04	23.02	0	1	1	5	0.5
T3	63.46	21.19	0	2	2	10	0.5
T4	65.27	21.74	1	0	1	10	0.5
T5	70.97	23.71	1	1	2	0	0
T6	68.18	22.68	1	2	0	5	0.5
T7	67.23	22.42	2	0	2	5	0.5
T8	64.37	21.44	2	1	0	10	0.5
T9	70.14	23.35	2	2	1	0	0

3.3. DMA Measurements

Dynamic mechanical analysis (DMA) was performed according to ASTM D4065-01 to determine storage modulus (E'), loss modulus (E', loss tangent ($\tan\delta$), and glass transition temperature (Tg) of the epoxy composites. The tests were conducted in the dual cantilever beam mode with a vibration frequency of 1 Hz and a displacement amplitude of 10 μm in a DMA analyzer (DMA 2980, TA Instruments). The temperature was ramped from 30 to 140 °C at a rate of 2 °C/min. At least three specimens of each type were tested and the data were analyzed. The DMA testing purpose was to simulate the loading conditions of the epoxy composite used in the packaging and subjected to periodic vibration from the surroundings.

4. Results and Discussion

4.1. The Dynamic Properties at Different Temperatures

Figure 4 presents the dynamic properties of Specimen T1 measured with the temperature sweeping from 30 °C to 140 °C. There are three results for each specimen: storage modulus, loss modulus, and loss tangent. As mentioned previously, there were three specimens denoted as T1-1, T1-2, and T1-3 that were tested for each specimen type of the design factors specified in Table 1. It is obviously observed that the test results had good repeatability. As observed in the figure, the storage modulus E' decreased sharply when the temperature reached near 60 °C while the loss modulus E'' increased to a maximum and dropped subsequently. According to these results, the loss tangent defined as $\tan\delta = E''/E'$ attained a maximum around 83 °C. The temperature at which the loss tangent has a maximum is usually defined as the glass transition temperature of the material, T_g. As far as the dynamic property is concerned, the loss tangent denotes the damping coefficient of the material. Larger material damping is always required for controlling the vibration of structure near resonance.

Similar test results are shown in Figure 5 for Specimen T2. The monotonous decreases in E' and a dome-shaped E'' with respect to the raise in temperature are observed, as shown in Figure 4. However, the loss tangent curve shows a plateau with two distinct peaks. Although the peak value of the loss tangent was not as high as the Specimen T1, the plateau became wider in the temperature range. In other words, Specimen T2 was more effective than Specimen T1 in reducing the vibration by material damping over a wider working temperature range. If the definition of glass transition temperature was recalled from the previous discussion, Specimen T2 had two T_g's at around 67 °C and 83 °C. The appearance of two loss tangent peaks is related to the HEMA toughener added in the epoxy composite, which will be discussed in more detail afterwards.

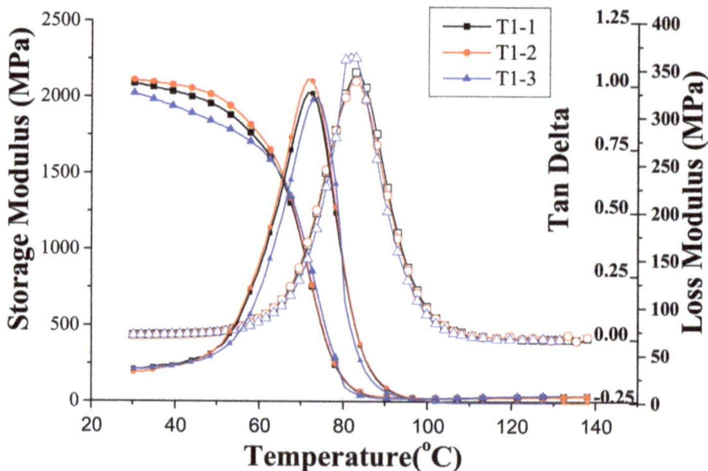

Figure 4. The dynamic properties of Specimen T1 measured with the temperature sweeping from 30 °C to 140 °C.

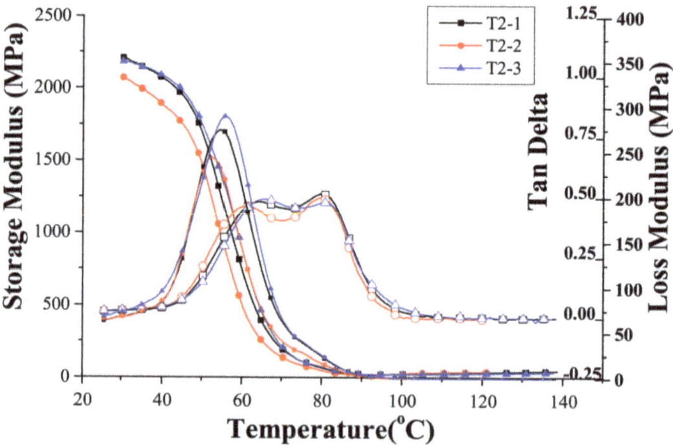

Figure 5. The dynamic properties of Specimen T2 measured with the temperature sweeping from 30 °C to 140 °C.

As observed from Table 2, Specimen T3 had an even higher concentration of HEMA toughener compared to the previous two specimens. The measured dynamical properties are similarly shown in Figure 6. Without too much surprise, there were also two peaks observed in the loss tangent curve with an even wider temperature range of the plateau or the lower T_g moving to lower temperatures. Similar measurements for the rest of the six specimens were also performed and the T_g's are summarized in Table 3.

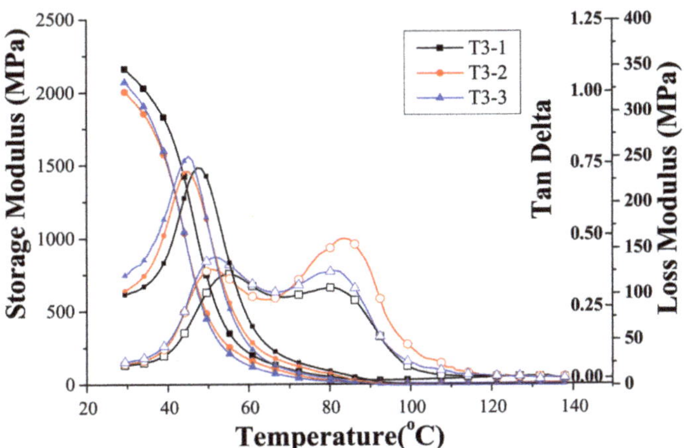

Figure 6. The dynamic properties of Specimen T3 measured with the temperature sweeping from 30 °C to 140 °C.

Table 3. Glass transition temperatures of the specimens.

Specimen	T_g1 (°C)	T_g2 (°C)
T1	-	83.08 ± 0.31
T2	67.42 ± 1.13	83.34 ± 0.80
T3	52.60 ± 2.28	81.34 ± 1.44
T4	55.93 ± 3.29	78.43 ± 1.76
T5	-	87.63 ± 1.13
T6	68.68 ± 2.29	83.86 ± 0.47
T7	65.92 ± 2.29	84.15 ± 0.42
T8	54.05 ± 1.69	78.77 ± 1.19
T9	-	90.26 ± 2.50

4.2. Glass Transition Temperatures

Table 3 lists the glass transition temperatures of nine specimens obtained from the previous measurements. By cross-referencing the compositions of the specimens reported in Table 3, those listed in Table 4 with only single T_g are specimens without adding the HEMA toughener. The specimens with the inclusion of HEMA toughener in the compositions revealed the lower T_g peak. In more detail, Specimens T2, T6, and T7, which all have a 5 wt.% of HEMA, had lower T_g's at 67.42, 68.68, and 65.92 °C, respectively. Moreover, with further increases in HEMA content to 10 wt.%, Specimens T3, T4, and T8 lowered their corresponding T_g's to 52.60, 55.93, and 54.05 °C, respectively. It is known that the glass transition temperature reflects the cross-linking density of its molecular chains inside the polymer material. Those specimens without HEMA inclusion in their compositions revealed only one T_g peak. The introduction of a thermoplastic HEMA into the microstructure somewhat hindered the cross-linking of epoxy during curing and the molecular chains of HEMA started to move locally at lower temperature. Minor or shoulder peaks may appear due to the phase separation between the epoxy resin and the toughening agent or inhomogeneous crosslinking density [14]. However, all specimens demonstrated the T_g peak at higher temperature around 78.77~90.26 °C, which reflected the movement of the molecular chains in the microstructural regions unaffected by the HEMA. With more HEMA contents, the regions affected by the HEMA enlarged and, consequently, further lowered the T_g. Of course, the movement of the molecular chain was still influenced by other nano-powders included in the composite. Therefore, the variation in T_g was measured for specimens with other different compositions but with less significance. The lowering of the resistance for the molecular chains to move at lower temperatures created more flexibility to the material and, therefore, alleviated the brittleness. That is the mechanism for using toughener in epoxy resin to raise its toughness. Since the main purpose of this study is to investigate the effects of the nano-powders on the dynamic property of the epoxy composite, only the second T_g is referred to specifically afterwards.

Table 4. Analysis on the measured results of the main glass transition temperature.

Specimen	Y_1 (°C)	Y_2 (°C)	Y_3 (°C)	Y_{Avg} (°C)	C.O.V. (%)	S/N (dB)
T1	82.84	82.94	83.45	83.08	0.39	38.39
T2	82.31	83.91	83.81	83.34	1.08	38.42
T3	80.14	83.01	80.87	81.34	1.83	38.20
T4	76.44	78.91	79.95	78.43	2.30	37.89
T5	86.71	88.96	87.21	87.63	1.35	38.85
T6	84.46	83.52	83.59	83.86	0.62	38.47
T7	83.88	83.87	84.71	84.15	0.57	38.50
T8	77.21	79.52	79.58	78.77	1.72	37.92
T9	91.54	87.12	92.11	90.26	3.03	39.10

Table 4 lists the main glass transition temperatures of all specimens and their calculated statistical properties. According to these measurements, the S/N ratio of the glass transition temperature for each control factor was summarized and presented in Table 5. It is observed that all nano-powders had higher effect (S/N ratio) on the glass transition temperature with higher level of doping concentration. Among them, SiO_2 had the most significant effect while Al_2O_3 had the least. It is worth mentioning that the amount of HEMA doping decreased the glass transition temperature drastically, which was mainly caused by the rotation and vibration of the organic molecules of HEMA toughener under the influence of heating. Moreover, the reaction from the added toughener in the epoxy also hindered its crosslinking and reduced the crosslinking density between the molecular chains. Therefore, the molecular chains had more space for moving and the T_g decreased accordingly. These nano-powders usually have higher stiffness than the epoxy matrix. The reinforcement demonstrates its effect with the raise in glass transition temperature. On the other hand, the HEMA toughener showed an opposite trend because of its interference with the crosslinking density, as discussed previously. Among the control factors, HEMA had the highest influence on the glass transition temperature followed by silica, alumina and carbon black, sequentially.

Table 5. Reaction table for the S/N ratio of glass transition temperature (unit: dB).

Control Factor	Al_2O_3	SiO_2	Carbon Black	HEMA
Level 1	38.34	38.26	38.26	38.78
Level 2	38.40	38.40	38.47	38.46
Level 3	38.51	38.59	38.52	38.00
Effect	0.17	0.33	0.26	0.78

With the obtained experimental data presented in Table 4, a multi-variable regression analysis by using SPSS software was performed on the relationship of the glass transition temperature of the epoxy composite with respect to the four control factors. The following polynomial equation was obtained for the prediction of the glass transition temperature Y of the epoxy composite.

$$Y = 83.08 + 0.537\,X_A + 0.183\,X_A^2 + 1.087\,X_B + 0.273\,X_B^2 + 2.978\,X_C - 0.832\,X_C^2 - 0.535\,X_D - 0.210\,X_D^2 \tag{8}$$

In the above equation, X_A, X_B, X_C, and X_D are the wt.% of the silica nano-powder, alumina nao-powder, carbon black nano-powder, and HEMA toughener, respectively. In order to check the validity or accuracy for the prediction equation, the specimen with the optimal combination of the control factors (A3, B3, C3, and D1 as revealed in Table 5) to maximize the glass transition temperature was fabricated and tested. The predicted 90.61 °C from

Equation (8) was close to the measured 90.91 °C. The 0.33% difference between the prediction and measurement showed good accuracy of the regression model in predicting the glass transition temperature of the epoxy composite.

4.3. Storage Modulus

The storage modulus of the material used in packaging denotes the material's measure to withstand deformation due to dynamic loading. The higher the storage modulus, the better the resistance to dynamic deformation. Table 6 presents the analysis on the measured results of the storage modulus at 30 °C for the nine specimens. Following the similar data processing of the previous glass transition temperature, the S/N ratios of storage modulus for each control factor at different levels were obtained and listed in Table 7. The results showed that the S/N ratio of the storage modulus increases with the doping content of each nano-powder. Among the three nano-powders studied, the carbon black is more effective in raising the storage modulus, while the Al_2O_3 has the least effect. However, the HEMA toughener played a different role of lowering the storage modulus with increasing doping content. The lowering effect was more pronounced than its nano-powder counterparts in raising the storage modulus of the epoxy composite as the S/N ratio was twice larger in magnitude. As mentioned previously, the incorporation of thermoplastic HEMA molecules into the epoxy polymer chains impedes the cross-linking and reduces the local cross-linking density. Thus, the storage modulus reduced with higher HEMA contents. On the other hand, the nano-powders filled in the epoxy matrix act as the reinforcement phase in the composite. Moreover, the dispersion of hard inorganic nano-powders in the microstructure of epoxy also serves as the pin-point sites to increase the resistance of chain movement. Thus, deformation reduced and the stiffness of the composite increased. It should also be mentioned that the insertion of HEMA molecules in blocking the crosslinking of the epoxy network and the incorporation of nano-powders in the microstructure could interfere with crack propagation in the epoxy composite. Therefore, the toughness of the epoxy composite can be enhanced.

Table 6. Analysis on the measured results of the storage modulus at 30 °C.

Specimen	E'_1 (MPa)	E'_2 (MPa)	E'_3 (MPa)	E'_{Avg} (MPa)	C.O.V. (%)	S/N (dB)
T1	2081	2107	2009	2066	3.21	66.30
T2	2184	2051	2178	2138	3.12	66.59
T3	2094	1969	2016	2026	3.26	66.13
T4	1688	1709	1764	1720	3.76	64.71
T5	2516	2485	2465	2489	2.73	67.92
T6	2166	2118	2173	2152	3.10	66.66
T7	2373	2285	2425	2361	2.86	67.45
T8	1963	1849	1818	1877	3.49	65.45
T9	2618	2557	2672	2616	2.61	68.35

Table 7. Reaction table for the S/N ratio of storage modulus (unit: dB).

Control Factor	Al_2O_3	SiO_2	Carbon Black	HEMA
Level 1	66.34	66.15	66.14	67.52
Level 2	66.43	66.65	66.55	66.90
Level 3	67.09	67.04	67.17	65.43
Effect	0.75	0.89	1.03	2.09

Similar regression fit using SPSS software was conducted relative to the relationship between the storage modulus and the control factors. The storage moduli measured from the nine specimens were employed in the regression fit with the content wt.% of the control factors. The obtained equation is listed as follows.

$$E' = 2066 - 16.667\,X_A + 60.333\,X_A^2 + 130.167\,X_B - 11.167\,X_B^2 + 122.500\,X_C + 3.833\,X_C^2 - 17.733\,X_D - 3.387\,X_D^2 \tag{9}$$

In order to check the accuracy of the above prediction function for the storage modulus, a specimen with the optimal combination of control factors (A3 B3 C3 D1 from Table 6 with maximum S/N ratio) was fabricated and tested. The prediction from Equation (9) was 2749.5 MPa while the measured one was 2770.3 MPa. A small deviation of 0.76% from prediction to measurement illustrates the model depicted in Equation (9). Equation (9). can be considered as a good model in the design of this epoxy composite for use in dynamic loading environments.

5. Conclusions

When material is used in vibrational environment, the capability to estimate its dynamic mechanical property is critical during the design to prevent the unexpected dynamic response from external excitation. The dynamic property such as the stiffness and damping coefficients are crucial for the packaging materials used in the delicate electronic devices. This study explored the feasibility to predict the storage modulus and glass transition temperature of an epoxy composite that incorporated the nano-powders of silica, alumina, and carbon black in a HEMA toughened epoxy matrix. The use of nano-powders as reinforcements was shown to increase both the T_g and the storage modulus, while the addition of HEMA toughener lowered both dynamic characteristics. With only the incorporation of each nano-powder of silica, alumina, and carbon black by 2 wt.%, the storage modulus of the epoxy composite at 30 °C showed a 34% increase from its pristine counterpart. On the other hand, the addition of 10 wt.% of HEMA into the epoxy could widen the damping plateau of the loss tangent spectrum from a temperature span around its T_g from 20 °C to 50 °C. This wider plateau denoted that the material damping could be operative in larger temperature span or frequency span. Finally, the prediction equations for the glass transition temperature and storage modulus were obtained from regression fits of the measured data, respectively. These equations could be used for design purpose and showed <1% deviation from the measurement within the range of control factors investigated in this study.

Author Contributions: Formal analysis, C.-M.C., H.-L.C. and C.-Y.L.; Investigation, C.-M.C., H.-L.C. and C.-Y.L.; Methodology, C.-M.C. and H.-L.C.; Project administration, H.-L.C. All authors have read and agreed to the published version of the manuscript.

Funding: This research received no external funding.

Institutional Review Board Statement: Not applicable.

Informed Consent Statement: Not applicable.

Data Availability Statement: The data underlying this article will be shared on reasonable request from the corresponding author.

Acknowledgments: The financial support from the Ministry of Science and Technology, Taiwan, for this study is gratefully acknowledged.

Conflicts of Interest: The authors declare no conflict of interest.

References

1. Montazeri, A.; Montazeri, N. Viscoelastic and mechanical properties of multi walled carbon nanotube/epoxy composites with different nanotube content. *Mater. Des.* **2011**, *32*, 2301–2307. [CrossRef]
2. Holck, O.; Dermitzaki, E.; Wunderle, B.; Bauer, J.; Michel, B. Basic thermo-mechanical property estimation of a 3D-crosslinked epoxy/SiO2 Interface using molecular modeling. *Microelectron. Reliab.* **2011**, *51*, 1027–1034. [CrossRef]
3. Manjunatha, C.M.; Taylor, A.C.; Kinloch, A.J.; Sprenger, S. The tensile fatigue behaviour of a silica nanoparticle-modified glass fibre reinforced epoxy composite. *Compos. Sci. Technol.* **2010**, *70*, 193–199. [CrossRef]
4. Chen, C.; Justice, R.S.; Schaefer, D.W.; Baur, J.W. Highly dispersed nanosilica–epoxy resins with enhanced mechanical properties. *Polymers* **2008**, *49*, 3805–3815. [CrossRef]
5. Liu, L.; Ying, G.; Sun, C.; Min, H.; Zhang, J.; Zhao, Y.; Wen, D.; Ji, Z.; Liu, X.; Zhang, C.; et al. MXene (Ti3C2Tx) Functionalized Short Carbon Fibers as Cross-Scale Mechanical Reinforcement for Epoxy Composites. *Polymers* **2021**, *13*, 1825. [CrossRef] [PubMed]

6. Hashim, U.R.; Jumahat, A.; Jawaid, M. Mechanical Properties of Hybrid Graphene Nanoplatelet-Nanosilica Filled Unidirectional Basalt Fibre Composites. *Nanomaterials* **2021**, *11*, 1468. [CrossRef] [PubMed]

7. Demirci, M.T. Low velocity impact and fracture characterization of SiO2 nanoparticles filled basalt fiber reinforced composite tubes. *J. Compos. Mater.* **2020**, *54*, 3415–3433. [CrossRef]

8. Jumahat, A.; Soutis, C.; Jones, F.R.; Hodzic, A. Effect of silica nanoparticles on compressive properties of an epoxy polymer. *J. Mater. Sci.* **2010**, *45*, 5973–5983. [CrossRef]

9. Sperling, L.H.; Taylor, D.W.; Kirkpatrick, M.L.; George, H.F.; Bardman, D.R. Glass-rubber transition behavior and compatibility of polymer pairs: Poly(ethyl acrylate) and poly(methyl methacrylate). *J. Appl. Polym. Sci.* **1970**, *14*, 73–78. [CrossRef]

10. Siegfried, D.L.; Thomas, D.A.; Sperling, L.H. Thermoplastic interpenetrating polymer networks of a triblock copolymer elastomer and an ionomeric plastic. I. Rheology and morphology. *J. Appl. Polym. Sci.* **1981**, *26*, 177–192. [CrossRef]

11. Hsieh, K.H.; Han, J.L. Graft interpenetrating polymer networks of polyurethane and epoxy. I. mechanical behavior. *J. Polym. Sci. Part. B Polym. Phys.* **1990**, *28*, 623–630. [CrossRef]

12. Lin, M.S.; Lee, S.T. Mechanical behaviours of fully and semi-interpenetrating polymer networks based on epoxy and acrylics. *Polymer* **1997**, *38*, 53–58. [CrossRef]

13. Mimura, K.; Ito, H.; Fujioka, H. Improvement of thermal and mechanical properties by control of morphologies in PES-modified epoxy resins. *Polymer* **2000**, *41*, 4451–4459. [CrossRef]

14. Kwon, W.; Han, M.; Kim, J.; Jeong, E. Comparative Study on Toughening Effect of PTS and PTK in Various Epoxy Resins. *Polymers* **2021**, *13*, 518. [CrossRef] [PubMed]

15. Silwal, B.; Mohamed, A.H.; Nonneman, J.; De Paepe, M.; Sergeant, P. Assessment of Different Cooling Techniques for Reduced Mechanical Stress in the Windings of Electrical Machines. *Energies* **2019**, *12*, 1967. [CrossRef]

16. Oh, D.H.; Kim, H.S.; Shim, J.H.; Jeon, Y.H.; Kang, D.W.; Lee, B.W. Characteristics of Gel Time and Dielectric Strength of Epoxy Composite According to the Mixing Ratio of Micro-Fillers. *Energies* **2020**, *13*, 5165. [CrossRef]

17. Khalil, N.Z.; Johanne, M.F.; Ishak, M. Influence of Al_2O_3 nanoreinforcement on the adhesion and thermomechanical properties for epoxy adhesive. *Compos. Part B Eng.* **2019**, *172*, 9–15. [CrossRef]

18. Chang, H.L.; Chen, C.M. Optimization of Epoxy Nanocomposite on Glass Transition Temperature Characteristics Utilizing Experimental Design Method. *Dig. J. Nanomater. Biostruct.* **2015**, *10*, 615–624.

Article

Thermally-Induced Deformations and Warpages of Flip-Chip and 2.5D IC Packages Measured by Strain Gauges

Ming-Yi Tsai *, Yu-Wen Wang and Chia-Ming Liu

Department of Mechanical Engineering, Chang Gung University, Kwei-Shan, Tao-Yuan 333, Taiwan; wanglovedog@gmail.com (Y.-W.W.); s3556947@gmail.com (C.-M.L.)
* Correspondence: mytsai@mail.cgu.edu.tw; Tel.: +886-3211-8800 (ext. 5743); Fax: +886-3211-834

Abstract: The thermal warpage problems in integrated circuit (IC) packaging exist in both flip-chip and two-and-a-half dimensional integrated circuits (2.5D IC) packages during manufacturing processes and thermal cycling service. This study proposes a simple and easy-to-use strain gauge measurement associated with a beam model theory to determine the thermally induced deformations and warpages of both packages. First, validation and limitations of the beam model theory are presented. Then, the thermally induced out-of-plane deformations for both packages are well described by the finite element method (FEM) simulation with a good consistency to full-field shadow moiré experimental results. The strain gauge measurements were implemented experimentally, and the thermal strain results were found to be well consistent with validated FEM ones. As a result, out-of-plane thermal deformations and warpages of the packages, calculated from the beam model theory with extracted curvature data from the strain gauge, were in reasonably good agreement with those from FEM analysis and shadow moiré measurements. Therefore, the strain gauge method of featuring point strain measurement combined with the beam model theory proved feasible in determining the thermal deformations and warpages of both IC packages.

Keywords: flip-chip package; 2.5D package; thermal warpage; strain gauge

Citation: Tsai, M.-Y.; Wang, Y.-W.; Liu, C.-M. Thermally-Induced Deformations and Warpages of Flip-Chip and 2.5D IC Packages Measured by Strain Gauges. *Materials* **2021**, *14*, 3723. https://doi.org/10.3390/ma14133723

Academic Editor: Kuo-Ning Chiang

Received: 24 May 2021
Accepted: 30 June 2021
Published: 2 July 2021

Publisher's Note: MDPI stays neutral with regard to jurisdictional claims in published maps and institutional affiliations.

1. Introduction

The flip-chip packages and the two-and-a-half dimensional integrated circuits (2.5D IC) packages are popularly used in advanced integrated circuits (IC) packaging [1–3]. However, those packages suffer from thermal warpage problems, which may cause solder joint defects, shown in Figure 1a, during manufacturing processes and thermal cycling service due to mismatches of coefficients of thermal expansion (CTE) between the inherent materials [3–7]. Unlike the conventional flip-chip packages, which use a silicon chip to directly bond to the substrate, the 2.5D IC packaging technology uses an additional silicon interposer as a platform to interconnect and integrate heterogeneous or homogeneous chips horizontally and vertically before flip-chip bonding to the substrate [1,3]. Such technology gained more attention in advanced IC packaging for heterogeneous integrations recently. For strain measurement, electrical resistance strain gauges were widely used for point (or local) mechanical strain measurements for more than a century in various engineering applications [8]. Its advantage over other full-field optical methods such as shadow moiré [4,9,10], Twyman–Green interferometry [9,11], or digital image correlation (DIC) [12,13] is that it is easy to use and can provide in-situ and real-time thermal strain measurements, especially for complex printed circuit boards (PCB) during the heating at a solder reflow oven [14]. Some preliminary results of thermal deformations of the 2.5D IC package were presented in an international conference [15]. In this study, the conventional strain gauges were employed for measuring the curvatures of the flip-chip package and the 2.5D IC package reinforced with a metal frame (as shown in Figure 1b,c) under thermal loads. Then, those curvature data were converted to deformations (or warpages) data using a beam model theory. Note that the thermal deformation of the package here is presented by its out-of-plane displacement referred to fixed center point, while the warpage is the out-of-plane displacement of a certain point. Feasibility and validity of this strain gauge method are thoroughly discussed herein by comparing the obtained thermal deformation and warpage results with those from the full-field shadow moiré and the FEM analysis in this study.

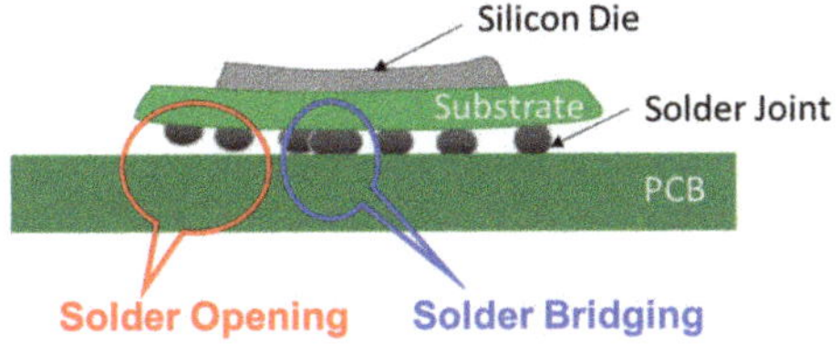

I. Concave Warpage of BGA Package

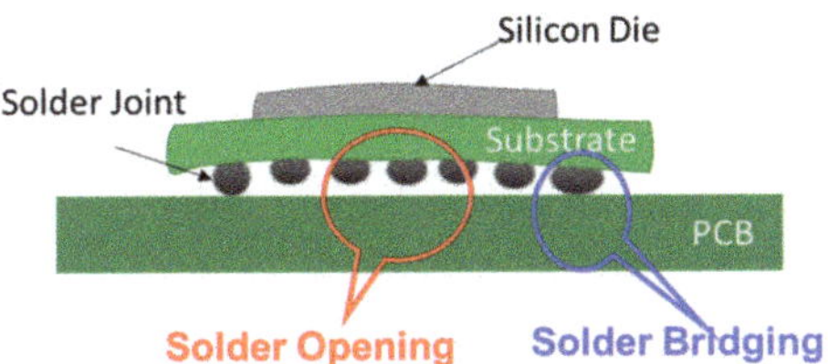

II. Convex Warpage of BGA Package

(**a**)

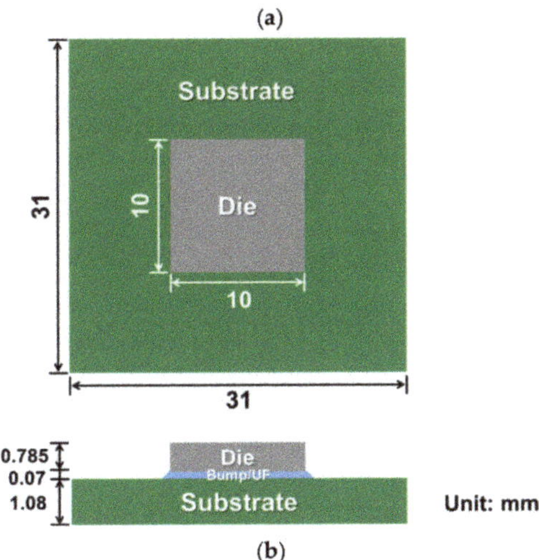

(**b**)

Figure 1. *Cont.*

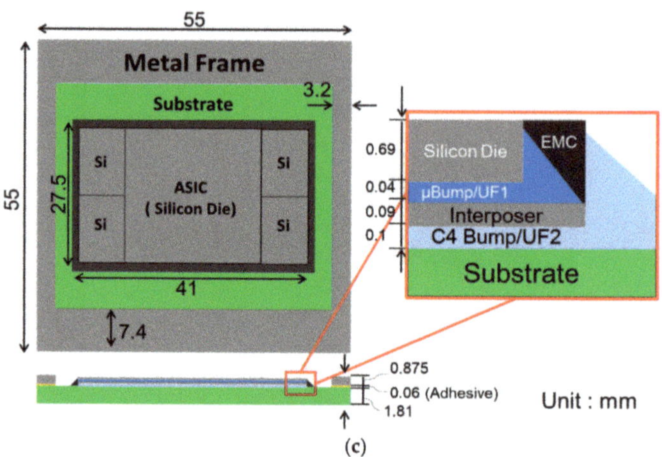

Figure 1. Schematics of (**a**) thermal warpage induced solder joint defects, (**b**) a flip-chip package and (**c**) a 2.5D IC package with detailed materials and dimensions.

2. Methodologies

The methods used in this study for determining the thermal deformations and warpages of the IC packages are briefly illustrated in this section, including the combination of strain gauge and a beam model theory, the finite element analysis, and a shadow moiré measurement.

2.1. Deformation Measurement by Stain Gauges Associated with A Beam Model

The typical strain gauge measurement [8] can be described as

$$\varepsilon_a = \varepsilon_t + \frac{(\gamma_g + \gamma_w)}{S_g}\Delta T \tag{1}$$

where ε_a is an apparent strain which is directly obtained from strain gauge measurement system, ε_t is a true strain which is an actual strain on the measured point of the specimen, S_g is a gauge factor, ΔT is thermal loading, and γ_g and γ_w are the temperature coefficients of resistivity of gauge-grid metal material and connected lead wire, respectively. Back-to-back strain gauges were adhered at a specific position on both top and bottom surfaces of the IC package for measuring the thermal strains on the package during heating and cooling processes. The apparent strains on the top and the bottom surfaces ($\varepsilon_{a,top}$ and $\varepsilon_{a,bot}$) of the package can be written individually in terms of the true strains on the top and the bottom surfaces ($\varepsilon_{t,top}$ and $\varepsilon_{t,bot}$) as

$$\varepsilon_{a,top} = \varepsilon_{t,top} + \frac{(\gamma_g + \gamma_w)}{S_g}\Delta T \tag{2}$$

$$\varepsilon_{a,bot} = \varepsilon_{t,bot} + \frac{(\gamma_g + \gamma_w)}{S_g}\Delta T \tag{3}$$

The strain data at various temperatures could be further converted into bending strains (ε_b) and bending curvatures (k) of the IC package by Equations (4) and (5), respectively. Then, the out-of-plane displacement (deformation) of the IC package under different temperatures could be further calculated by Equation (6) with a given curvature and distance x from the center.

$$\varepsilon_b = (\varepsilon_{a,bot} - \varepsilon_{a,top})/2 = (\varepsilon_{t,bot} - \varepsilon_{t,top})/2 \tag{4}$$

$$k = 2\varepsilon_b/t = (\varepsilon_{t,bot} - \varepsilon_{t,top})/t \qquad (5)$$

$$W = kx^2/2 \qquad (6)$$

where t is the package thickness, and W is the out-of-plane displacement (deformation) at the given distance x with a constant curvature k.

Since the curvature of the package may not be constant across the entire surface of the package, a beam model theory, as shown in Figure 2, with multiple curvatures for the out-of-plane displacement calculation is proposed in this study. This model is based on a beam theory with an assumption of a small deflection. The k_i is a constant curvature on the segment i between the length l_{i-1} and l_i. The k_1 represents the curvature of the center segment. The associated equations of an out-of-plane displacement (or deflection) W_i within the segment i are listed [14] as follows:

$$W_0 = 0, x = 0 \qquad (7a)$$

$$W_1(x) = \frac{1}{2}k_1x^2, \; 0 \leq x \leq l_1 \qquad (7b)$$

$$W_2(x) = W_1(l_1) + k_1l_1(x - l_1) + \frac{1}{2}k_2(x - l_1)^2, \; l_1 \leq x \leq l_2 \qquad (7c)$$

$$W_3(x) = W_2(l_2) + [k_1l_1 + k_2(l_2 - l_1)](x - l_2) + \frac{1}{2}k_3(x - l_2)^2, \; l_2 \leq x \leq l_3 \qquad (7d)$$

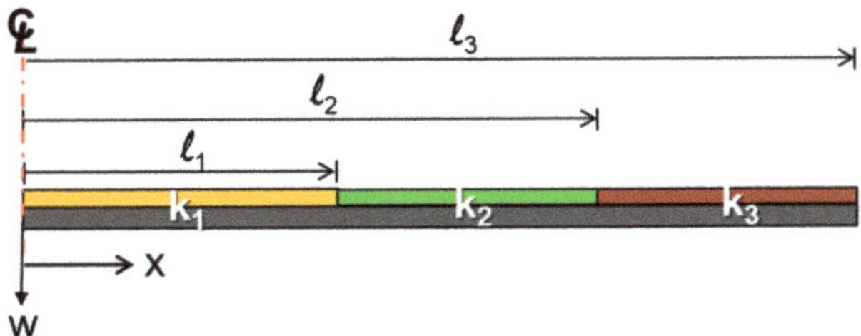

Figure 2. Schematic of the beam with constant curvatures k_1, k_2, and k_3 at different segments.

2.2. FEM Simulation

An isothermal linear FEM analysis was performed to calculate the thermally induced package deformation and warpage due to the CTE mismatch between the inherent materials properties. The schematics of the FEM models are also shown in Figure 3a,b for the flip-chip and the 2.5D IC packages, respectively, with related boundary conditions and meshes in a quarter model. Material mechanical properties used in the finite element analysis are shown in Table 1, in which most data are provided by material vendors. Since the obtained strain data from the FEM analysis were in the x–y coordinate, the transformation of strain vector $\{\varepsilon\}_x$ from the x-y coordinate to $\{\varepsilon\}_{1,\theta}$ in the 1–2 coordinate with a rotation angle of θ was needed during the data process and is illustrated in Figure 4, in which [T] is a transformation matrix as a function of θ. It is noted that this strain transformation was used for calculating a normal (or axial) strain in the 1-axis direction, such as the diagonal direction of the package or the strain gauge direction during the data process. Then, the FEM simulation results were further used to compare with those from strain gauge measurements in terms of curvatures and from moiré measurement in terms of deformations for both packages. In addition, there were two 2D FEM models—plane stress and axisymmetric models—implemented in this study for validating the beam model theory.

2.3. Shadow Moiré Measurements

The shadow moiré method [4,9,10] is widely used to measure the out-of-plane displacement of the specimens. The systems with the sensitivity of 12.7 µm/fringe and

25.4 µm/fringe were used for the thermal deformation measurement of the flip-chip and the 2.5D IC packages, respectively. An oven or a hot plate were used to heat the specimens and provided thermal loading from room temperature to 260 °C. The present thermal cycling test plus data recording in general took about one hour from room temperature to 260 °C and longer for cooling by natural cooling. Two test samples, a flip-chip package with a size of $31 \times 31 \times 1.94$ mm^3 and a 2.5D IC package with a size of $55 \times 55 \times 2.73$ mm^3 (shown in Figure 1b,c), were tested in the moiré experiments by measuring surfaces on their substrates in this study.

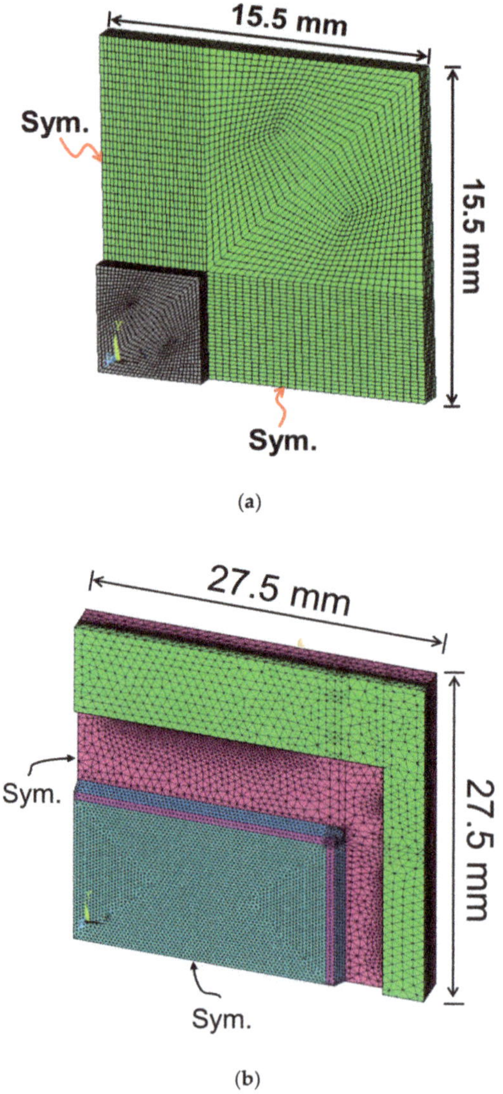

Figure 3. A quarter model and meshes of (**a**) a flip-chip package and (**b**) a 2.5D IC package used in the finite element analysis.

Table 1. Material mechanical properties for flip-chip and 2.5D IC packages used in the finite element analysis and provided by material vendors (* note that the effective composite material properties were used based on the rule of mixture).

Package Type	Material	E (MPa)	ν	α (ppm/°C)	Tg (°C)
	Si	131,000	0.28	2.8	-
Flip-chip Package	Bump/UF *	7600/100	0.32	29/98	120
	Substrate	26,000	0.39	14.4	-
	Si	169,000	0.28	2.3	-
	EMC	23,000/2300	0.69	10.8/24	140
	μBump/UF1 *	7200/72	0.28	28/97	140
2.5D IC Package	C4 Bump/UF2 *	7600/76	0.28	29/98	140
	Substrate	20,000	0.42	13.2	
	Metal Frame	117,000	0.3	16.9	
	Adhesive	100/20	0.3	125/170	50

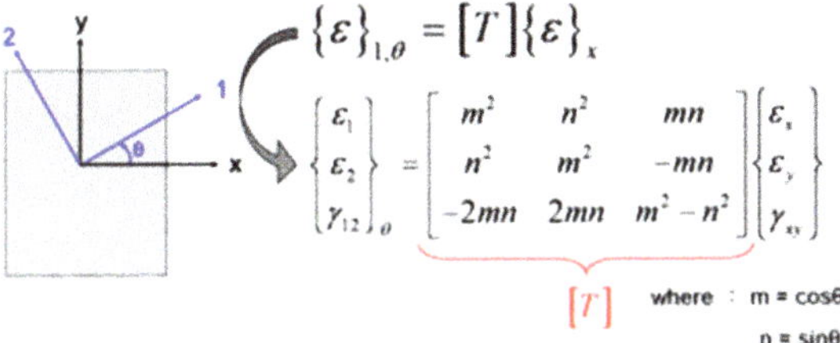

Figure 4. Transformation of strain vector $\{\varepsilon\}_x$ from the x–y coordinate to $\{\varepsilon\}_{1,\theta}$ in the 1–2 coordinate with a rotation angle of θ.

3. Results and Discussion

The obtained results of the flip-chip package and the 2.5D IC package are extensively discussed individually in this section, including the beam model verification, the strain gauge measurement, the moiré measurement, and the FEM analysis.

The Case of Flip-Chip Package

(a)   Validation of beam model with FEM simulation

A beam with two curvatures (k_1 and k_2) over the entire length was considered. A schematic of the out-of-plane displacements of the beam with given constant curvatures k_1 and various k_2 is shown in Figure 5. The normalized curvature was $\lambda = k_2/k_1$ and the normalized beam length was $\beta = l_2/l_1$. From the beam theory (Equations (7b) and (7c)), $W_2(l_2)_{\lambda=1}$ and $W_2(l_2)_\lambda$, representing the out-of-displacements (warpages) for $\lambda = 1$ and any value, respectively, at $x = l_2$, could be described as follows:

$$W_2(l_2)_{\lambda=1} = \frac{1}{2}k_1 l_2^2 \tag{8}$$

$$W_2(l_2)_\lambda = \frac{1}{2}k_1 l_1^2 + k_1 l_1 (l_2 - l_1) + \frac{1}{2}k_2(l_2 - l_1)^2 \tag{9}$$

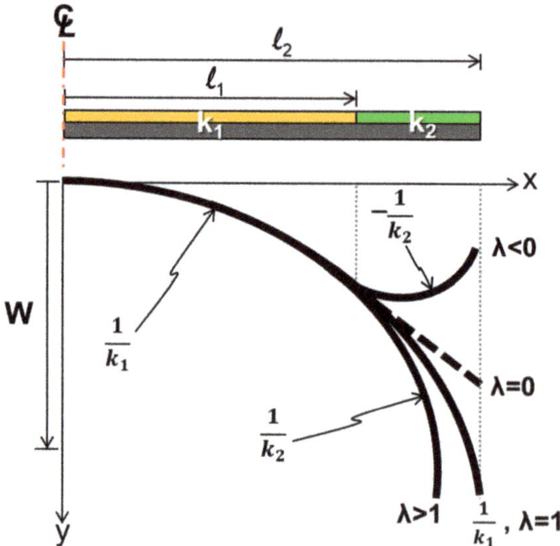

Figure 5. Schematic of the out-of-plane displacement of the beam with given constant curvatures k_1 and various k_2 (or various $\lambda = k_2/k_1$).

It can be also seen from Figure 5 that $W_2(l_2)_\lambda$, the out-of-displacement at $x = l_2$, was larger than $W_2(l_2)_{\lambda\,=\,1}$ when $\lambda > 1$, otherwise it was smaller. Thus, in order to reduce the warpage of the beam, the $\lambda < 1$ had to be selected. That meant the beam with $k_2 < k_1$ was preferred. In addition, from Equation (9), the curvature k_1 played a more important (or dominant) role than k_2 in the value of $W_2(l_2)_\lambda$ because it had an additional slope term $k_1 l_1(l_2 - l_1)$ affecting $W_2(l_2)_\lambda$. For parametric demonstrations, the normalized $W_2(l_2)_\lambda$ by the value of $W_2(l_2)_{\lambda\,=\,1}$ is shown as

$$\frac{W_2(l_2)_\lambda}{W_2(l_2)_{\lambda=1}} = \left(\frac{l_1}{l_2}\right)^2 + \frac{2l_1(l_2 - l_1)}{l_2^2} + \frac{k_2}{k_1}\frac{(l_2 - l_1)^2}{l_2^2} \tag{10}$$

The above equation could be further expressed in terms of λ and β, as described by

$$\frac{W_2(l_2)_\lambda}{W_2(l_2)_{\lambda=1}} = \frac{2}{\beta} - \frac{1}{\beta^2} + \lambda\left(1 - \frac{1}{\beta}\right)^2 \tag{11}$$

For validation of this beam theory, Equation (11) is plotted against β and λ in Figure 6 and was compared with the results from the plane stress and the axisymmetric models of 2D FEM analyses. It is shown that the results from the beam theory and the plane stress model were very consistent, but they were slightly different from the axisymmetric model, which represented a plate (with a Poisson effect) rather than a beam (without a Poisson effect). Moreover, for $\lambda = 1$, the results from the beam theory and the plane stress model were close to those from the axisymmetric model; for $\lambda < 1$, they were overestimated, but for $\lambda > 1$, they were underestimated. In order to understand this mechanism, curvature distributions for the $\beta = 2$ case from the beam theory are plotted in Figure 7 along the entire length of the beam with various values of normalized curvature (λ) in comparison with those from the 2D plane stress and the axisymmetric models of the FEM analyses. It was found that, unlike consistent and constant curvatures (k_1 and k_2) over each segment between the beam theory and the plane stress model, the axisymmetric model (representing the plate case) gave smaller curvature k_1 and non-uniform k_2 values as $\lambda < 1$. In other words, the out-of-plane displacement (or warpage) caused by smaller curvature k_1 and non-uniform k_2 resulting from the Poisson effect in the axisymmetric model was lower than that in the beam theory and the plane stress model. As a result, both beam and plane stress models gave an overestimated value of the displacement or the warpage as $\lambda < 1$. However, for $\lambda > 1$, this was reversed. Furthermore, to examine the detailed warpages, the out-of-displacement of the beam along the entire length of the beam for the $\beta = 2$ case is plotted for those three models in Figure 8. It can be also seen that the out-of-plane displacement curves were very consistent for the three models for $\lambda = 1$, but the curves from both beam theory and plane stress model started deviating from those in the axisymmetric model with increasing values for $\lambda < 1$ and decreasing values for $\lambda > 1$. Based on the above observations, it was proven that the beam theory described in Equation (7) coincided with the 2D plane stress model but had some deviations from the 2D axisymmetric model due to the Poisson effect. The question is whether the beam theory can be used for calculating the out-of-displacement of the plate-like IC packages with the curvature data obtained from the strain gauge measurements. That is answered later in this paper.

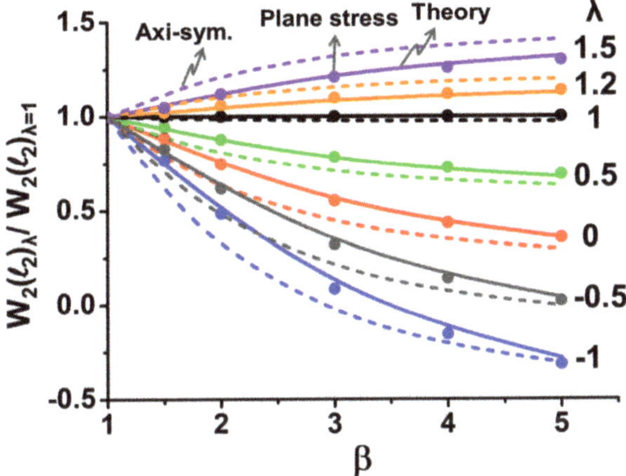

Figure 6. Normalized warpage $(W_2(l_2)_\lambda / W_2(l_2)_{\lambda = 1})$ vs. normalized beam length $(\beta = l_2/l_1)$ for the beam theory (in solid lines) with various values of normalized curvature $(\lambda = k_2/k_1)$ compared with the results from the plane stress (in dot points) and the axisymmetric models (in dash lines) of 2D FEM analyses.

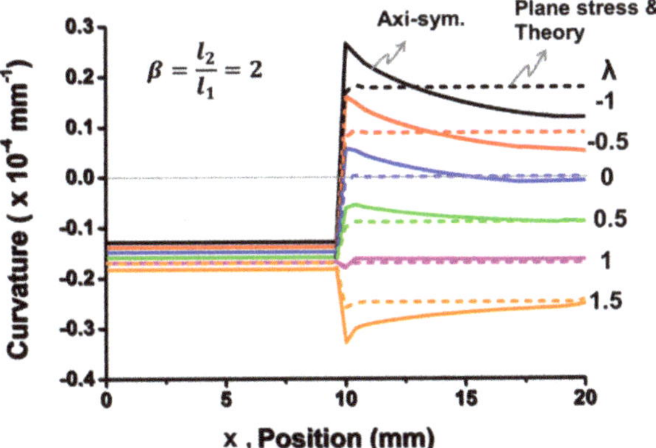

Figure 7. Curvature distributions along the length of the beam for the $\beta = 2$ case from the beam theory (in dash lines) with various values of normalized curvature (λ), compared with the results from plane stress (in dash lines) and axisymmetric models (in solid lines) of 2D FEM analyses.

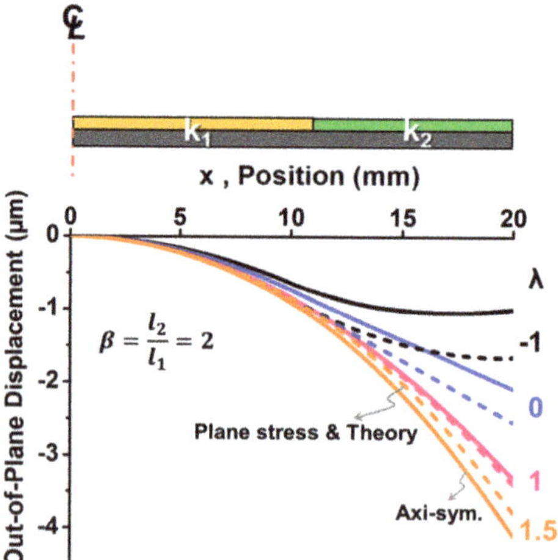

Figure 8. Out-of-plane displacement of the beam along the length of the beam for the $\beta = 2$ case from the beam theory (in dash lines) with various values of normalized curvature (λ) compared with the results from 2D plane stress (in dash lines) and axisymmetric (in solid lines) models of FEM.

(b)   Validation of FEM simulation with shadow moiré

To verify the result from the 3D FEM analysis, the out-of-plane displacements of the flip-chip package along the diagonal line *ac* at room temperature (T = 25 °C) and T = 260 °C are shown in Figure 9 from the 3D FEM analysis and the moiré measurement. It can be seen that the 3D FEM model effectively and precisely described the thermal deformation of the flip-chip package based on those consistent results. However, whether the 2D FEM (with axisymmetric model) or the strain gauge measurement associated with the beam model theory can do the same is discussed later.

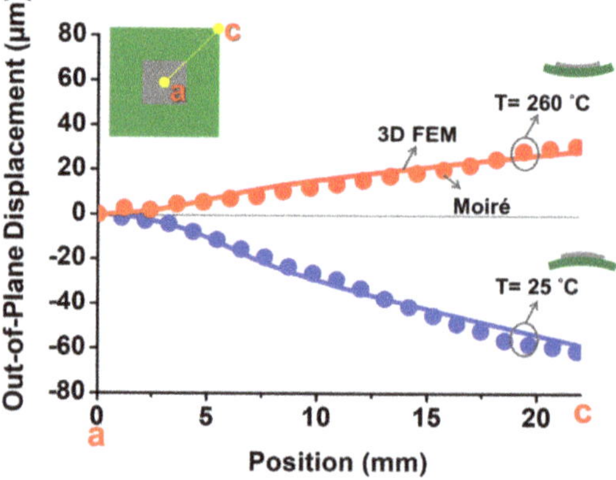

Figure 9. Out-of-plane displacement of the package along the diagonal line *ac* and with a full field at room temperature (T = 25 °C) and T = 260 °C from the 3D FEM analysis compared with moiré results.

(c)   Implementation of gauge measurement

The back-to-back gauges $G_1/G_{1'}$, $G_2/G_{2'}$, and $G_3/G_{3'}$ were attached on the top and the bottom surfaces of the flip-chip package at the points *a*, *b*, and *c* (representing center point, near die corner, and substrate corner points), respectively, for the strain measurement. The measured strain data with various temperatures are shown in Figure 10. The data indicate that the difference of the strain data between the bottom and the top gauges was large at point a (the center point), while it was small at point c (near substrate corner). That meant there existed the bending strains (ε_b) with a large value at the center point, an intermediate value at the near die corner, and a small value near the substrate corner based on Equation (4). Furthermore, by Equation (5), those bending strain data could be converted to the curvature data, which are shown in Figure 11a, for various temperatures with curvatures k_a, k_b, and k_c at the measured points a, b, and c, respectively. It was found that there existed a kinked point on the curve of k_a but not on other curves near T = 120 °C, which was the glass transition temperature (T_g) of the underfill material. This resulted from the elastic modulus (E) of the underfill material with about two orders of magnitude decrease at a temperature above its T_g [5]. Additionally, at this T_g, the curvature on every point on the flip-chip package was almost zero based on the observation of flatness of the package in moiré experiments. Thus, all curvature curves could be further shifted to T = 120 °C with the predefined zero curvature (shown in Figure 11b). It was evident that there existed the apparent value of k_b (near the corner of the chip) but with the close zero value of k_c near the substrate corner. This was due to the Poisson effect, which was consistent with the case of $\lambda = 0$ with the axisymmetric model in Figure 7. Moreover, the curvature data from T = 120 °C cooling to 25 °C in Figure 11b were used to calculate the out-of-plane displacement of the package at room temperature (T = 25 °C) along the diagonal line ac using Equation (7), and this displacement result obtained from the strain gauge was compared with those from moiré and FEM (with 2D axisymmetric and 3D models), as shown in Figure 12. It was shown that, due to an inherent limitation of the beam theory (Equation (7)), the displacements obtained from strain gauge were closer to those of the 2D axisymmetric FEM model than those from moiré and 3D FEM models. Furthermore, the thermally induced warpages of the package at the point c (the corner of the package) from the strain gauge measurement are plotted in Figure 13 against temperature in comparison with those from moiré data and FEM results (with 2D axisymmetric and 3D models). Results also indicated that the gauge result was consistent with the 2D axisymmetric result but was slightly off from results of moiré and 3D FEM models. However, in the engineering applications, the strain gauge measurement associated with the beam model theory was accordingly proved to be feasible and good enough to determine the thermally induced out-of-plane deformations and warpage of the flip-chip packages.

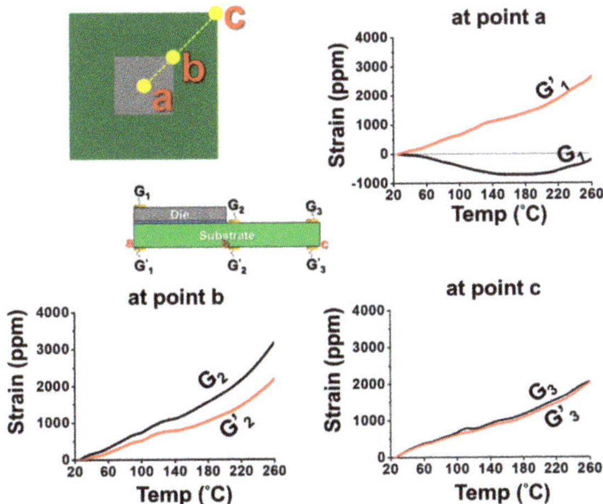

Figure 10. The strain data from the back-to-back gauges with various temperatures and with gauges $G_1/G_{1'}$, $G_2/G_{2'}$, and $G_3/G_{3'}$ at the points a, b, and c (representing center point, near die corner, and substrate corner points), respectively, during the strain measurement under thermal loading from 25 °C to 260 °C in heating.

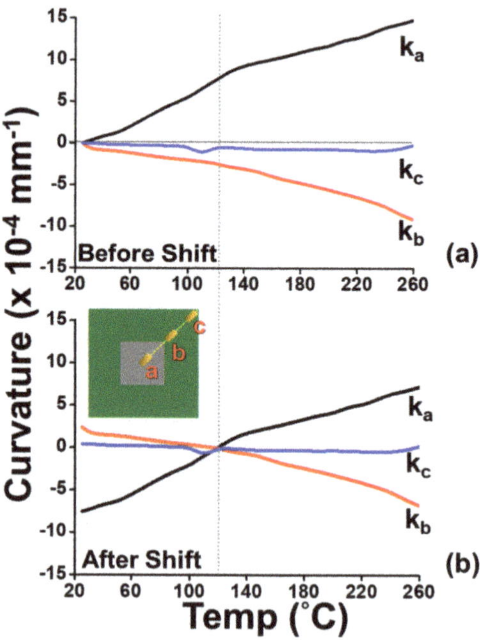

Figure 11. The curvature data at the points *a*, *b*, and *c* (representing center point, near die corner, and substrate corner points, respectively) with various temperatures from the strain gauge measurement (**a**) before shift and (**b**) after shift to 120 °C with the zero curvature.

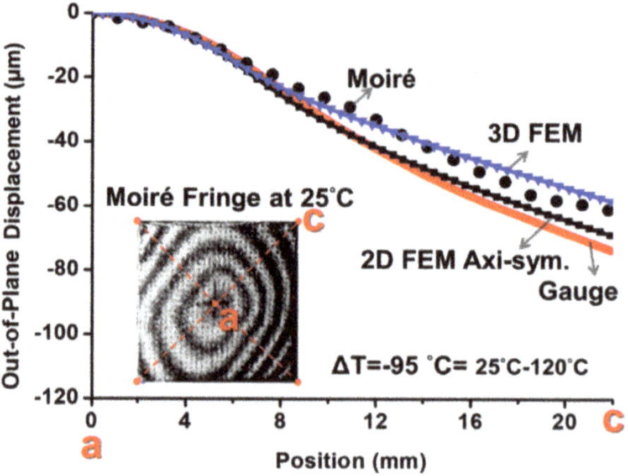

Figure 12. Out-of-plane displacement of the package at room temperature (T = 25 °C) along the diagonal line *ac* from the strain gauge measurement compared with the results from moiré data (with a sensitivity of 12.7 μm/fringe) and FEM results (with 2D axisymmetric and 3D models).

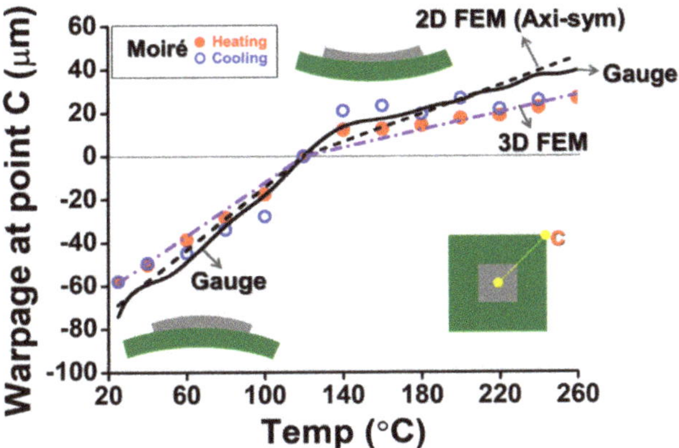

Figure 13. The thermally induced warpage of the package at the point c (the corner of the package) with various temperatures from the strain gauge measurement compared with the results from moiré data and FEM results (with 2D axisymmetric and 3D models).

4. The Case of 2.5D IC Package

(a)   Validation of FEM simulation

Moiré measurement of the 2.5D IC package with a metal frame under thermal loading from 25 °C to 260 °C in heating and cooling was performed, and its moiré fringe patterns (out-of-plane displacement contours with a sensitivity of 25.4 µm/fringe) are shown in Figure 14 [16]. It was obvious that, during the heating process, the spherically convex shape of the specimen at 25 °C became less warped as the temperature increased and then turned out to be flat at 140 °C. Upon continuous heating, the specimen became concave above 140 °C and increasingly up to 260 °C. On the other hand, the specimen deformed in the reverse way during the cooling process and then came back to the same deformed shape at 25 °C. Such thermal elastic deformations of the package along the diagonal line *oa* at 25 °C and 260 °C are plotted and shown in Figure 15 in detail in comparison with those from the FEM analysis. The consistent results between moiré and FEM analysis indicated that the FEM model was valid and precise enough to describe the thermal deformations of the 2.5D IC package.

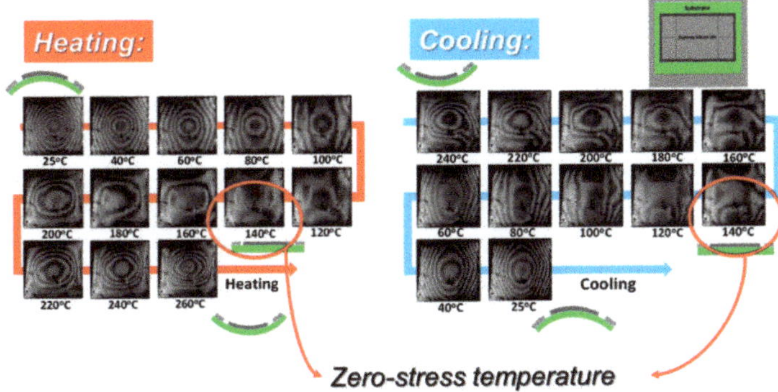

Figure 14. Moiré fringe patterns of the 2.5D IC package with a metal frame under thermal loading from 25 °C to 260 °C in heating and then from 260 °C to 25 °C in cooling (with a sensitivity of 25.4 µm/fringe) [16].

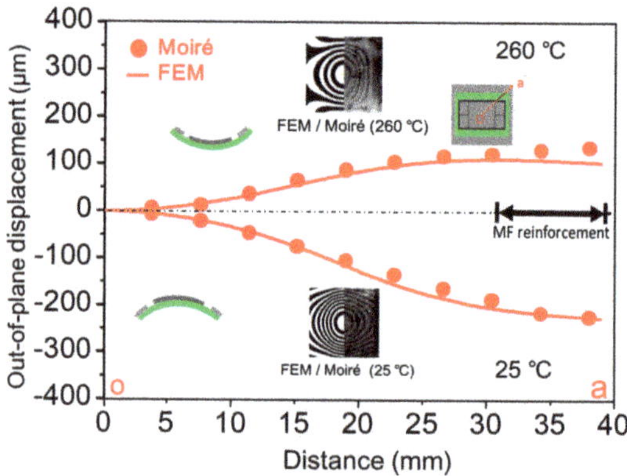

Figure 15. Out-of-plane deformation along the diagonal line *oa* for 2.5D IC package at the temperatures of 25 °C and 260 °C from moiré and FEM results.

(b)　Beam model verification

After the FEM model was validated, this model was then employed for verifying the beam model theory, which was applied to calculate the thermal deformations of the 2.5D IC package with the strain gauge data later. Curvature distribution along the diagonal line *oa* of the package under thermal loading $\Delta T = -115\ °C$ (from 140 °C cooling down to 25 °C) from the FEM analysis is shown in Figure 16a with the average curvatures of k_1, k_2, and k_3 over the gauge length at the corresponding segment. These gauge-assumed curvature data extracted from the FEM model were put into the beam model theory as described in Equation (7) to calculate the out-of-plane displacement. The thermally induced out-of-plane displacement of the 2.5D IC package obtained is shown in Figure 16b compared with that from the FEM analysis. The almost identical results of both methods revealed that the method of using the point-wise data of strain gauge associated with the beam model theory was feasible for measuring the thermal deformation of the 2.5D IC package.

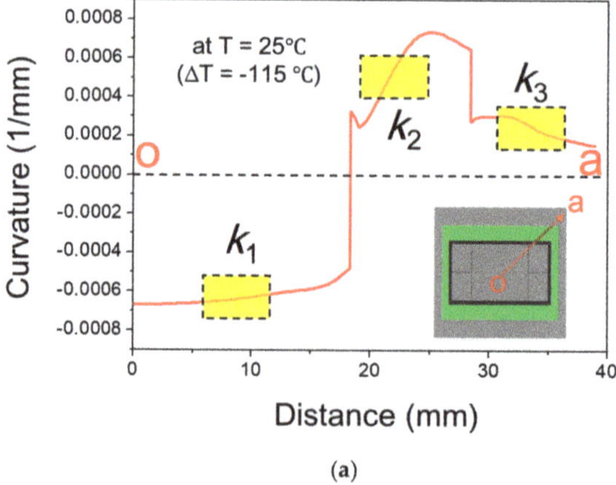

(a)

Figure 16. *Cont.*

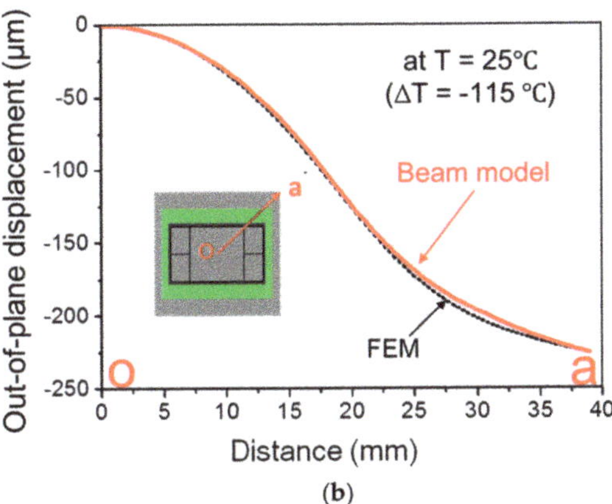

(b)

Figure 16. (a) Curvature distribution along the diagonal line *oa* of the 2.5D IC package and the average curvatures k_1, k_2, and k_3 over the gauge length at each segment from FEM analysis, and (b) comparison of out-of-plane displacements of the 2.5D package from FEM and beam model associated with the average curvatures k_1, k_2, and k_3.

(c)   Implementation of strain gauge measurement

The strain gauge measurement was further carried out for determining the bending curvatures and deformations of the 2.5D IC package under thermal loading from 140 °C to 25 °C in the cooling process. Note that the strain gauge readings were reset at zero at 140 °C. In Figure 17, the back-to-back strain gauges were attached at the centers of lines *AB*, *BC*, and *CD* along half of a diagonal line with the gauge pairs of G1/G2, G3/G4, and G5/G6, respectively. The axial thermal strains of each reading from those six gauges are also shown in Figure 17 in comparison with those from the FEM simulation. The consistency between both data indicated that the strain gauges could measure the thermal strains of the package. The bending curvature data under this thermal loading were further extracted using Equations (4) and (5) and are shown in Figure 18 in a detailed comparison with the FEM results. Almost identical results revealed that the curvature data for k_1, k_2, and k_3, corresponding to those on lines *AB*, *BC*, and *CD*, increased positively for k_2 and k_3 but increased negatively for k_1 in the cooling process. Their maximum curvatures occurred at 25 °C. The differences of the maximum curvatures between strain gauge measurement and FEM simulation were within 8% (listed in the left table). Moreover, the out-of-plane displacement at 25 °C was plotted using Equation (7) of the beam model theory associated with the curvature data (k_1, k_2, and k_3) from the gauge measurement and is shown in Figure 19 in comparison with those from shadow moiré measurement and FEM simulation. The results indicated that the out-of-plane displacement of the 2.5D IC package from gauge measurement reasonably agreed with the other two with a minor difference. The warpages of the 2.5D IC package at various temperatures are further plotted in Figure 20 from the strain gauge measurement, the shadow moiré measurement, and the FEM simulation. Those results also showed a reasonably good agreement between each other. Overall, this study demonstrated that the strain gauge measurement associated with the beam model theory can be used for characterizing the out-of-plane thermal deformation and warpage of the 2.5D IC package.

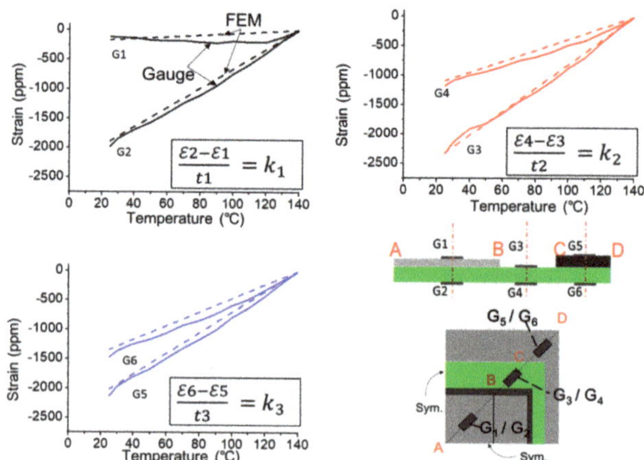

Figure 17. Axial strains from gauge measurements and FEM simulation along the diagonal line of the 2.5D IC package under thermal loading from 140 °C to 25 °C in cooling.

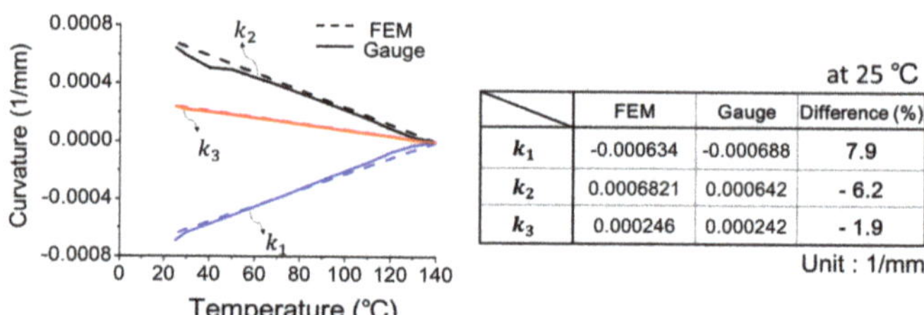

	FEM	Gauge	Difference (%)
k_1	-0.000634	-0.000688	7.9
k_2	0.0006821	0.000642	- 6.2
k_3	0.000246	0.000242	- 1.9

Unit : 1/mm

Figure 18. Comparison of the curvature data (k_1, k_2, and k_3) extracted from strain gauge measurement and FEM simulation for the 2.5D IC package at various temperatures.

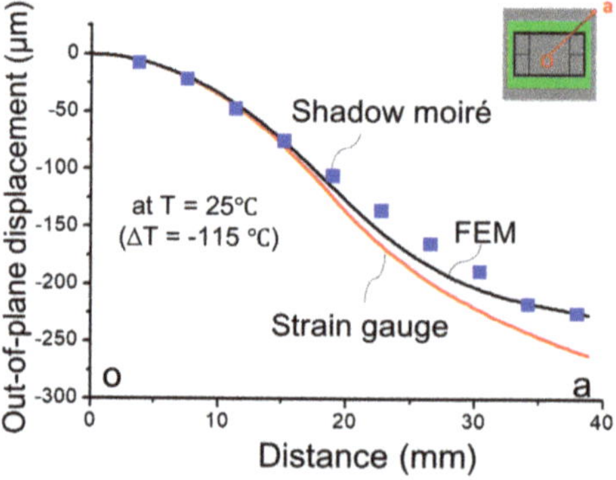

Figure 19. Comparison of the out-of-plane displacement of the 2.5D IC package at temperature of 25 °C ($\Delta T = -115$ °C) from strain gauge measurement, shadow moiré, and FEM simulation.

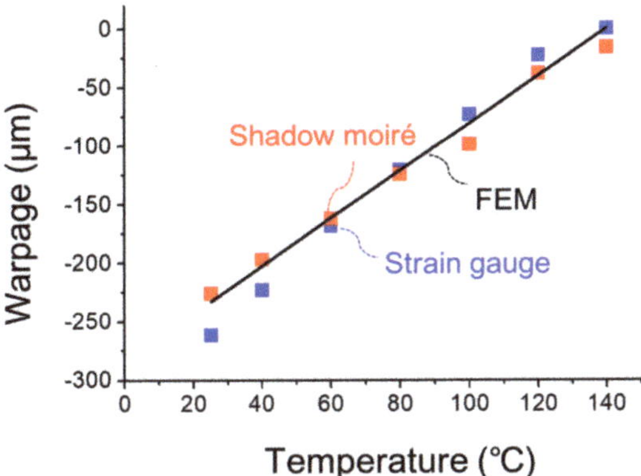

Figure 20. Comparison of warpages of the 2.5D IC package at various temperatures from strain gauge measurement, shadow moiré, and FEM simulation.

(d) Effect of segment curvature variation on warpage

It is interesting to know how the curvature data k_1, k_2, and k_3 at each segment affected the warpage of the 2.5D IC package. The plus and the minus 10% variations of each curvature data in Figure 16a used in plotting displacement using the beam model were analyzed, and the obtained warpage results and their difference are listed in Table 2. Note that k_1, k_2, and k_3 were normalized values and defined as unity for reference. It could be seen that the warpage was the most sensitive to k_1, rather than k_2 and k_3, with increases or decreases of ~15% of warpage value by changing $\pm$10% of k_1 values, respectively. This was consistent with the above-mentioned finding that k_1, the curvature of the center segment, was a dominant value in the warpage of the packages. Therefore, the precisely defined or measured k_1 was more important than the other two (k_2 and k_3) for the warpage calculation or the measurement of the 2.5D IC package.

Table 2. The warpage difference of the 2.5D IC package with a variation of segment curvatures at each segment.

k_1	1	1.1	1	1	0.9	1	1
k_2	1	1	1.1	1	1	0.9	1
k_3	1	1	1	1.1	1	1	0.9
Warpage (μm)	-226	-260	-215	-224	-191	-236	-227
Difference (%)	0	15.34	-4.75	-0.60	-15.35	4.75	0.60

(e) Effect of gauge misalignment on warpage

Since the precise measurement of k_1 was critical for determining the warpage of the 2.5D IC package, it was desired to understand how the misalignment of k_1 gauges affected the warpage. Various angle misalignments (with $\pm$5° and $\pm$10°) of top and bottom strain gauges in the 45° direction in segment AB (the center segment) were analyzed using a beam model for calculating the package warpage through the measurement of k_1 curvatures. The results are shown in Table 3, in which the warpage determined from the top and the bottom gauges in the angle of 45° was used as a based (or reference) value. It could be seen that, for $\pm$5° misalignments, the obvious warpage differences occurring at the angle pairs of 50°/50° and 40°/40° were 3.57% and -3.76%, while the other pairs were below 3%. However, for $\pm$10° misalignments, the warpage difference increased by more than double. As a result, as long as the gauge alignment on the top and the bottom surfaces was under control with less than $\pm 5°$ misalignments, the results of the warpage measurement were acceptable with less than 5% error.

Table 3. The warpage (with a unit of μm) and the warpage difference of the 2.5D IC package with various angle misalignments ($\pm 5°$ and $\pm 10°$) of top and bottom strain gauges for measuring k_1 curvatures in segment AB (the center segment).

$\pm 5°$		Top Strain Gauge		
	Degree	50°	45°	40°
Bottom strain gauge	50°	−234 (3.57%)	−231 (2.38%)	−229 (1.3%)
	45°	−228 (0.96%)	−226 (Reference)	−223 (−1.31%)
	40°	−222 (1.49%)	−220 (−2.68%)	−217 (−3.76%)
$\pm 10°$		Top Strain Gauge		
	Degree	55°	45°	35°
Bottom strain gauge	55°	−243 (7.49%)	−237 (5.06%)	−233 (3.02%)
	45°	−231 (2.21%)	−226 (Reference)	−221 (−2.26%)
	35°	−220 (−2.48%)	−215 (−4.91%)	−210 (−6.95%)

5. Conclusions

This study proposed the strain gauge method associated with a beam model theory for determining the thermally induced deformations and warpages of the flip-chip and the 2.5D IC packages. The beam model theory was thoroughly evaluated in this study, and it was found that the theory coincided with the 2D plane stress model but had some deviations from the 2D axisymmetric model due to the Poisson effect. The finite element method (FEM) resulted with good consistency with the full-field shadow moiré experiment, showing good prediction of thermally induced out-of-plane deformations for both packages. Furthermore, the strain gauge measurement with the beam model theory was actually implemented for both packages. It was found that the obtained thermal strain data were in good agreement with the FEM data. The gauge-determined thermal deformations and warpages of both packages also showed reasonably good agreement with those from the FEM analyses and the shadow moiré measurements. Moreover, the curvature in the center segment of the 2.5D IC package was found to be a dominant value in control of thermal warpages of the package. The gauge misalignment effect was also evaluated and discussed in detail. Overall, it was proven that the strain gauge method of featuring point strain measurement associated with the beam model theory can be feasible for measuring the thermal deformations and warpages of both packages.

Author Contributions: Conceptualization, M.-Y.T., Y.-W.W. and C.-M.L.; Methodology, M.-Y.T., Y.-W.W. and C.-M.L.; Software, Y.-W.W. and C.-M.L.; Validation, M.-Y.T., Y.-W.W. and C.-M.L.; Formal Analysis, Y.-W.W. and C.-M.L.; Investigation, M.-Y.T., Y.-W.W. and C.-M.L.; Resources, M.-Y.T.; Data Curation, Y.-W.W. and C.-M.L.; Writing—Original Draft Preparation, M.-Y.T.; Writing—Review & Editing, M.-Y.T.; Visualization, M.-Y.T.; Supervision, M.-Y.T.; Project Administration, M.-Y.T.; Funding Acquisition, M.-Y.T. All authors have read and agreed to the published version of the manuscript.

Funding: This study was supported by Ministry of Science and Technology, Taiwan under project MOST-108-2221-E-182-045-MY3.

Institutional Review Board Statement: Not applicable.

Informed Consent Statement: Not applicable.

Data Availability Statement: On inquiry, the data presented in this study is available from the authors.

Acknowledgments: Financial support from Ministry of Science and Technology, Taiwan is gratefully acknowledged.

Conflicts of Interest: The authors declare no conflict of interest.

References

1. Lau, J.H. *Low Cost Flip Chip Technologies*; McGraw-Hill: New York, NY, USA, 2000.
2. Garrou, P. *Introduction to 3D Integration*; Wiley: Hoboken, NJ, USA, 2008; pp. 1–11.
3. Lau, J.H. Critical Issues of 3D IC Integrations. In Proceedings of the IMAPS International Symposium on Microelectronics, San Jose, CA, USA, 1–5 November 2009; pp. 585–592.
4. Tsai, M.-Y.; Chang, H.-Y.; Pecht, M. Warpage Analysis of Flip-Chip PBGA Packages Subject to Thermal Loading. *IEEE Trans. Device Mater. Reliab.* **2009**, *9*, 419–424. [CrossRef]
5. Tsai, M.-Y.; Wang, Y.-W. A Theoretical Solution for Thermal Warpage of Flip-Chip Packages. *IEEE Trans. Compon. Packag. Manuf. Technol.* **2020**, *10*, 72–78. [CrossRef]
6. Zhang, X.; Chai, T.; Lau, J.H.; Selvanayagam, C.S.; Biswas, K.; Liu, S.; Pinjala, D.; Tang, G.; Ong, Y.; Vempati, S.; et al. Development of through silicon via (TSV) interposer technology for large die (21 × 21 mm) fine-pitch Cu/low-k FCBGA package. In Proceedings of the 2009 59th Electronic Components and Technology Conference, San Diego, CA, USA, 26–29 May 2009; pp. 305–312.
7. Lee, C.-C.; Hung, C.; Cheung, C.; Yang, P.-F.; Kao, C.-L.; Chen, D.-L.; Shih, M.-K.; Chien, C.-L.C.; Hsiao, Y.-H.; Chen, L.-C.; et al. An Overview of the Development of a GPU with Integrated HBM on Silicon Interposer. In Proceedings of the 2016 IEEE 66th Electronic Components and Technology Conference (ECTC), Las Vegas, NV, USA, 31 May–3 June 2016; pp. 1439–1444.
8. Dally, J.W.; Riley, W.F. *Experimental Stress Analysis*, 14th ed.; College House Enterprises: Knoxville, TN, USA, July 2006.
9. Post, D.; Han, B.; Ifju, P. *High Sensitivity Moiré Experimental Analysis for Mechanics and Materials*; Springer: New York, NY, USA, 1993.
10. Tsai, M.-Y.; Chen, Y.-C.; Lee, S. Correlation Between Measurement and Simulation of Thermal Warpage in PBGA With Consideration of Molding Compound Residual Strain. *IEEE Trans. Compon. Packag. Technol.* **2008**, *31*, 683–690. [CrossRef]
11. Tsai, M.-Y.; Hsu, C.; Wang, C. Investigation of Thermomechanical Behaviors of Flip Chip BGA Packages During Manufacturing Process and Thermal Cycling. *IEEE Trans. Compon. Packag. Technol.* **2004**, *27*, 568–576. [CrossRef]
12. Park, S.; Dhakal, R.; Joshi, R. Comparative Analysis of BGA Deformations and Strains Using Digital Image Correlation and Moiré Interferometry. In Proceedings of the Sem Annual Conference and Exposition on Experimental and Applied Mechanics, Portland, OR, USA, 7–9 June 2005.
13. Kim, Y.; Park, A.-Y.; Kao, C.-L.; Su, M.; Black, B.; Park, S. Prediction of deformation during manufacturing processes of silicon interposer package with TSVs. *Microelectron. Reliab.* **2016**, *65*, 234–242. [CrossRef]
14. Liao, M.C.; Huang, P.S.; Lin, Y.H.; Tsai, M.Y.; Huang, C.Y.; Huang, T.C. Measurements of Thermally-induced Cur-vatures and Warpages of Printed Circuit Board During a Solder Reflow Process Using Strain Gauges. *Appl. Sci.* **2017**, *7*, 739. [CrossRef]
15. Tsai, M.Y.; Liu, G.M.; Wang, Y.W. Measurement of Thermally-Induced Curvatures and Warpages of 2.5D Packages by Using Strain Gauges. In Proceedings of the IMPACT 2019, Taipei, Taiwan, 23–25 October 2019.
16. Tsai, M.-Y.; Liu, C.-M.; Wang, Y.-W.; Liu, H.-Y.; Liu, G.M. Effects of Metal Frame and Adhesive on Thermally-Induced Warpage and Stress of 2.5D Packages: Experimental and Numerical Studies. *IEEE Trans. Device Mater. Reliab.* **2018**, *18*, 450–455. [CrossRef]

 *materials*

Article

Highly Robust Ti Adhesion Layer during Terminal Reaction in Micro-Bumps

Chen-Wei Kao [1], Po-Yu Kung [1], Chih-Chia Chang [1], Wei-Chen Huang [1], Fu-Ling Chang [1] and C. R. Kao [1,2,*]

[1] Department of Materials Science and Engineering, National Taiwan University, Taipei 10617, Taiwan; r09527a09@ntu.edu.tw (C.-W.K.); a0987700171@gmail.com (P.-Y.K.); rroo1115@gmail.com (C.-C.C.); f08527014@ntu.edu.tw (W.-C.H.); f09527008@ntu.edu.tw (F.-L.C.)

[2] Advanced Research Center for Green Materials Science & Technology, National Taiwan University, Taipei 10617, Taiwan

* Correspondence: crkao@ntu.edu.tw; Tel.: +886-2-33663745

Abstract: The use of scaled-down micro-bumps in miniaturized consumer electronic products has led to the easy realization of full intermetallic solder bumps owing to the completion of the wetting layer. However, the direct contact of the intermetallic compounds (IMCs) with the adhesion layer may pose serious reliability concerns. In this study, the terminal reaction of the Ti adhesion layer with Cu–Sn IMCs was investigated by aging the micro-bumps at 200 °C. Although all of the micro-bumps transformed into intermetallic structures after aging, they exhibited a strong attachment to the Ti adhesion layer, which differs significantly from the Cr system where spalling of IMCs occurred during the solid-state reaction. Moreover, the difference in the diffusion rates between Cu and Sn might have induced void formation during aging. These voids progressed to the center of the bump through the depleting Cu layer. However, they neither affected the attachment between the IMCs and the adhesion layer nor reduced the strength of the bumps. In conclusion, the IMCs demonstrated better adhesive behavior with the Ti adhesion layer when compared to Cr, which has been used in previous studies.

Keywords: micro-joints; solid-state reaction; intermetallic; adhesion layer

Citation: Kao, C.-W.; Kung, P.-Y.; Chang, C.-C.; Huang, W.-C.; Chang, F.-L.; Kao, C.R. Highly Robust Ti Adhesion Layer during Terminal Reaction in Micro-Bumps. *Materials* **2022**, *15*, 4297. https://doi.org/10.3390/ma15124297

Academic Editor: Matthias Kern

Received: 17 May 2022
Accepted: 15 June 2022
Published: 17 June 2022

Publisher's Note: MDPI stays neutral with regard to jurisdictional claims in published maps and institutional affiliations.

1. Introduction

Micro-joints play an important role in three-dimensional integrated circuit (3D IC) technology in the realization of vertical chip stacking. However, the micro-joint volume is significantly little, nearly six orders of magnitude less than that of the conventional ball grid array joints or flip-chip joints and is a critical issue. With the miniaturization of solder joints and thickness reduction of under-bump metallurgy (UBM), it is observed that both Sn and Cu present limited amounts in micro-joints. This results in a large portion, or even the entire micro-joint, being occupied by intermetallic compounds (IMCs) after assembly [1]. Hence, this phenomenon introduces new reliability concerns.

The adhesion layer, such as Cr and Ti, is the main component of the UBM used in the electronics industry. Owing to the rapidly consumed solder joints and wetting layer, the IMCs would attach to the adhesion layer with limited Cu during the solid–liquid reaction. This condition causes spalling, which is reported in numerous studies [2–7]. Spalling of the IMCs causes a significant reduction in bonding strength and affects the reliability of electronic products. In addition, the formation of IMCs in Cu–Sn reactions with limited Sn has been reported previously [1,8–12]. These studies only focused on the micro-structural evolution in the early stages of interfacial reactions, which were abundant with either Cu or Sn during the experiment. However, fewer studies have investigated the same when the constraint volume of Cu and Sn is simultaneously converted into the full IMC joints in the terminal solid-state reaction. In a previous study, Tsai et al. [13] found that when the IMCs were attached to the Cr adhesion layer, Cu_6Sn_5 and Cu_3Sn detached and left the substrate during the terminal solid-state reaction. In addition, this spalling process did not dominate the ripening or gravity effects of Cu_6Sn_5 compared to regular spalling. It was confirmed that the spalling was due to the high interfacial energy between the Cu and Sn intermetallic and Cr. Furthermore, void formation during the reaction was also observed, which was attributed to Cu diffusion. In contrast, the widespread usage of the Ti adhesion layer in the electronics industry is well-known, but no study has discussed the phenomenon when the IMCs contact the Ti adhesion layer during the solid-state reaction.

This study aims to investigate the micro-structure evolution of ultrathin Sn/Cu micro-bumps at the terminal reaction and observe the phenomenon when the IMCs directly contact the Ti layer during the solid-state reaction. In addition, the relationship between the voids and the reactive layer thickness are proposed and illustrated. Furthermore, since IMC formation and void nucleation deteriorate the reliability of micro-bumps [14–19], a die shear test was conducted to study the mechanical properties of the micro-bumps with different aging times. Finally, the correlations between the shear strength and evolution of IMCs during the aging process are discussed.

2. Experimental

Figure 1 presents a schematic of the dimensions of the Sn/Cu/Ti structure used in this study. The detailed procedure for sample preparation has been presented previously [13]. In brief, the samples were first fabricated by electroplating Cu and Sn, sequentially on the Ti adhesion layer, to form micro-bumps and then aged at 200 °C, which could accelerate the IMCs formation to reach the terminal stage of interfacial reaction, for 0, 24, 36, and 42 h to observe the evolution between layers.

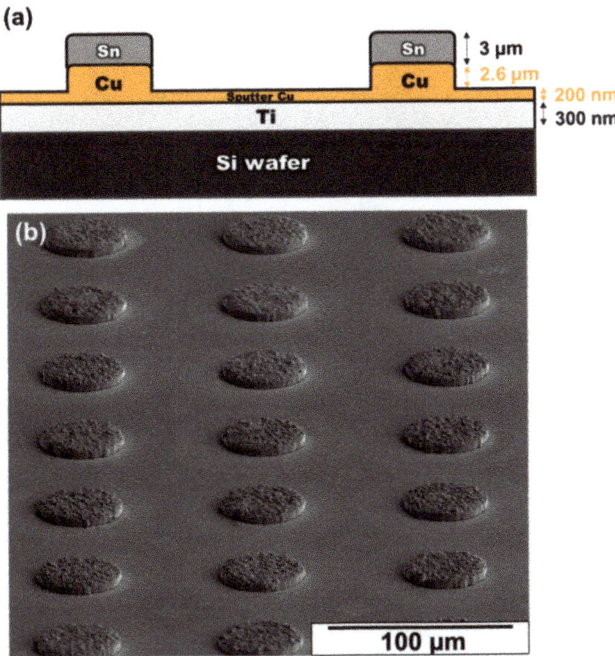

Figure 1. (a) Schematic drawing of the configuration of the Sn/Cu/Ti structure in this study; (b) Micrograph showing the top view of micro-bumps.

After aging, the samples were cross-sectioned and polished using an ion-milling system (Hitachi IM4000Plus) with an Ar^+ ion beam for subsequent analysis. These samples were examined using a scanning electron microscope (SEM, Hitachi SU5000) by BSE detector and transmission electron microscope (TEM, FEI Tecnai G2 F20) to observe the sample morphology and micro-structural evolution of the Cu–Sn IMCs. The chemical composition of the IMCs was analyzed by energy-dispersive X-ray spectrometry (EDX). The shear strengths at different aging times were determined by the die shear tests (Xyztec sigma) using a bonding tester with a 1 µm shear tool height from the substrate and a 10 µm/s shear speed. For each aging time, the corresponding shear strengths for 15 specimens were averaged. After the shear test, the fracture surfaces were analyzed using optical microscopy, SEM and EDX.

3. Results and Discussion

3.1. Micro-Structure Evolution of Sn/Cu Micro-Bumps before and after Aging at 200 °C

Figure 2a shows a backscattered image of the as-fabricated Sn/Cu micro-bump after the electroplating process. The average thicknesses of the Cu and Sn layers were 2.6 µm and 3 µm, respectively. In this stage, a thin Cu_6Sn_5 layer (0.63 µm) formed immediately at the Sn/Cu interface after the deposition process. The formation of the thin Cu_6Sn_5 layer is similar to that reported in the literature [20]. The study indicated that the Cu_6Sn_5 formed during or immediately after the deposition of Sn on Cu in bimetallic Cu–Sn films [20].

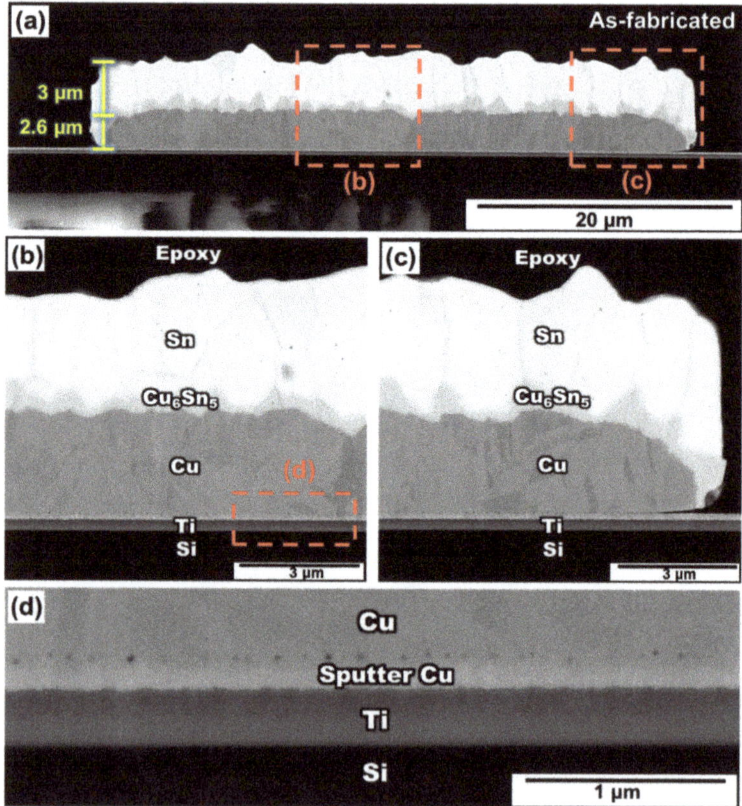

Figure 2. (**a**) Cross-section SEM image of the as-fabricated Sn/Cu micro-bump, and the zoomed-in image in a region near the (**b**) middle region, (**c**) right edge, and (**d**) interface between Cu and Ti layer.

Figure 2b,c shows the zoomed-in images of the Sn/Cu micro-bump at different places, as represented by the red rectangle in Figure 2a. It was evident that Cu, Ti and Si were in contact with each other and no defects, such as cracks, were observed along their interfaces. However, apparent voids existed at the interface of the sputtered Cu and the electroplated Cu, as shown in Figure 2d. The formation of these voids would be discussed in Section 3.3.

Figure 3 shows the micro-structure of the Sn/Cu micro-bumps aged at 200 °C for 24 h. At this stage, Sn is completely converted into Cu_6Sn_5 and Cu_3Sn. As shown in Figure 3b, since the Cu layer was still in contact with the IMC and Ti layer, it demonstrated good adhesion properties in the middle of the bump. Furthermore, because of the surface diffusion [21–23], the bump edge developed IMCs more rapidly than the middle. In addition, the Cu layer was completely consumed at the edge of the bump. Therefore, the IMCs touched the Ti adhesion layer effortlessly at the bump edge. As shown in Figure 3c, even when all the Cu layers were fully depleted, the right edge of the bump exhibited good adhesion between the IMCs and the Ti layer, without any obvious cracks. Similarly, the left edge of the bump shows the same phenomenon. This result indicates that no spalling effect occurred between the IMCs and the Ti layer, implying that the IMCs adhered to the Ti layer more stably on mutual contact.

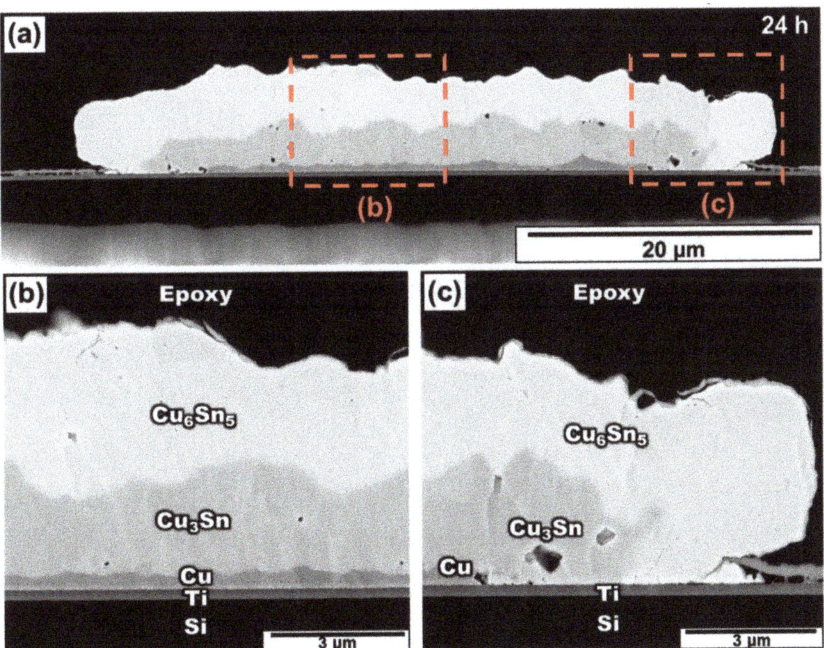

Figure 3. (**a**) Cross-section SEM image of the Sn/Cu micro-bump after aging at 200 °C for 24 h and the zoomed-in image in a region near the (**b**) middle region and (**c**) right edge.

Figure 4 shows the micro-structure of the Sn/Cu bump aged for 36 h. The Cu_3Sn layer enlarged because of the reaction between Cu and Cu_6Sn_5. At this stage, as most of the Cu converted into Cu_3Sn, the Cu layer became discontinuous and formed a series of isolated regions. Compared to the result of aging for 24 h, the interface of $Cu_3Sn/Cu/Ti$ exhibited more voids in the middle region after aging for 36 h. As shown in Figure 4b, although many voids appeared, no continuous cracks formed. In addition, Figure 4c shows a zoomed-in image on the right edge of the bump. Although the contact area between the IMC and the Ti layer grew owing to further Cu depletion, it still exhibited good adhesion without any cracks. Therefore, spalling did not occur between the Cu_3Sn and Ti layers when accompanied by the formation of voids.

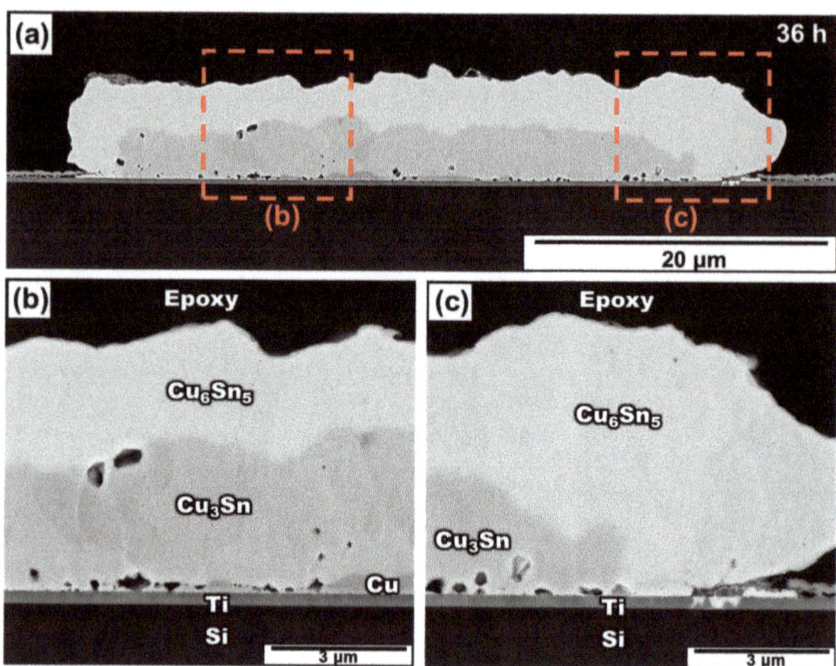

Figure 4. (**a**) Cross-section SEM image of the Sn/Cu micro-bump after aging at 200 °C for 36 h and the zoomed-in image in a region near the (**b**) middle region and (**c**) right edge.

Figure 5 shows the micro-structure of the Sn/Cu bump aged at 200 °C for 42 h. In this stage, Cu was completely consumed and reacted with Cu_6Sn_5 to form Cu_3Sn and the IMCs came in direct contact with the Ti layer through all the bumps. Although many voids formed at the Cu_3Sn/Ti interface, no continuous gaps or detachments between the IMCs and the substrate were observed. Besides, Figure 6 exhibits the TEM observation at the interface between Cu–Sn IMCs and Ti layer. After the higher resolution was examined, the full IMC structure of micro-bump still had excellent adhesion on the Ti adhesion layer. This demonstrated a significant difference from the system of the Cr adhesion layer, which spalled IMCs from the substrate at this stage. This implies that using the Ti adhesion layer succeeded in preventing the spalling of the IMC from the substrate. This phenomenon is discussed in detail in the next section.

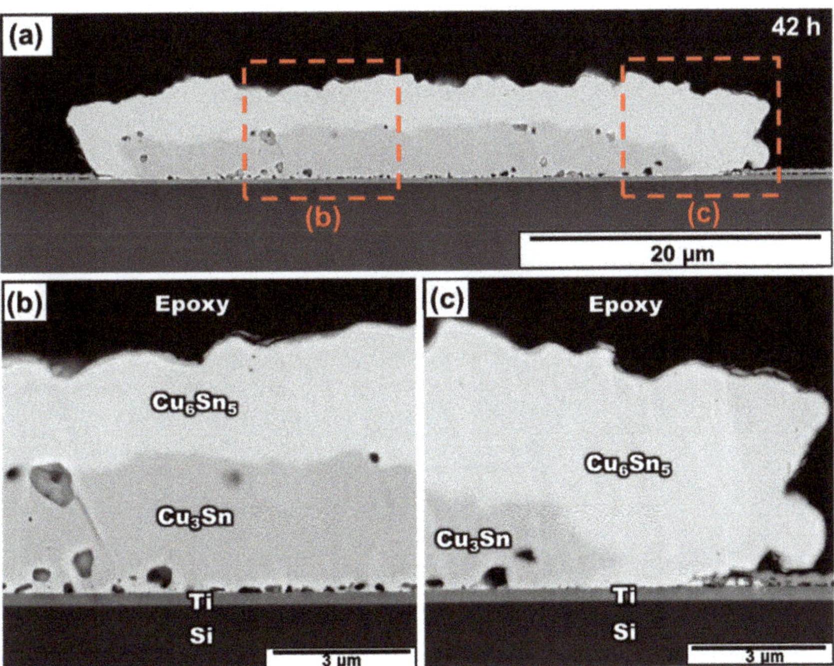

Figure 5. (**a**) Cross-section SEM image of the Sn/Cu micro-bump after aging at 200 °C for 42 h and the zoomed-in image in a region near the (**b**) middle region and (**c**) right edge.

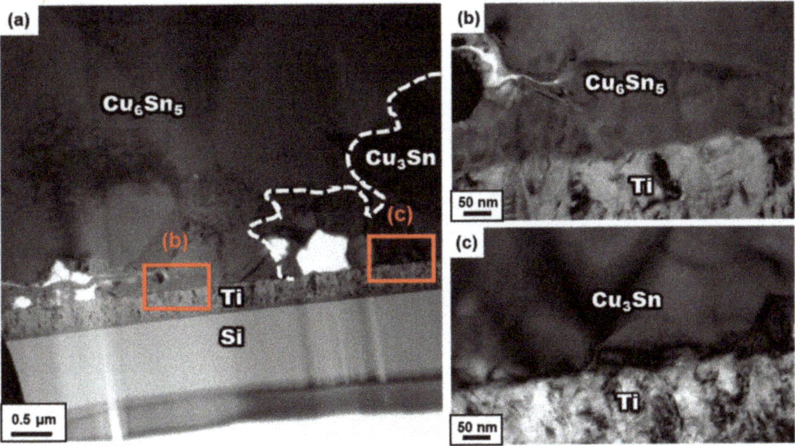

Figure 6. The TEM observation of the micro-bump aging after 42 h in (**a**) the region at the interface between Cu–Sn IMCs and Ti layer, and the enlarged view into the (**b**) Cu_6Sn_5 interface and (**c**) Cu_3Sn interface from (**a**).

3.2. Phenomenon between Cu_3Sn/Ti Interface

Cu–Sn IMCs may spall from the Cr and Ti surfaces when the Cu film is consumed in solid–liquid reactions [2–7]. In addition, the spalling phenomenon between the IMCs and Cr adhesion layer has also been observed during the solid-state reaction, which was confirmed by the high interfacial energy between the IMCs and Cr [13]. However, in this study, the SEM cross-sectional results illustrate that the Cu_6Sn_5 and Cu_3Sn compounds stably adhered to the Ti surface, as shown in Figure 7. The IMCs attached to the Ti layer did not spall in the same manner as the Cr system. It is assumed that this phenomenon might result from the lower interfacial energy between the Cu and Sn intermetallic and Ti in the solid-state reaction. The relationship between interfacial energies can be described by the following inequality:

$$\gamma_{Cu-Sn\ IMC} + \gamma_{Ti} > \gamma_{Cu-Sn\ IMC/Ti} \tag{1}$$

where $\gamma_{Cu-Sn\ IMC}$ is the surface free energy of the intermetallic, γ_{Ti} is the surface free energy of Ti and $\gamma_{Cu-Sn\ IMCs/Ti}$ represents the interfacial energy between the Cu and Sn intermetallic and Ti. When Cu was depleted and the IMC came in contact with the Ti adhesion layer, the interfacial energy between the Cu and Sn intermetallics and Ti was still lower than that of the individuals. Therefore, the IMC layers did not leach from the Ti adhesion layer. In other words, spontaneous spalling did not occur in the Ti system. However, further studies are required to elucidate this mechanism.

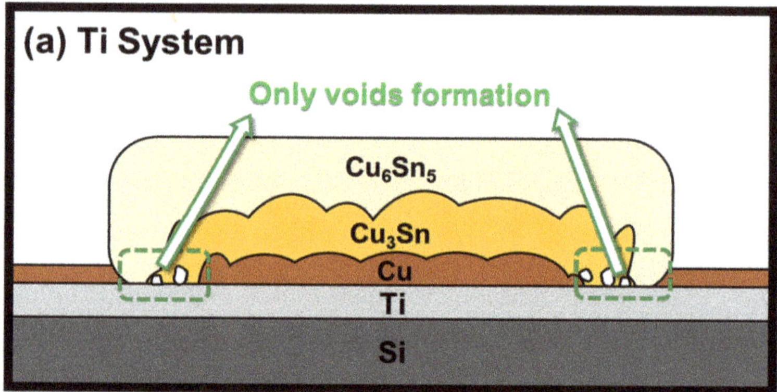

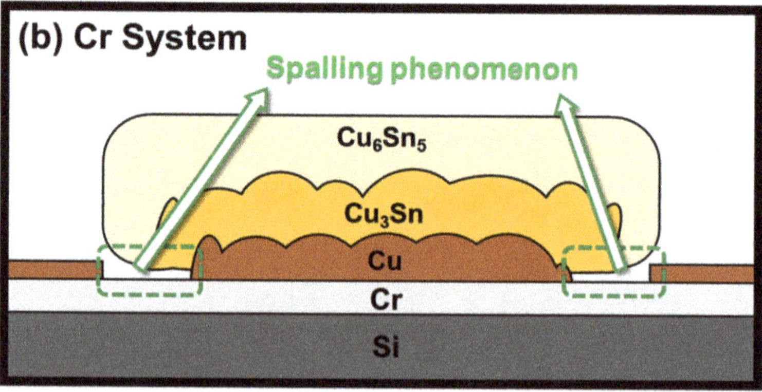

Figure 7. Schematic drawing of a Sn/Cu micro-bump (**a**) using the Ti adhesion layer; (**b**) using the Cr adhesion layer after aging at 200 °C with Cu existence.

3.3. Correlation of Void Formation with Reactive Metal Layers

According to the above results, micro-voids are often observed at the Cu_3Sn/Ti interface after isothermal aging tests. In addition, many micro-voids coalesce into larger voids, as shown in Figure 8. This phenomenon could strongly weaken the properties of micro-bumps, such as their electrical and mechanical properties. Therefore, these voids are key factors that threaten the reliability of electronic packages and it is important to realize the formation of these voids.

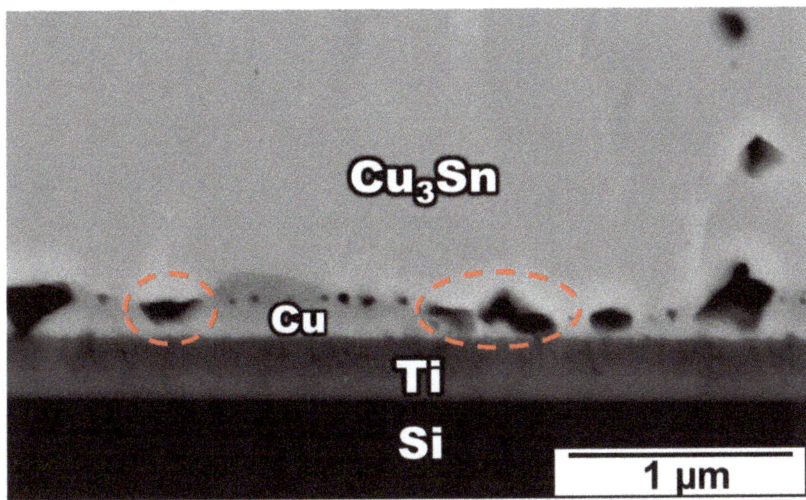

Figure 8. The zoomed-in image of voids formation and voids growth at the Cu_3Sn/Cu interface after aging at 200 °C for 36 h.

There are two common ways of void formation in the Sn/Cu system. One is a series of voids that were formed between the sputtered Cu and the electroplated Cu, as shown in Figure 2. This phenomenon is described as the introduction of several impurities into the deposited Cu film during the electroplating. Organic additives in the electroplating solution may be the main reason for the void formation [24–27]. However, the total volume of this voiding would not change with aging time [25,28]. The other method of void formation is the Kirkendall voids, which is observed at the Cu_3Sn layer. This voiding could be related to the Kirkendall effect, which was attributed to the unequal diffusion rates of Cu and Sn in the IMC [29] and impurity segregation [30–34]. It is generally believed that Cu is the dominant diffusing species in Cu_3Sn, a larger diffusion flux of Cu would occur from the Cu substrate [35–37]. If the vacancies, left by the diffusing-out of Cu atoms, cannot be occupied, they would gather to form new micro-voids with impurity [30–34]. Consequently, if the unequal diffusion rate between Cu and Sn exists, the voids would increase continuously.

From Figures 2–4, it is obvious that the voids became larger with the increase in aging time. Based on previous research [38,39], the voids tended to agglomerate together and became large, since agglomeration could reduce the surface energy. However, the voids becoming larger did not symbolize the increase in voids. Therefore, the relationship between the void quantities and the aging time is required to be determined. Two methods were used to calculate the void percentage in the experiment.

$$\text{Porosity 1 (\%)} = \frac{\text{Pores Area } (\mu m^2)}{\text{Total Micro} - \text{bump Area } (\mu m^2)} \tag{2}$$

$$\text{Porosity 2 (\%)} = \frac{\text{Pores Area } (\mu m^2)}{Cu_3Sn \text{ Area } (\mu m^2) + \text{Pores Area } (\mu m^2)} \tag{3}$$

where Equation (3) is used to calculate the relationship between Cu_3Sn layer and pores because the voids are all found in the Cu_3Sn/Ti interface.

Based on Equations (2) and (3), the blue and black curves, respectively, exhibit a similar trend in Figure 9a,b. First, it is evident that the curves increased rapidly in region II (Sn depleted/Cu remaining region), which is the region where Sn was completely consumed and only the Cu remained. This phenomenon demonstrated that the voids kept growing due to the unequal diffusion flux, forming more Kirkendall voids. Second, in region III (Cu-depleted region), which is the region that Sn and Cu were all completely consumed, because the entire Cu was depleted, the porosity did not change with time. Consequently, no more Cu diffused to form voids.

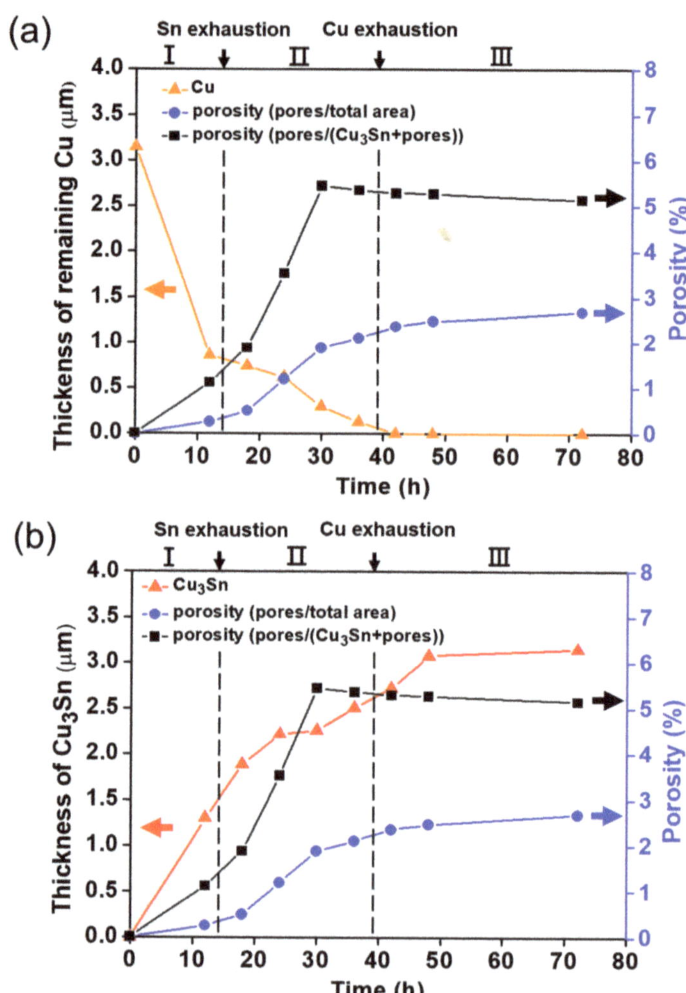

Figure 9. The correlation of porosity and (**a**) Cu thickness, (**b**) Cu_3Sn thickness with different aging times.

Figure 9 also shows the relationship between the remaining Cu thickness (orange line) and Cu_3Sn growth thickness (red line) versus the two porosities mentioned in Equations (2) and (3). Figure 9a shows that the porosity increased rapidly when the remaining Cu thickness decreased steeply. This strongly indicated again that the void formation was caused by the Cu diffusion, which is referred to as Kirkendall voids. In Figure 9b, although the Cu_3Sn growth thickness did not fully correspond to the trend of the porosity, it showed a similar change with the porosity curve.

3.4. Shear Strength and Fracture Surface of Sn/Cu Micro-Bumps

Although the interfacial IMCs and void formation at the interface typically have a significant effect on the solder joint reliability [14–19], a die shear test was conducted to evaluate the effect of interfacial reactions of Sn/Cu micro-bumps in a solid-state reaction.

Figure 10 shows the SEM images and EDX mapping analysis of the fracture surface of the bump at various aging times. Shear tests were conducted from left to right. Based on previous studies, the Sn/Cu_6Sn_5 interface is the main factor affecting the fracture location in the as-fabricated sample [40]. Therefore, as shown in Figure 10a, the bump was broken at the Sn/Cu_6Sn_5 interface without aging. However, as shown in Figure 2, there were a series of voids and the formation of small cracks between the sputtered Cu and the electroplated Cu without aging. These defects do not affect the mechanical properties at this stage. From the SEM images, the fracture surface did not pass through these voids, so the mechanical strength was determined by the Sn/Cu_6Sn_5 interface.

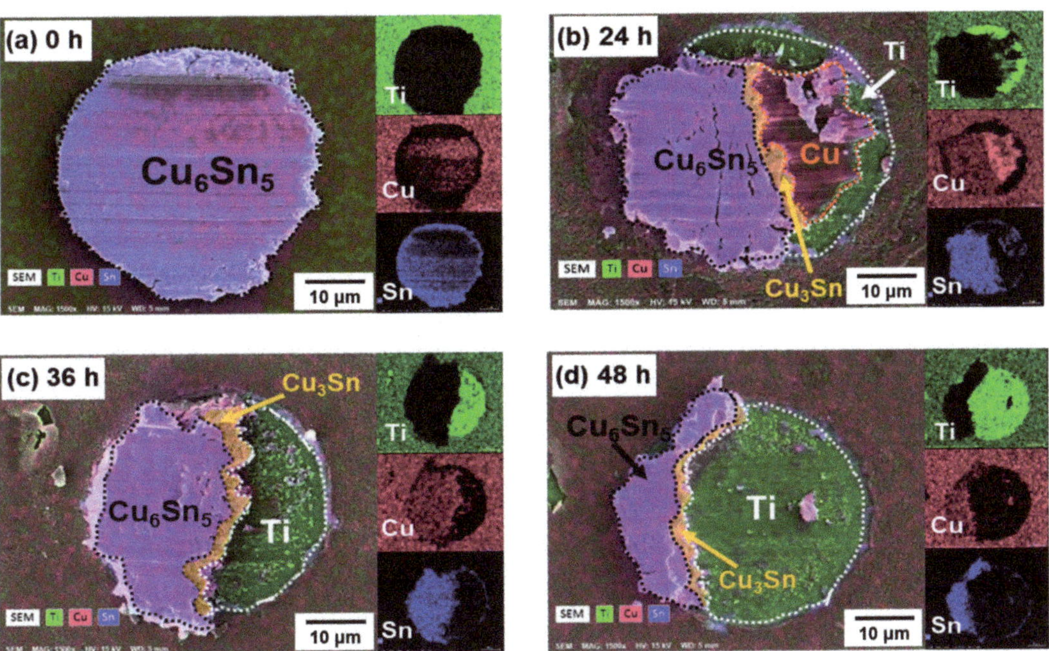

Figure 10. The fracture surfaces of solder joints after aging at 200 °C for: (**a**) 0 h, (**b**) 24 h, (**c**) 36 h and (**d**) 48 h with their elemental area distribution of Ti (green), Cu (red), Sn (blue).

Figure 10b–d shows the fracture surface after aging for 24–48 h and Figure 11 shows the phase percentages of the fracture surface with different aging times. To ensure the full IMC structure in micro-bumps, the sample aging after 48 h was used in the die shear test. In the analysis of the fracture surface, four main layers, Cu_6Sn_5, Cu_3Sn, Cu and Ti, were exposed at different aging times. The Cu_3Sn layer was only slightly exposed in these SEM images. Interestingly, the exposed Ti layer increased with decreasing Cu_6Sn_5 layer thickness as the aging time increased.

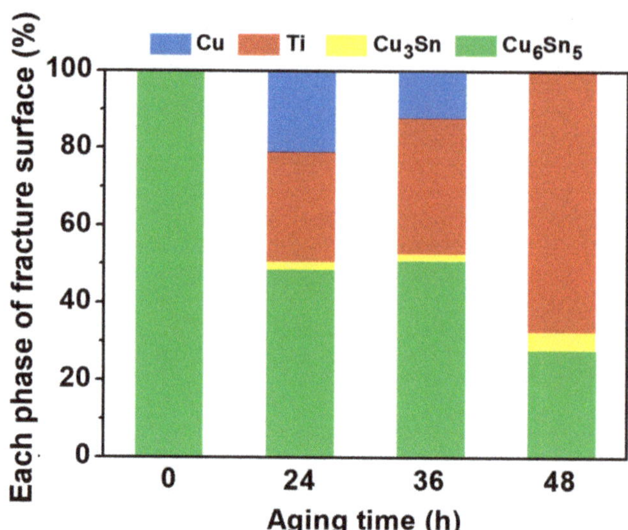

Figure 11. The phase percentages of fracture surface against aging time.

Figure 12 shows the correlation between the aging time and the shear strength. The strength was approximately 9 MPa without aging. This means that the Sn/Cu_6Sn_5 would be the main weak interface in the as-fabricated micro-bump. However, the shear strength steeply increased to 32 MPa after aging for 24 h and then remained constant after aging for 36 h and 48 h. During this period, the Cu_3Sn layer thickened and was the major component of IMCs attached to the Ti layer through aging. In addition, the number of voids increased with the growth of Cu_3Sn. However, we found that the number of voids has little effect on the strength. Thus, the shear strength was dominated by Cu_3Sn/Ti adhesive interface but not by voids. This phenomenon is finally discussed below.

By combining the findings from the cross-sectional observation, fracture surface observation and strength measurement, a schematic summarizing the results of the fracture analysis is presented in Figure 13. After electroplating, a thin Cu_6Sn_5 layer was formed between the Cu and Sn layers; thus, the fracture occurred through the Sn/Cu_6Sn_5 interface with a very smooth fractured surface. When the aging time reached 24 h, the shear strength increased with the growth of Cu_3Sn. In addition, according to the phenomenon mentioned above, the outer edge of the micro-bump formed some voids at the Cu_3Sn/Ti interface. The stress concentration would be developed around these voids, which could more easily cause the crack growth. Therefore, the crack propagated along with the Cu_6Sn_5 layer and extended to the Cu_3Sn/Ti interface. Furthermore, when the aging time increased to 36 and 48 h, the Cu_3Sn layer grew increasingly with the depletion of the Cu layer, and the void formations also increased at the Cu_3Sn/Ti interface. Thus, the cracks propagated more easily and extended to the Cu_3Sn/Ti interface. This led to an increase in the percentage of the exposed Ti area on the fracture surface with increasing aging time. However, since the fracture surface mainly occurred at the Cu_3Sn/Ti interface, the adhesion between Cu_3Sn and Ti was the main factor controlling the shear strength. Therefore, the strength did not change with increasing aging time and the number of micro-voids. The micro-structural characterization suggested that the key reason for determination of the micro-bump strength was not the void formation but the types of the IMC interface. Therefore, it is important to realize the interfacial reaction and IMC formation during the miniaturization of micro-bumps to achieve better joint strength.

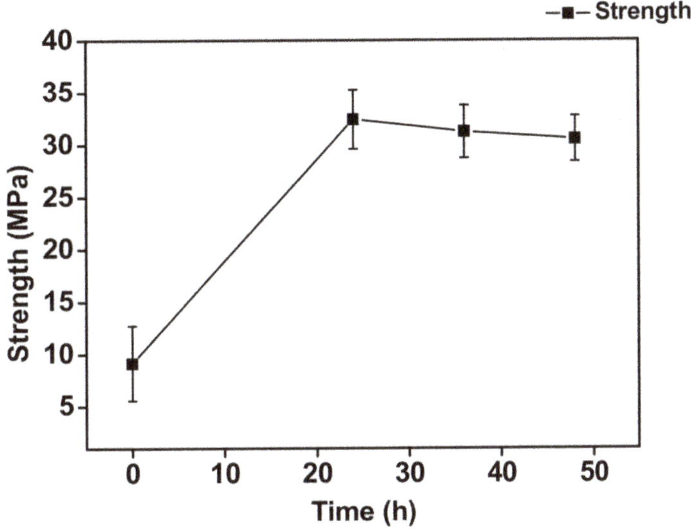

Figure 12. Variation of the shear strength with the aging time at 200 °C.

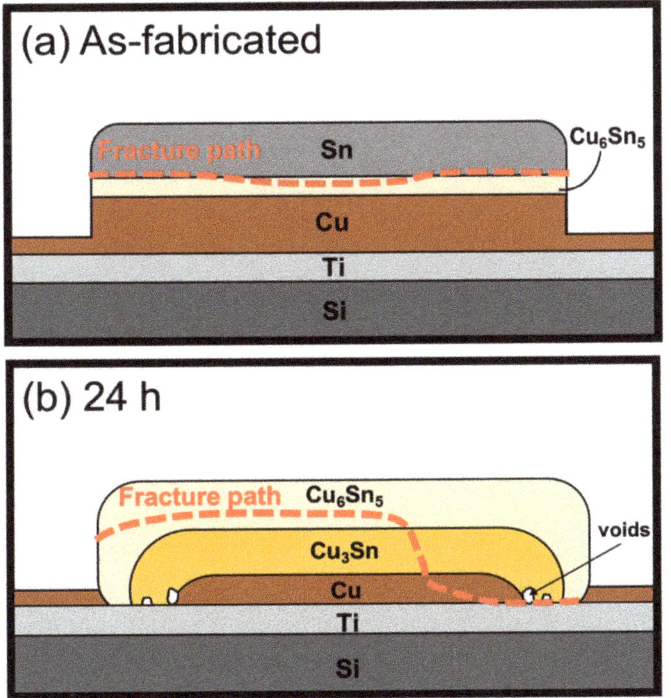

Figure 13. *Cont.*

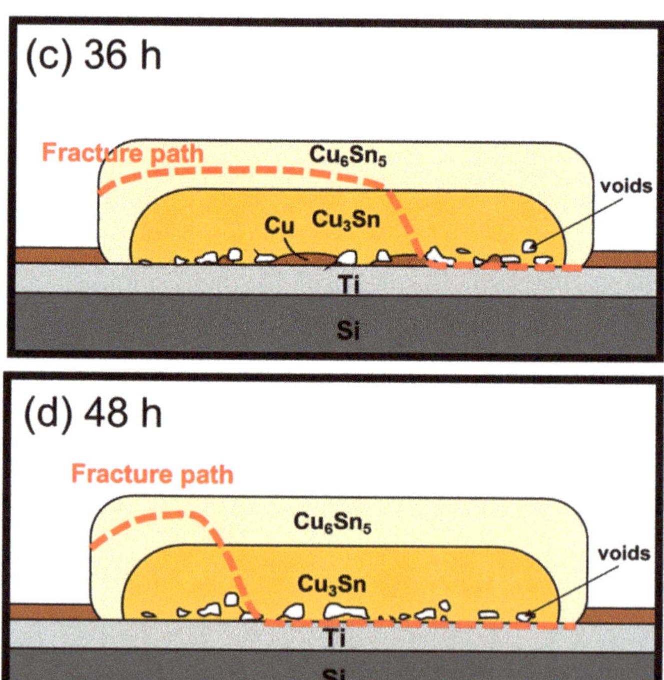

Figure 13. Schematic drawing showing the crack propagation of Sn/Cu micro-bumps (**a**) as-fabricated, and after (**b**) 24 h, (**c**) 36 h, and (**d**) 48 h of aging.

4. Conclusions

In this study, the micro-structural evolution of Sn/Cu micro-bumps during the solid-state reaction with the Ti adhesion layer at 200 °C was investigated. Based on the results, the following conclusions are drawn.

1. Under the as-fabricated condition, a thin layer of Cu_6Sn_5 was formed at the Cu and Sn interfaces. In addition, after the aging time increased from 24 to 42 h, the Cu_6Sn_5 and Cu_3Sn phases primarily thickened and extended towards the substrate.
2. With the complete consumption of Sn and Cu, voids could be found and extended to the entire bump at the Cu_3Sn/Ti interface; however, these voids could not induce cracks or gaps.
3. After aging for 42 h, the micro-bumps transformed into intermetallic (IMC) structures. Owing to the lower interfacial energy between the Cu and Sn IMCs and the Ti layer, they all attached well to each other and the micro-bump did not exhibit spontaneous spalling.
4. The correlation of void formation is related to the unbalanced diffusion rate in Cu_3Sn. Since the trend of porosity had highly corresponding with Cu thickness consumption, these voids could be indicated as the Kirkendall voids.
5. In the die shear test analysis, the strength of the bumps was determined by the type of the IMC interface. In addition, after aging, the voids extended to the entire bump, and the fracture surface exposed more Ti area. This phenomenon could be attributed to the fact that the voids would only influence the fracture path.

Finally, compared to the Cr system, the Ti adhesion layer imparted excellent properties to the full IMC joints. Therefore, the Ti system can be used to obtain reliable micro-joints as an UBM layer, as it prevents spalling of the IMCs produced during the terminal solid-state reaction.

Author Contributions: Conceptualization, C.-W.K., P.-Y.K., C.-C.C., W.-C.H., F.-L.C. and C.R.K.; Data curation, C.-W.K. and P.-Y.K.; Formal analysis, C.-W.K., P.-Y.K. and C.-C.C.; Funding acquisition, C.R.K.; Investigation, C.-W.K. and P.-Y.K.; Methodology, C.-W.K., P.-Y.K., C.-C.C., W.-C.H., F.-L.C. and C.R.K.; Project administration, C.R.K.; Supervision, C.R.K.; Validation, C.-W.K.; Writing—original draft, C.-W.K.; Writing—review & editing, C.-W.K., P.-Y.K., C.-C.C., W.-C.H., F.-L.C. and C.R.K. All authors have read and agreed to the published version of the manuscript.

Funding: This work was supported by the "Advanced Research Center for Green Materials Science and Technology" from the Featured Area Research Center Program within the framework of the Higher Education Sprout Project by the Ministry of Education (109006) and the Ministry of Science and Technology (MOST 109-2634-F-002-042) of Taiwan.

Institutional Review Board Statement: Not applicable.

Informed Consent Statement: Not applicable.

Data Availability Statement: Not applicable.

Acknowledgments: The authors thank Ya-Yun Yang and Ching-Yen Lin from the Precious Instrument Center of the College of Science at National Taiwan University for their assistance with ion-milling.

Conflicts of Interest: The authors declare no conflict of interest.

References

1. Yang, T.L.; Yu, J.J.; Shih, W.L.; Hsueh, C.H.; Kao, C.R. Effects of silver addition on Cu–Sn microjoints for chip–stacking applications. *J. Alloys Compd.* **2014**, *605*, 193–198. [CrossRef]
2. Berry, B.S.; Ames, I. Studies of the SLT chip terminal metallurgy. *IBM J. Res. Dev.* **1969**, *13*, 286–296. [CrossRef]
3. Pan, G.Z.; Liu, A.A.; Kim, H.K.; Tu, K.N.; Totta, P.A. Microstructures of phased–in Cr–Cu/Cu/Au bump–limiting metallization and its soldering behavior with high Pb content and eutectic PbSn solders. *Appl. Phys. Lett.* **1997**, *71*, 2946–2948. [CrossRef]
4. Liu, C.Y.; Kim, H.K.; Tu, K.N.; Totta, P.A. Dewetting of molten Sn on Au/Cu/Cr thinfilm metallization. *Appl. Phys. Lett.* **1996**, *69*, 4014–4016. [CrossRef]
5. Liu, A.A.; Kim, H.K.; Tu, K.N.; Totta, P.A. Spalling of Cu_6Sn_5 spheroids in the soldering reaction of eutectic SnPb on Cr/Cu/Au thin films. *J. Appl. Phys.* **1996**, *80*, 2774–2780. [CrossRef]
6. Kim, P.G.; Jang, J.W.; Lee, T.Y.; Tu, K.N. Interfacial reaction and wetting behavior in eutectic SnPb solder on Ni/Ti thin films and Ni foils. *J. Appl. Phys.* **1999**, *86*, 6746–6751. [CrossRef]
7. Kim, H.K.; Tu, K.N.; Totta, P.A. Ripening–assisted asymmetric spalling of Cu–Sn compound spheroids in solder joints on Si wafers. *Appl. Phys. Lett.* **1996**, *68*, 2204–2206. [CrossRef]
8. Zhang, R.; Tian, Y.; Hang, C.; Liu, B.; Wang, C. Formation mechanism and orientation of Cu_3Sn grains in Cu–Sn intermetallic compound joints. *Mater. Lett.* **2013**, *110*, 137–140. [CrossRef]
9. Yao, P.; Li, X.; Liang, X.; Yu, B.; Jin, F.; Li, Y. A study on interfacial phase evolution during Cu/Sn/Cu soldering with a micro interconnected height. *Mater. Char.* **2017**, *131*, 49–63. [CrossRef]
10. Mo, L.; Wu, F.; Liu, C. Growth kinetics of IMCs in Cu–Sn intermetallic joints during isothermal soldering process. In Proceedings of the 2015 IEEE 65th Electronic Components and Technology Conference (ECTC), San Diego, CA, USA, 26–29 May 2015; pp. 1854–1858. [CrossRef]
11. Li, J.F.; Agyakwa, P.A.; Johnson, C.M. Interfacial reaction in Cu/Sn/Cu system during the transient liquid phase soldering process. *Acta Mater.* **2011**, *59*, 1198–1211. [CrossRef]
12. Kao, C.R.; Chuang, H.Y.; Chen, W.M.; Yang, T.L.; Kuo, M.S.; Chen, Y.J.; Yu, J.J.; Li, C.C. Soldering reactions under space confinement for 3D IC applications. In Proceedings of the 2012 IEEE 62nd Electronic Components and Technology Conference (ECTC), San Diego, CA, USA, 29 May–1 June 2012; pp. 724–728. [CrossRef]
13. Tsai, C.H.; Lin, S.Y.; Lee, P.T.; Kao, C.R. A new spalling mechanism of intermetallics from the adhesion layer in the terminal–stage reaction between Cu and Sn. *Intermetallics* **2021**, *138*, 107342. [CrossRef]
14. Yoon, J.W.; Jung, S.B. Phase analysis and kinetics of solid–state ageing of Pb–free Sn–3.5Ag solder on electroless Ni–P substrate. *Surf. Interface Anal.* **2004**, *36*, 963–965. [CrossRef]
15. Hu, X.; Li, Y.; Liu, Y.; Liu, Y.; Min, Z. Microstructure and shear strength of Sn37Pb/Cu solder joints subjected to isothermal aging. *Microelectron. Reliab.* **2014**, *54*, 1575–1582. [CrossRef]
16. Yang, C.; Le, F.; Lee, S.R. Experimental investigation of the failure mechanism of Cu–Sn intermetallic compounds in SAC solder joints. *Microelectron. Reliab.* **2016**, *62*, 130–140. [CrossRef]

17. Jeong, M.H.; Lim, G.T.; Kim, B.J.; Lee, K.W.; Kim, J.D.; Joo, Y.C.; Park, Y.B. Interfacial reaction effect on electrical reliability of Cu pillar/Sn bumps. *J. Electron. Mater.* **2010**, *39*, 2368–2374. [CrossRef]

18. Peng, W.; Monlevade, E.; Marques, M.E. Effect of thermal aging on the interfacial structure of SnAgCu solder joints on Cu. *Microelectron. Reliab.* **2007**, *47*, 2161–2168. [CrossRef]

19. Tu, K.N.; Thompson, R.D. Kinetics of interfacial reaction in bimetallic Cu–Sn thin films. *Acta Metall.* **1982**, *30*, 947–952. [CrossRef]

20. Tu, K.N. Interdiffusion and reaction in bimetallic Cu–Sn thin films. *Acta Metall.* **1973**, *21*, 347–354. [CrossRef]

21. Chiu, W.L.; Liu, C.M.; Haung, Y.S.; Chen, C. Formation of nearly void–free Cu_3Sn intermetallic joints using nanotwinned Cu metallization. *Appl. Phys. Lett.* **2014**, *104*, 171902. [CrossRef]

22. Panchenko, I.; Croes, K.; de Wolf, I.; de Messemaeker, J.; Beyne, E.; Wolter, K.J. Degradation of Cu_6Sn_5 intermetallic compound by pore formation in solid–liquid interdiffusion Cu/Sn microbump interconnects. *Microelectron. Eng.* **2014**, *117*, 26–34. [CrossRef]

23. Chu, D.T.; Chu, Y.C.; Lin, J.A.; Chen, Y.T.; Wang, C.C.; Song, Y.F.; Chiang, C.C.; Chen, C.; Tu, K.N. Growth competition between layer–type and porous–type Cu_3Sn in microbumps. *Microelectron. Reliab.* **2017**, *79*, 32–37. [CrossRef]

24. Andricacos, P.C.; Uzoh, C.; Dukovic, J.O.; Horkans, J.; Deligianni, H. Damascene copper electroplating for chip interconnections. *IBM J. Res. Dev.* **1998**, *42*, 567–574. [CrossRef]

25. Moriyama, M.; Konishi, S.; Tsukimoto, S.; Murakami, M. Effect of organic additives on formation and growth behavior of micro–void in electroplating copper films. *Mater. Trans.* **2004**, *45*, 3172–3176. [CrossRef]

26. Kelly, J.J.; West, A.C. Copper deposition in the presence of polyethylene glycol: II. Electrochemical impedance spectroscopy. *J. Electrochem. Soc.* **1998**, *145*, 3477–3481. [CrossRef]

27. Nawafune, H.; Kitamura, H.; Mizumoto, S.; Uchida, E.; Okada, T. Electrodeposition of void–free copper from ethylenediamine complex bath for ULSI metallization. *J. Surf. Finish. Soc. Jpn.* **2000**, *51*, 1142–1147. [CrossRef]

28. Konishi, S.; Moriyama, M.; Murakami, M. Effect of annealing atmosphere on void formation in copper interconnects. *Mater. Trans.* **2002**, *43*, 1624–1628. [CrossRef]

29. Strandlund, H.; Larsson, H. Prediction of Kirkendall shift and porosity in binary and ternary diffusion couples. *Acta Mater.* **2004**, *52*, 4695–4703. [CrossRef]

30. Yin, L.; Borgesen, P. On the root cause of Kirkendall voiding in Cu_3Sn. *J. Mater. Res.* **2011**, *26*, 455–466. [CrossRef]

31. Stangl, M.; Dittel, V.; Acker, J.; Hoffmann, V.; Gruner, W.; Strehle, S.; Wetzig, K. Investigation of organic impurities adsorbed on and incorporated into electroplated copper layers. *Appl. Surf. Sci.* **2005**, *252*, 158–161. [CrossRef]

32. Uedono, A.; Mori, K.; Ito, K.; Imamizu, K.; Hachiya, T.; Kamijo, H.; Suzuki, R. Impact of residual impurities on annealing properties of vacancies in electroplated Cu studied using monoenergetic positron beams. *Jpn. J. Appl. Phys.* **2007**, *46*, L483. [CrossRef]

33. Yu, J.; Kim, J.Y. Effects of residual S on Kirkendall void formation at Cu/Sn–3.5 Ag solder joints. *Acta Mater.* **2008**, *56*, 5514–5523. [CrossRef]

34. Kim, J.Y.; Yu, J.; Kim, S.H. Effects of sulfide–forming element additions on the Kirkendall void formation and drop impact reliability of Cu/Sn–3.5 Ag solder joints. *Acta Mater.* **2009**, *57*, 5001–5012. [CrossRef]

35. Oh, M. Growth Kinetics of Intermetallic Phases in the Copper-Tin Binary and the Copper-Nickel-Tin Ternary Systems at Low Temperatures. Ph.D. Thesis, Lehigh University, Bethlehem, PA, USA, 1994.

36. Oberndorff, P. Lead-Free Solder Systems: Phase Relations and Microstructures. Ph.D. Thesis, Technical University of Eindhoven, Eindhoven, The Netherlands, 2001.

37. Paul, A. The Kirkendall Effect in Solid State Diffusion. Ph.D. Thesis, Technical University of Eindhoven, Eindhoven, The Netherlands, 2004.

38. Wen, J.; Huang, Y.; Hwang, K.C.; Liu, C.; Li, M. The modified Gurson model accounting for the void size effect. *Int. J. Plast.* **2005**, *21*, 381–395. [CrossRef]

39. Goodhew, P.J. Shapes of pores in metals. *Metal. Sci.* **1981**, *15*, 377–385. [CrossRef]

40. Chen, Y.J.; Chung, C.K.; Yang, C.R.; Kao, C.R. Single–joint shear strength of micro Cu pillar solder bumps with different amounts of intermetallics. *Microelectron Reliab.* **2013**, *53*, 47–52. [CrossRef]

materials

Article

Study on the Strip Warpage Issues Encountered in the Flip-Chip Process

Wan-Chun Chuang * and Wei-Long Chen

Department of Mechanical and Electromechanical Engineering, Engineering Technology Research & Promotion Center, National Sun Yat-sen University, Kaohsiung 804, Taiwan; d043020007@student.nsysu.edu.tw
* Correspondence: wcchuang@mail.nsysu.edu.tw; Tel.: +886-7525-2000 (ext. 4228)

Abstract: This study successfully established a strip warpage simulation model of the flip-chip process and investigated the effects of structural design and process (molding, post-mold curing, pretreatment, and ball mounting) on strip warpage. The errors between simulated and experimental values were found to be less than 8%. Taguchi analysis was employed to identify the key factors affecting strip warpage, which were discovered to be die thickness and substrate thickness, followed by mold compound thickness and molding temperature. Although a greater die thickness and mold compound thickness reduce the strip warpage, they also substantially increase the overall strip thickness. To overcome this problem, design criteria are proposed, with the neutral axis of the strip structure located on the bump. The results obtained using the criteria revealed that the strip warpage and overall strip thickness are effectively reduced. In summary, the proposed model can be used to evaluate the effect of structural design and process parameters on strip warpage and can provide strip design guidelines for reducing the amount of strip warpage and meeting the requirements for light, thin, and short chips on the production line. In addition, the proposed guidelines can accelerate the product development cycle and improve product quality with reduced development costs.

Keywords: flip-chip process; strip warpage; bump

Citation: Chuang, W.-C.; Chen, W.-L. Study on the Strip Warpage Issues Encountered in the Flip-Chip Process. *Materials* **2022**, *15*, 323. https://doi.org/10.3390/ma15010323

Academic Editors: Kuo-Ning Chiang, Sandra Maria Fernandes Carvalho and Cédric Delattre

Received: 29 July 2021
Accepted: 30 December 2021
Published: 3 January 2022

Publisher's Note: MDPI stays neutral with regard to jurisdictional claims in published maps and institutional affiliations.

1. Introduction

Integrated circuit (IC) packaging technology is continually innovating, with chips becoming lighter, thinner, and shorter. Due to the reduced size and an unmatched coefficient of thermal expansion (CTE) between the materials of the package, warpage occurring after completion of the thermal process can cause numerous problems. Large deformation results in the weak attachment of bumps or copper rods to the circuit board, which could even damage the structure and signal. Therefore, many studies have addressed warpage-related issues, and finite element (FE) simulation methods have been employed to analyze warpage behavior.

In the related literature on warpage theory because IC packages are composed of composite materials, the warpage behavior generated by thermal processes has been concluded to be complicated. Timoshenko [1] proposed a theory to explain the warpage caused by temperature changes in two bonded materials. Chen [2] discussed the effect of two bonding materials on warpage at different temperatures, theoretically analyzed a bimaterial structure, and sketched a multilateral structure. Garrett [3], of the technical department of Akrometrix, used Timoshenko's biomaterial warpage theory [1] to derive the variables of warpage after an IC molding process: material properties (Young's modulus and CTE), mold compound, substrate thickness, and temperature of the molding process. Wu et al. [4,5] discussed how the material properties of epoxy composites affected the performance of electronic devices and discovered the desirable dielectric and thermal properties for their design; they also studied the microwave absorption for nanorod and spinel structures [6,7]. Although some studies have analyzed package warpage theoretically, they simplified the structure into only the mold compound and substrate. A package also contains complex structures such as a die, wire bonding, copper rods, and bump solder balls. Thus, this simplification cannot fully reflect the actual situation.

Some scholars have recently employed FE simulation to analyze package warpage. Dudek et al. [8] used FE simulation to analyze the effects of the mechanical and thermal properties of the material body on warpage. In addition, Hu et al. [9] employed FE simulation to investigate plastic ball grid array packaging technology and developed a bimaterial warping model, where the substrate and mold compound in the package were used to discuss the effect of the molding process on warpage. Moreover, Huang et al. [10] used FE simulation to explore the effect of geometric die and substrate thicknesses on warpage. For packaging technology, they employed a single-sample window ball grid array, and they obtained simulated values consistent with the trend described by Timoshenko's bimaterial theory [1] for the effect of the reflow process on warpage. Chae and Ouyang [11] discussed the effect of molding temperature on strip warpage for flip-chip strip packaging technology. They discovered that a high molding temperature will cause large strip warpage. In addition, they proposed the use of mechanics of composite materials theory for calculating the CTE of a substrate and mold compound at 25, 150, and 260 °C. Huber et al. [12] employed an FE simulation to determine the effect of mold compound on warpage after a long period of thermal aging. Bin et al. [13] applied FE simulation to fine pitch ball grid array packaging technology; they also discussed the effect of the geometric thickness of the mold compound and die on warpage when a strip was subjected to the molding process. The results revealed a negative correlation between the strip warpage and geometric thickness of the mold compound and die. Zheng et al. [14] proposed a reference temperature calibration of the flip-chip warping simulation model, obtaining consistency and small errors between the simulated and experimental warpage values. Chen et al. [15] employed FE simulation to investigate embedded silicon fan-out wafer-level package technology. The technique does not require EMC materials, and its structure is relatively simple. Therefore, the proposed simulation model achieved relatively good agreement with experimental and theoretical value, and the error between the experimental and the simulation results was only approximately 9%. However, this method is not feasible for a more complex simulation model with an EMC structure. Tsai et al. [16] proposed a new Suhir-solution-based theory for predicting the thermal deformation of flip-chip packages with the capillary underfill process and discussed the effect of the temperature of the reflow process on warpage. The deviation between the model and experimental results was approximately 25%. They also proposed a strain gauge measurement associated with a beam model theory to determine the thermally induced warpages of packages. The thermal strain results were consistent with those of validated FEM. Therefore, the strain gauge method proved feasible in determining the thermal warpages of packages [17]. Yao et al. [18] proposed an analytical model to evaluate the pore and superficial permeability of an underfill porous medium in a flip-chip packaging; they also presented an approach to predict the flow front and the filling time [19]. Chiang et al. [20] proposed an overview of artificial intelligence assisted design on simulation technology for reliability life prediction of advanced packaging. Developers only need to input geometric data of the package structures, and then the reliability life cycle can be obtained by this AI-trained model. Lin et al. [21] presented a finite element method to predict the final warpage of an ultra-thin flip chip scale package based on chemical shrinkage and cure-dependent viscoelasticity of molded underfill. Errors between the experimental and simulation results were approximately 10%. Cheng et al. [22] investigated the warpage behavior of a flip chip package-on-package (FCPoP) assembly during the fabrication process. They took some effects into account, such as the viscoelastic behavior and cure shrinkage of the epoxy molding compound. The results showed that simulation data fell within the ranges of the measured data.

The flip-chip is a commonly used packaging technique. Figure 1 shows the flip-chip process flow investigated in this study, and Figure 2 shows a schematic of a single flip-chip unit. As shown in Figure 1, the bonding process involves 12 steps:

1. Wafer Grinding: the wafer is first processed by grinding before any fabrication procedure is conducted.
2. Wafer Saw: the wafer is diced into small dies of target size.
3. Flip-chip Bond: the cut die obtained from Step 2 is placed on the substrate with the help of bumps and soldering flux, but these are not completely fused together.
4. Reflow: through the reflow process, the bumps and soldering flux on the substrate are fused so that the die can be fixed on the substrate.
5. Flux Cleaning: plasma cleaning can remove contaminants formed during the production process, thereby effectively enhancing the strength of the bond between the die and substrate.
6. Pre-MD Baking: this refers to the baking before molding, where water moisture subsequently formed in the die, substrate, and bumps must be completely removed to ensure that the mold compound fits tightly to protect the die, substrate, and bumps.
7. Pre-MD Plasma Cleaning: the plasma surface is cleaned to remove impurities on the surface so that gaps between the internal components can be filled during molding.
8. Molding: a mold compound is injected into the package to seal all the components, protecting the die and bumps inside the device.
9. Post-molding cure (PMC): the sealed device is cured again to enhance its structural stability.
10. Pre-treatment: pre-heat treatment before implantation of solder balls.
11. Ball Mounting and Reflow: solder balls are implanted underneath the base substrate for future signal connection with the external circuit.
12. Package Saw: the strip is diced into single wafers for packaging and shipping.

Because of the CTE mismatch between the packaging material, strip warpage often occurs during processes that require heating. Among the procedures, Steps 3–4, 6, and 8–11 are all performed at high temperatures. However, based on manufacturing experience, the strip is almost completely flat during Steps 3–7. No strip warpage occurs, even during the molding process in Step 8, when the strip is sealed by the mold compound and other materials at 175 °C. At this point, the strip remains almost flat. As discussed in several papers [3,9,11,14], strip warpage mainly occurs when the device is cooled to room temperature at 25 °C after the molding process. Severe strip warpage occurring during the post-molding period reduces product yield. The literature [8–14,20–22] indicates that FE simulation is frequently used to solve packaging warpage problems. Material parameters are crucial for determining whether consistency between simulated and experimental values is achieved. In particular, the mold compound is a high molecular polymer with a Young's modulus and CTE that exhibit large temperature-based variations. In addition, few scholars have simultaneously simulated the effects of different geometric structures and process temperatures on warpage.

The present study input the temperature variation curve of the material properties into the simulation model and simulated the effects of different steps in the continuous process on warpage. The effects of the process temperature and geometric thickness of the mold compound, substrate, and die on warpage were also determined. Therefore, this study simulated a more realistic situation than previous studies. The proposed model and warpage analysis method can be used by designers to predict warpage under a continuous process and identify the optimal parameter design conditions for reducing strip warpage.

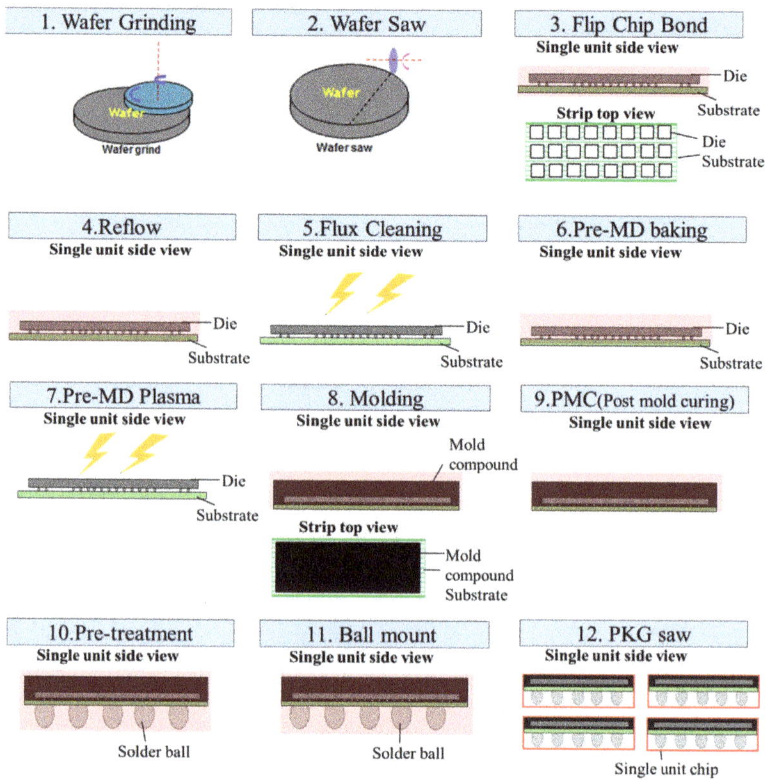

Figure 1. Flip-chip process flow.

Material	Area ratio (%)
Mold compound	45.76
Die	20.69
Bump	4.98
Substrate	28.57

Figure 2. Structural schematic of a single unit inside a strip.

2. Research Method

The research method was divided into three steps: strip model, Taguchi method, and structural design.

2.1. Strip Simulation Model

2.1.1. Model Establishment

This study used the COMSOL Multiphysics software to establish a strip model. Table 1 lists the specifications of the strip structure. The shape of the strip included a long side (x-direction) and a short side (y-direction). When the molding process was cooled to room temperature (25 °C), the amount of warpage on the short side of the strip was relatively small; only the direction of the long axis exhibited severe warpage (Figure 3). Table 2 lists the experimental values of the strip warpage (with compound 1). As shown in the table, no warpage occurred on either the long or short side during the molding process (Step 8) at 175 °C. However, as the strip was cooled to 25 °C, the CTE mismatch between the packaging materials caused a severe warpage of 7 mm but no measurable warpage on the short side. Until the end of Step 11 (ball mount), only minor warpage was observed on the short side. Therefore, strip warpage mainly occurred on the long side. To identify the main cause of warpage and reduce the simulation time required, this study simplified the 3D strip model to a 2D strip model. Subsequently, because a strip is a symmetrical structure, a quarter of the 2D strip model was used in this study. Figure 4 illustrates the quarter 3D strip model and 2D strip models. Figure 2 shows a structural diagram of a single unit in the strip, where each strip contained a total of 119 single-unit chips, and the structures included a mold compound, die, bump, and substrate. As shown in Table 3, we also validated the feasibility of a 2D simulation model. The results obtained from the 2D simulation model were found to be consistent with the experimental values, suggesting that it is possible to simplify the 3D model to a 2D model.

Table 1. Specifications of the strip structure.

PKG Information	
PKG size (mm^2)	7×7
Mold compound thickness (μm)	450
Die size (mm^2)	6.3×6.34
Die thickness (μm)	150
Bump type	SAC405
Bump pitch (μm)	190

Figure 3. Actual strip warpage on the production line.

Table 2. Experimental values of strip warpage.

Process Flow		Strip Warpage (mm)	
		Long Side (x-Direction)	Short Side (y-Direction)
8. Molding	175 °C	0	0
	25 °C	7	0
9. PMC		5	N/A
10. Pre-treatment		7	N/A
11. Ball mount		7.5	N/A

Table 3. Experimental values vs. simulation values of strip warpage.

Process Flow		Strip Warpage (mm)	
		Experimental Value	Simulation Value
8. Molding	175 °C	0	0
	25 °C	7	7.36
9. PMC		5	5.14
10. Pre-treatment		7	6.73
11. Ball mount		7.5	6.90

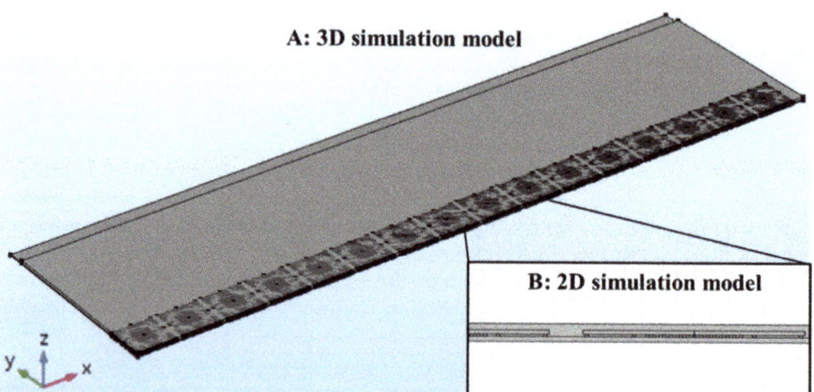

Figure 4. One quarter of the (**A**) 3D model and (**B**) 2D model.

2.1.2. Establishing Material Parameters

Material parameters are crucial to ensure that the simulation results of a model match the actual situation. As illustrated in Figure 2, four materials were present in the structure: bump, die, substrate, and mold compound. To match the materials on the production line, the bump was simulated as SAC405 (95.5Sn:4.0Ag:0.5Cu) and the die as being silicon. As shown in Tables 4 and 5, the most important issue was based on [8–10,12]. The mold compound is a polymer material, whereas the substrate is composed of different materials, so it has varying mechanical and thermal properties due to differences in ambient temperature. The mold compounds employed in this study were the mold compounds 1 and 2, which are used on the production line. A dynamic mechanical analyzer (DMA) was employed to measure the Young's modulus of the mold compounds and substrate. The range of temperatures used was 25–260 °C, and the Young's modulus curves (E(T) curves) of the substrate and mold compounds are presented in Figures 5a and 6a, respectively. The Young's modulus curve is steep and has a negative slope at ambient temperatures of both mold compounds 1 and 2, close to its glass transition temperature (Tg; ca. 165 °C and 130 °C for mold compounds 1 and 2, respectively). A thermal mechanical analyzer (TMA) was employed to measure the CTE of the substrate and mold compounds 1 and 2 over the temperature range of 25–260 °C (Figures 5b and 6b, respectively). The CTE curves of mold compounds 1 and 2 are steep and have a positive slope at ambient temperatures close to Tg. The CTE measurement results obtained for the substrate agree with the IPC-4101 specification [23]. Unlike related studies [8–14], this study input the temperature variation curves E(T) and α(T) of mold compounds and the substrate into the simulation model, which ensured that the material parameters of the simulation model were close to the actual situation.

Table 4. Material parameters of bump, silicon, and substrate.

	Bump-SAC405	Die (Silicon100)	Substrate
Young's modulus (GPa)	53	131	Figure 5a
Poisson's ratio	0.40805	0.27	0.2
Density (kg/m^3)	7445.45	2330	1938
CTE (ppm/°C)	20	2.8	Figure 5b

Table 5. Material parameters of mold compounds.

	Mold Compound 1	Mold Compound 2
Young's modulus (GPa)	Figure 6a	Figure 6a
Poisson's ratio	0.3	0.3
Density (kg/m^3)	2010	1990
CTE (ppm/°C)	Figure 6b	Figure 6b

(a)

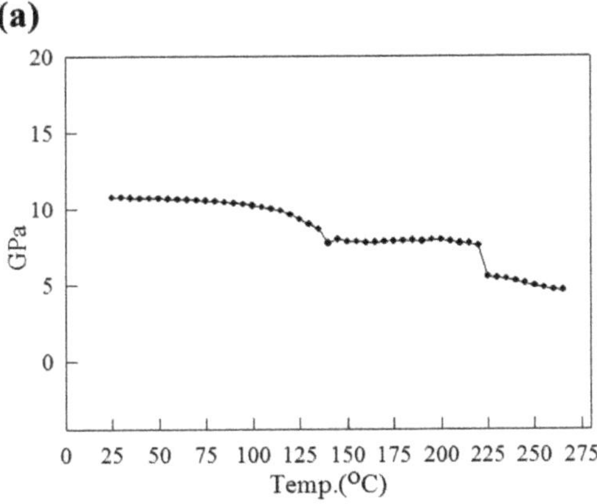

(b)

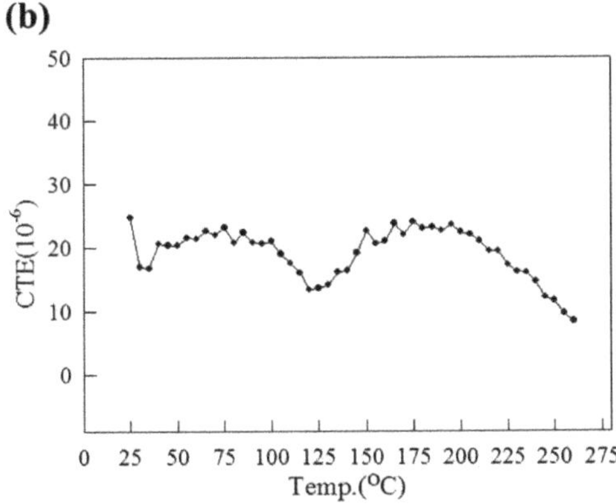

Figure 5. (a) Young's modulus E(T) and (b) CTE α(T) of substrate.

As illustrated in Figure 2, the bump was located between the die and the substrate and was used for signal connection. According to the mechanics of materials [24], when the stress of a material exceeds the yield stress, the material is no longer a linear elastic material, and it undergoes plastic deformation. The stress–strain diagram of an SAC405 bump was presented in [25]. The stress–strain diagram obtained in the present study (Figure 7) indicates a yield stress of 26 MPa and was input into the simulation model. According to [26], a creep effect can occur in a metal when the ambient temperature exceeds one-third of its melting point, and one-third of the melting point of an SAC405 bump is 72.28 °C. The flip-chip considered in this study had a process temperature of ≥175 °C. This study was different from other related studies [8–14] as the plastic effect of a bump [25] and the creep effect [26] were considered in this model. Both effects were input into the simulation model to match a realistic situation.

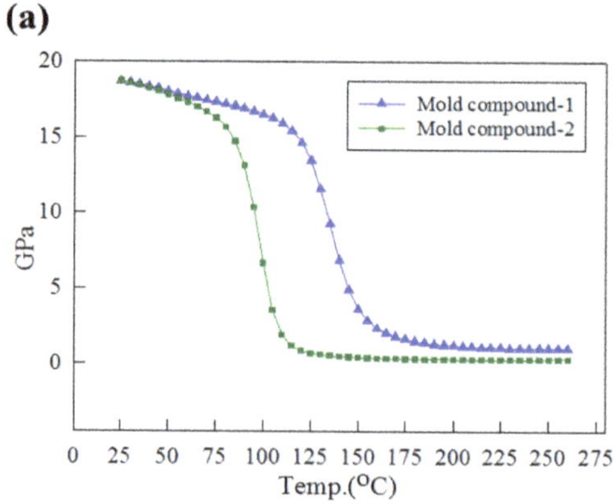

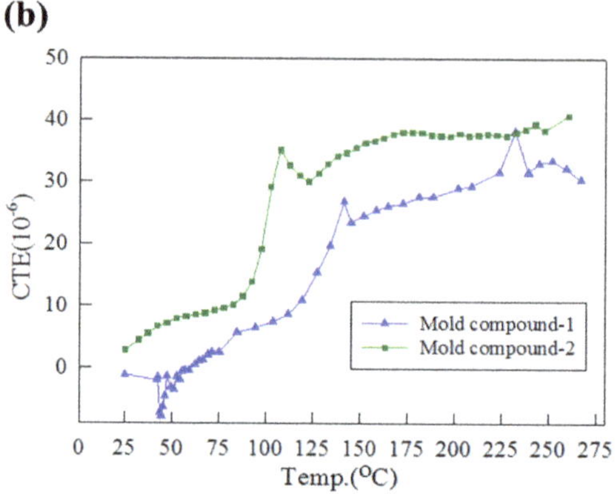

Figure 6. (**a**) Young's modulus E(T) and (**b**) CTE α(T) of mold compounds.

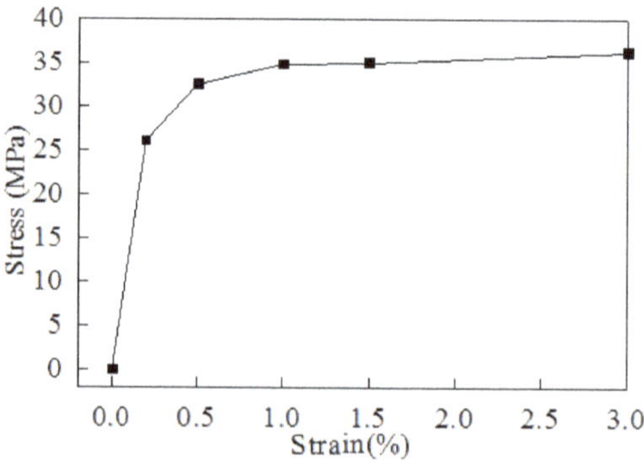

Figure 7. Stress–strain curve of the SAC405 bump [25].

2.1.3. Boundary Condition Settings

The process flow of the flip-chip on the production line consisted of twelve steps, as illustrated in Figure 1. The issue was warpage generation due to the unmatched CTEs of the package materials. Warpage mostly occurred after thermal processes. The six steps involving thermal processes were reflow, pre-MD (molding) baking, molding, PMC, pretreatment, and ball mounting. However, according to experience from production line workers, strips are almost flat after Steps 3–7, as illustrated in Figure 8a. Even when the mold compound is combined with the other materials into the strip package at 175 °C during molding, warpage had not yet occurred, and the strip was almost flat. However, severe warpage occurred when the strip was cooled to a room temperature of 25 °C, as shown in Figure 8b. In related studies [8–14], the simulation model was set for a stress-free state at a molding temperature of 175 °C. As illustrated in Figure 8, the stage and strip at both ends of the strip model were defined as contact points at position A, and the stage was a simple support. For any end point A located between $y = 0$ and w (where w is the width of the strip), the following 2D boundary condition can be imposed: point A is allowed to move freely along the x-direction with an arbitrary displacement (displacement along the x-direction is u_x = constant) but fixed along the z-direction (displacement along the z-direction is $u_z = 0$).

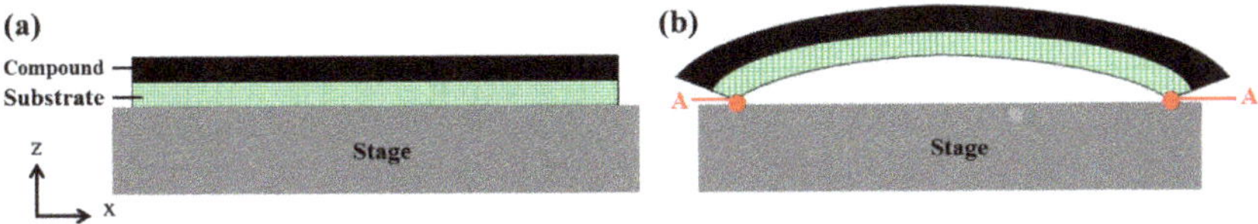

Figure 8. Strip placed on the stage: (**a**) at 175 °C after molding and (**b**) after cooling to 25 °C.

According to [8,9,11,13,14], warpage mainly occurs during molding. Therefore, the simulation model starts from the molding process (Step 8). Continuous simulation calculations of different processes were conducted to simulate Steps 8–11 of the thermal process. Table 6 details the process temperature ranges and times for these steps. First, the molding process was simulated (Step 8) by cooling the system from 175 °C to a room temperature of 25 °C. Subsequently, the temperature was increased to and then maintained at 175 °C for 240 min during the PMC process (Step 9) to completely cure the mold compound and eliminate internal stress. Finally, in Steps 10 and 11, the reflow process for pretreatment of the solder ball and implantation of the ball into the substrate was simulated.

Table 6. Process temperature ranges and times for flip-chip.

Process Flow	Temperature Range (°C)	Process Time (s)
8. Molding	175 → 25	30
9. PMC	25 → 175 → 25	20,040
10. Pre-treatment	25 → 238 → 25	935
11. Ball mount	25 → 238 → 25	935

2.2. Taguchi Method

Various variables, such as geometric structure and process temperature, can affect the warpage during the process. Few scholars have simultaneously studied the effects of different geometric structures and process temperatures on strip warpage. Therefore, this study explored the effects of process temperature and geometric thickness of the mold compound, substrate, and die on strip warpage. The process temperature was changed mostly in Steps 8 and 9. Hence, this study used Taguchi's orthogonal arrays to establish an L16 variable combination. Table 7 lists the control factors and their settings. This study explored five factors: three related to structural thickness (thickness of the mold compound, die, and substrate) and two related to process temperature (molding and PMC temperatures).

Table 7. Control factors and their settings.

Name of Control Factors	Level 1	Level 2	Level 3	Level 4
Mold compound thickness(μm)	150	450	750	1100
Die thickness (μm)	75	150	250	400
Substrate thickness (μm)	100	180	300	500
Molding temp. (°C)	150	165	175	185
PMC temp. (°C)	175	185		

2.3. Structural Design

IC packages are composed of numerous different materials and geometrical shapes. This study used the neutral axis theory of composite materials [24] as the structural design criteria. Severe strip warpage occurs when the strip is cooled to room temperature (25 °C). Therefore, this study investigated the relationship between the z_n-coordinate of the neutral axis and warpage at 25 °C. Equation (1) was used to calculate the neutral axis z-coordinate formula of the composite material, where $A_{Mold\ compound}$, A_{Die}, A_{Bump}, and $A_{Substrate}$ are the areas of the mold compound, die, bump, and substrate, respectively. Table 8 lists the Young's modulus of each material at 25 °C. The Young's moduli were normalized by dividing each by the minimum. n represents the transformation factor, and $n_{Mold\ compound}$, n_{Die}, n_{Bump}, and $n_{Substrate}$ represent the proportional Young's moduli of the materials. Furthermore, $z_{Mold\ compound}$, z_{Die}, z_{Bump}, and $z_{Substrate}$ are the centroid z-coordinates of the mold compound, die, bump, and substrate, respectively. Substituting the values in Table 8 into Equation (1), the z-coordinate z_n of the neutral axis of the 2D strip in Figure 3 was obtained as 280.61 μm.

$$z_n = \frac{A_1 \times n_1 \times z_1 + A_2 \times n_2 \times z_2 + A_3 \times n_3 \times z_3 + A_4 \times n_4 \times z_4}{A_1 \times n_1 + A_2 \times n_2 + A_3 \times n_3 + A_4 \times n_4} \tag{1}$$

were

$A_1 = A_{Mold\ compound}$, $n_1 = n_{Mold\ compound}$, $z_1 = z_{Mold\ compound}$.
$A_2 = A_{Die}$, $n_2 = n_{Die}$, $z_2 = z_{Die}$;
$A_3 = A_{Bump}$, $n_3 = n_{Bump}$, $z_3 = z_{Bump}$;
$A_4 = A_{Substrate}$, $n_4 = n_{Substrate}$, $z_4 = z_{Substrate}$.

Table 8. Transformation factor n for each material.

Material	Young's Modulus (GPa)	n = Material (E)/Substrate (E)
Substrate	10.75	1
Bump	53	4.93
Die	131	12.18
Mold compound 1	18.66	1.73

3. Results

3.1. Experimental Results

Figure 9 indicates eight points (D1–D8) as the measuring positions along the strip. D4 and D8 were in the middle of the long side of the strip, while D2 and D6 were in the middle of the short side of the strip. A ruler was used to measure the warpage at the start of the molding process at 175 °C and the end of the process at 25 °C. Table 9 lists the experimental values of strip warpage at the reference points during the molding process. The results show that D4 and D8 are the positions at which the maximum warpage occurred. A similar process was applied to obtain the maximum warpage of the strip during the post-mold curing, pre-treatment, and ball mount processes. These results are listed in Table 10.

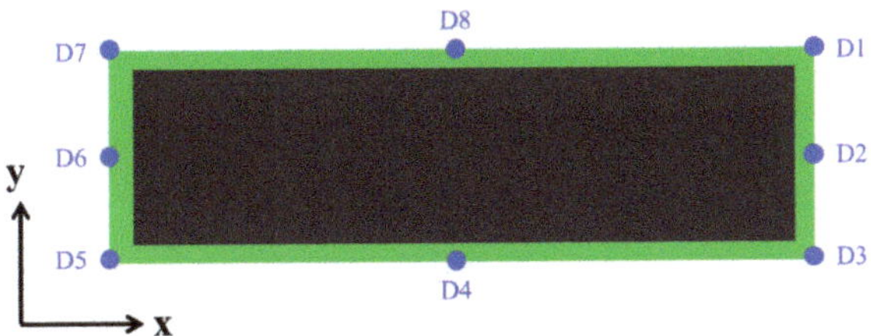

Figure 9. Measuring positions along the strip (D1–D8).

Table 9. Experimental values of strip warpage at reference points.

Process Flow		Strip Warpage (mm)							
		Long Side (x-Direction)		Short Side (y-Direction)		Corner			
		D4	D8	D2	D6	D1	D3	D5	D7
8. Molding	175 °C	0	0	0	0	0	0	0	0
	25 °C	7	7	0	0	0	0	0	0

Table 10. Comparison of strip warpage simulation and experimental values using mold compound 1.

Process Flow	Strip Warpage (mm)		
	Experimental Value	Simulation Value	Error (%)
8. Molding	7	7.36	5.14
9. PMC	5	5.14	2.8
10. Pre-treatment	7	6.73	3.85
11. Ball mount	7.5	6.90	8.00

3.2. Simulation Results

3.2.1. Experimental and Simulation Results

Most related studies only considered a single process step (molding and reflow) [8–14]; in contrast, this study successfully simulated the continuous flip-chip process from molding to ball mounting (Steps 8–11). Table 10 presents a comparison of the simulation results and experimental values, and Figure 10 illustrates the simulation results of strip warpage after each process for mold compound 1. The warpage trends are consistent, revealing a concave shape facing downwards. The simulated and experimental values were similar, with differences all lower than 8%. Therefore, the model established in this study is feasible for simulating the flip-chip process.

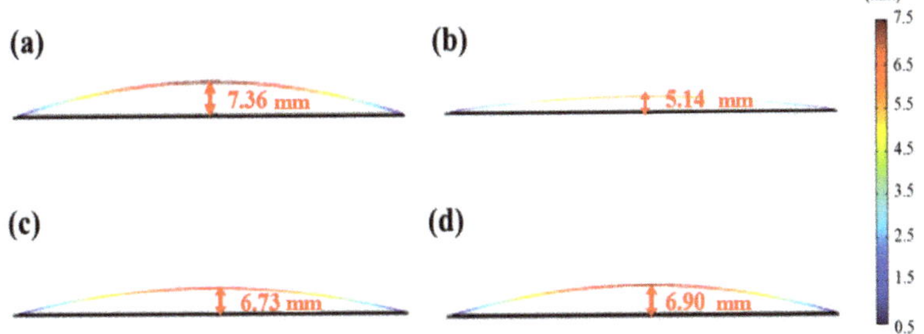

Figure 10. Strip warpage simulation results for mold compound 1: (**a**) Step 8—molding, (**b**) Step 9—PMC, (**c**) Step 10—pretreatment, and (**d**) Step 11—ball mounting.

3.2.2. Comparison of Mold Compounds

To investigate the effect of mold compounds for strip warpage on the production line, the strip warpage of different mold compounds (mold compounds 1 and 2) was compared under the same geometry and process conditions. Table 11 presents a comparison of strip warpage for mold compounds 1 and 2, and the results demonstrate that the strip containing mold compound 2 warped less than mold compound 1, regardless of the specific process step. The two main factors affecting warpage under the same geometric structure and process conditions are as follows:

1. The CTE difference between the mold compound and substrate had a greater impact, the larger the difference, the larger the strip warpage.
2. The Young's modulus of the mold compound exerted an effect, the larger the young's modulus, the greater the structural rigidity of the strip and the lower the warpage.

Table 11. Comparison of strip warpage simulation values for mold compounds 1 and 2.

Process Flow	Strip Warpage (mm)	
	Mold Compound 1	Mold Compound 2
8. Molding	7.36	5.02
9. PMC	5.14	3.48
10. Pre-treatment	6.73	4.65
11. Ball mount	6.90	4.71

As shown in Figures 5b and 6b, the CTE difference between mold compound 2 and the substrate was smaller than that between mold compound 1 and the substrate. In addition, Figure 6a indicates that the Young's modulus of mold compound 2 was slightly larger than that of mold compound 1. The use of mold compound 2 resulted in less strip warpage for each process than the use of mold compound 1.

4. Discussion

4.1. Taguchi Analysis

Since the degree of warpage that occurs during ball mounting (Step 11) directly affects the single-chip yield in the singulation step (Step 12) of the flip-chip process, this study focused on the strip warpage that occurred during the ball mounting process. Figure 11 displays the Taguchi analysis' main effect diagram of the ball mounting warpage. According to Figure 11, the main factors affecting strip warpage are the die and substrate thickness, followed by mold compound thickness and molding temperature. The least influential factor is PMC temperature. This study investigated the effects of these five factors sequentially. First, the effect of die thickness was evaluated. The Young's modulus of the die was relatively high (131 GPa), which indicates that the die was the most rigid material in the strip. Therefore, when the die thickness was increased, the structural rigidity of the strip notably increased, reducing the amount of warpage. Second, the effect of substrate thickness factor was determined. When the substrate thickness was increased to 300–500 µm, the amount of warpage rose sharply to >7 mm; thus, substrate thickness was positively correlated with warpage (Figure 11). Third, the effect of mold compound thickness was investigated. A higher mold compound thickness resulted in greater strip rigidity. However, the warpage was considerably reduced only when the thickness was increased to 1100 µm; thicknesses of 150, 450, and 750 µm were unable to cause large warpage reduction. Fourth, the effect of molding temperature factor was evaluated. The smaller difference between the molding process temperature and room temperature of 25 °C resulted in the lower warpage. Nonetheless, the molding process temperature must be higher than the Tg of the mold compound. Finally, the effect of PMC temperature factor was determined. The results clearly demonstrate that a change in the PMC process temperature has no strong effect on warpage.

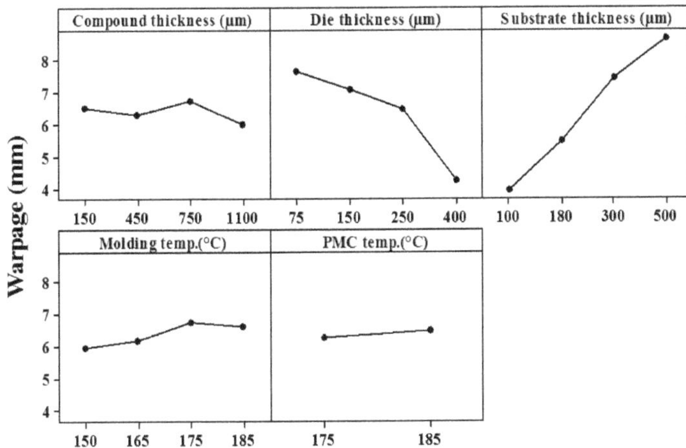

Figure 11. Main effect diagram of factors affecting warpage during ball mounting.

For the current flip-chip structure design, the following four process condition designs can reduce the amount of strip warpage:

1. greater die thickness (>150 μm).
2. greater mold compound thickness (>1100 μm).
3. smaller substrate thickness (<100 μm); and
4. lower molding temperature, although it should not be lower than the Tg of the mold compound.

4.2. Structural Design Criteria

According to the Taguchi analysis presented in Section 4.1, a greater die thickness, greater mold compound thickness, smaller substrate thickness, and lower molding temperature are ideal for reducing warpage. However, the trend in manufacturing is for thinner, lighter, and shorter chips, and increasing the die and mold compound thicknesses to reduce warpage is not optimal. Therefore, this study constructed a structural strip design based on composite material neutral axis theory [24,27] and investigated the relationship between z_n and warpage. Table 12 lists the ball mounting warpage simulation results, where the original parameters reflect the original conditions on the production line, and z_n is the neutral axis z-coordinate of each structural condition. The positions of the structural neutral axis z_n-coordinates were divided into three categories: (1) neutral axis on the mold compound (No. 13), (2) neutral axis on the die (Nos. 2, 3, 4, 5–12, and 14–16), and (3) neutral axis on the bump (No. 1).

First, the neutral axes on the mold compound (No. 13) and die (Nos. 2, 3, 4, 5–12, and 14–16) are discussed. Three characteristics were identified after comparing with the main effect analysis diagram in Figure 11:

1. less warpage occurred when the substrate thickness was <180 μm (e.g., Nos. 16, 12, 6, 11, and 15) and the warpage was <5.15 mm.
2. greater warpage occurred when the substrate thickness was ≥300 μm (e.g., Nos. 14, 7, 9, and 10) and the warpage was >7.92 mm; and
3. Nos. 5 and 8 were special because the substrate thickness was only 180 μm (in No. 5), and the die thickness was only 75 μm, which caused insufficient structural rigidity, resulting in a warpage of 7.63 mm.

Table 12. Warpage simulation results obtained with production-line and L16 parameters.

No.	Mold Compound Thickness (μm)	Die Thickness (μm)	Substrate Thickness (μm)	Molding Temp. ($^\circ$C)	PMC Temp. ($^\circ$C)	Ball Mount Warpage (mm)	z_n (μm)
Original	450	150	180	175	175	6.36	280.61
1	150	75	100	150	175	4.69	156.18
2	150	150	180	165	175	6.21	258.21
3	150	250	300	175	185	7.71	410.69
4	150	400	500	185	185	7.47	661.20
5	450	75	180	175	185	7.63	264.51
6	450	150	100	185	185	4.47	205.25
7	450	250	500	150	175	8.31	602.85
8	450	400	300	165	175	4.75	479.12
9	750	75	300	185	175	9.36	434.78
10	750	150	500	175	175	9.79	620.90
11	750	250	100	165	185	4.81	276.04
12	750	400	180	150	185	2.94	394.61
13	1100	75	500	165	185	8.98	703.36
14	1100	150	300	150	185	7.92	524.47
15	1100	250	180	185	175	5.15	421.30
16	1100	400	100	175	175	1.91	373.81

In contrast, although the substrate and die thicknesses of No. 8 were 300 and 400 μm, respectively, the structural rigidity was sufficient, and thus the warpage was only 4.75 mm. For No. 1 (bump on the neutral axis), even if the mold compound and die thicknesses were only 150 and 75 μm, respectively, with the substrate thickness being only 100 μm, a relatively small warpage of 4.69 mm occurred. Therefore, the group with neutral axis coordinates below the die and above the bump resulted in relatively small warpage.

In summary, the die and substrate thicknesses have a strong effect on warpage in the structural design. This study discovered that in the flip-chip process design, in addition to meeting the production requirement that the mold compound should nearly completely cover the die, two conditions must be fulfilled to minimize the amount of strip warpage: (1) neutral axis on the bump and (2) neutral axis on the die with a die thickness of >150 μm and substrate thickness of <180 μm. If one of these conditions is met, the strip will have a smaller warpage of <5.15 mm. The authors suggest that designers set the neutral axis of the strip structure on the bump, which is more suitable for the current production trend of thin, light, and short design.

5. Conclusions

This study successfully established a strip warpage simulation model of the flip-chip process and investigated the effects of structure design and process (molding, PMC, pretreatment, and ball mounting) on strip warpage. The errors between model and experimental values were less than 8%, indicating that the simulation method can be applied to the flip-chip process steps and can be extended to strip warpage analysis of different mold compounds in the future. In addition, Taguchi analysis was employed to identify the key factors affecting strip warpage, which were discovered to be die thickness and substrate thickness, followed by mold compound thickness and molding temperature. Although greater die and mold compound thicknesses result in less warpage, they cause a substantially greater overall strip thickness, which does not comply with the current trend towards thin, light, and short chips. To overcome this problem, this study proposed the design concept of setting the neutral axis of the strip structure on the bump, which reduces the amount of strip warpage and the overall strip thickness. In summary, the model proposed in this study can be used to evaluate the effect of structural design and process parameters on strip warpage and can provide strip design guidelines for minimizing strip warpage to requirements of the production line. Moreover, the guidelines can accelerate the product development cycle and improve product quality with reduced development costs.

Author Contributions: Conceptualization, W.-C.C.; methodology, W.-C.C.; software, W.-L.C.; validation, W.-C.C. and W.-L.C.; formal analysis, W.-C.C.; investigation, W.-C.C. and W.-L.C.; resources, W.-C.C.; data curation, W.-C.C., W.-L.C., and flip chip assembly process engineering div. Advanced Semiconductor Engineering Inc.; writing—original draft preparation, W.-C.C. and W.-L.C.; writing—review and editing, W.-C.C.; supervision, W.-C.C.; project administration, W.-C.C. All authors have read and agreed to the published version of the manuscript.

Funding: This study was supported by the Ministry of Science and Technology, Taiwan under project MOST 110-2221-E-110-035.

Institutional Review Board Statement: Not applicable.

Informed Consent Statement: Not applicable.

Data Availability Statement: Not applicable.

Acknowledgments: Financial support from the Ministry of Science and Technology, Taiwan is gratefully acknowledged. Support for experimental observations from flip-chip assembly process engineering div. Advanced Semiconductor Engineering Inc. is gratefully acknowledged.

Conflicts of Interest: The authors declare no conflict of interest.

References

1. Timoshenko, S. Analysis of Bi-Metal Thermostats. *J. Opt. Soc. Am.* **1925**, *11*, 233–255. [CrossRef]
2. Chen, C.S. Different Conservation Laws Constructed on Warpage Analyses for Bimaterial Plates With Temperature-Dependent Properties. *J. Electron. Packag.* **2011**, *133*, 1–14. [CrossRef]
3. Elevated Temperature Measurements of Warpage of BGA Package. Available online: https://akrometrix.com/wp-content/uploads/2016/02/Elevated-Temp-BGA-Warpage-Measurements.pdf (accessed on 1 July 2021).
4. Wang, Z.; Wang, X.; Zhao, N.; He, J.; Wang, S.; Wu, G.; Cheng, Y. The desirable dielectric properties and high thermal conductivity of epoxy composites with the cobweb-structured SiCnw–SiO$_2$–NH$_2$ hybrids. *J. Mater. Sci. Mater. Electron.* **2021**, *32*, 20973–20984. [CrossRef]
5. Wang, Z.; Wang, X.; Wang, S.; He, J.; Zhang, T.; Wang, J.; Wu, G. Simultaneously Enhanced Thermal Conductivity and Dielectric Breakdown Strength in Sandwich AlN/Epoxy Composites. *Nanomaterials* **2021**, *11*, 1898. [CrossRef] [PubMed]
6. Wang, J.; Jia, Z.; Liu, X.; Dou, J.; Xu, B.; Wang, B.; Wu, G. Construction of 1D Heterostructure NiCo@C/ZnO Nanorod with Enhanced Microwave Absorption. *Nanomicro Lett.* **2021**, *13*, 175. [CrossRef] [PubMed]
7. Zhou, X.; Jia, Z.; Zhang, X.; Wang, B.; Liu, X.; Xu, B.; Bi, L.; Wu, G. Electromagnetic wave absorption performance of NiCo2X4 (X = O, S, Se, Te) spinel structures. *Chem. Eng. J.* **2021**, *420*, 129907. [CrossRef]
8. Dudek, R.; Walter, H.; Auersperg, J.; Michel, B. Numerical Analysis for Thermo-Mechanical Reliability of Polymers in Electronic Packaging. In Proceedings of the Polytronic 6th International Conference on Polymers and Adhesives in Microelectronics and Photonics, Tokyo, Japan, 15–18 January 2007; pp. 220–227.
9. Hu, G.; Chew, S.; Singh, B. Cure Shrinkage Analysis of Green Epoxy Molding Compound with Application to Warpage Analysis in a Plastic IC Package. In Proceedings of the 8th International Conference on Electronic Packaging Technology, Shanghai, China, 14–17 August 2007; pp. 1–5.
10. Huang, C.Y.; Li, T.D.; Tsai, M.Y. Warpage measurement and design of wBGA package under thermal loading. In Proceedings of the 4th International Microsystems, Packaging, Assembly and Circuits Technology Conference, Taipei, Taiwan, 21–23 October 2009; pp. 415–418.
11. Chae, M.; Ouyang, E. Strip warpage analysis of a flip chip package considering the mold compound processing parameters. In Proceedings of the IEEE 63rd Electronic Components and Technology Conference, Las Vegas, NV, USA, 28–31 May 2013; pp. 441–448.
12. Huber, S.; von Dijk, M.; Walter, H.; Wittler, O.; Thomas, T.; Lang, K.D. Improving the FE simulation of molded packages using warpage measurements. *Microelectron. Reliab.* **2014**, *54*, 1862–1866. [CrossRef]
13. Bin, G.; Dimaano, J.; Chen, R.; Bool, E.; Shi, S.F.; Ang, C.G.; Suthiwongsunthorn, N. Unit warpage control with universal die thickness. In Proceedings of the IEEE 16th Electronics Packaging Technology Conference, Singapore, 3–5 December 2014; pp. 303–306.
14. Zheng, J.; Zhou, E.; Wang, L.; Aldrete, M.; Kumar, R.; Syed, A. Relative and absolute warpage modeling on molded packages. In Proceedings of the IEEE 65th Electronic Components and Technology Conference, San Diego, CA, USA, 26–29 May 2015; pp. 1538–1545.

15. Chen, C.; Yu, D.Q.; Wang, T.; Xiao, Z.Y.; Wan, L.X. Warpage Prediction and Optimization for Embedded Silicon Fan-Out Wafer-Level Packaging Based on an Extended Theoretical Model. *IEEE Trans. Compon. Packag. Manuf. Technol.* **2019**, *9*, 845–853. [CrossRef]
16. Tsai, M.Y.; Wang, Y.W. A Theoretical Solution for Thermal Warpage of Flip-Chip Packages. *IEEE Trans. Compon. Packag. Manuf. Technol.* **2020**, *10*, 72–78. [CrossRef]
17. Tsai, M.Y.; Wang, Y.W.; Liu, C.M. Thermally Induced Deformations and Warpages of Flip-Chip and 2.5D IC Packages Measured by Strain Gauges. *Materials* **2021**, *14*, 3723. [CrossRef] [PubMed]
18. Yao, X.J.; Fang, J.J.; Zhang, W. A Further Study on the Analytical Model for the Permeability in Flip-Chip Packaging. *J. Electron. Packag.* **2018**, *140*, 011001–011006. [CrossRef]
19. Yao, X.; Jiang, W.; Yang, J.; Fang, J.; Zhang, W. A Surface Energy Approach to Developing an Analytical Model for the Underfill Flow Process in Flip-Chip Packaging. *J. Electron. Packag.* **2021**, *144*, 041003–041014. [CrossRef]
20. Panigrahy, S.K.; Tseng, Y.C.; Lai, B.R.; Chiang, K.N. An Overview of AI-Assisted Design-on-Simulation Technology for Reliability Life Prediction of Advanced Packaging. *Materials* **2021**, *14*, 5342. [CrossRef] [PubMed]
21. Lin, P.Y.; Lee, S. Warpage Modeling of Ultra-Thin Packages Based on Chemical Shrinkage and Cure-Dependent Viscoelasticity of Molded Underfill. *IEEE Trans. Device Mater. Reliab.* **2020**, *20*, 67–73. [CrossRef]
22. Cheng, H.C.; Tai, L.C.; Liu, Y.C. Theoretical and Experimental Investigation of Warpage Evolution of Flip Chip Package on Packaging during Fabrication. *Materials* **2021**, *14*, 4816. [CrossRef] [PubMed]
23. IPC-4101: Specification for Base Materials for Rigid and Multilayer Printed Boards. Available online: https://www.ipc.org/TOC/TOC-IPC-4101E.pdf (accessed on 1 July 2021).
24. Hibbeler, R.C. *Mechanics of Materials*, 9th ed.; Pearson: London, UK, 2014.
25. Yeh, C.; Lai, Y. A numerical approach towards the correlation between ball impact test and drop reliability. In Proceedings of the 8th Electronics Packaging Technology Conference, Singapore, 6–8 December 2006; pp. 161–167.
26. Syed, A. Accumulated creep strain and energy density based thermal fatigue life prediction models for SnAgCu solder joints. In Proceedings of the 54th Electronic Components and Technology Conference, Las Vegas, NV, USA, 4 June 2004; pp. 737–746.
27. Chuang, W.C.; Chen, W.L. Investigation of Strip Warpage Behavior in Wire Bonding Process. *J. Electron. Packag.* **2020**, *142*, 021002–021014. [CrossRef]

Article

The Effect of the Crucible on the Temperature Distribution for the Growth of a Large Size AlN Single Crystal

Yue Yu, **Botao Liu, Xia Tang**, **Botao Song, Pengfei Han, Sheng Liu and Bing Gao ***

The Institute of Technological Sciences, Wuhan University, Wuhan 430072, China; 2019106520025@whu.edu.cn (Y.Y.); 2018106520020@whu.edu.cn (B.L.); 2018106520022@whu.edu.cn (X.T.); 2020106520027@whu.edu.cn (B.S.); 2020106520028@whu.edu.cn (P.H.); victor_liu63@vip.126.com (S.L.)
* Correspondence: gaobing@whu.edu.cn

Abstract: The appropriate distribution of temperature in the growth system is critical for obtaining a large size high quality aluminum nitride (AlN) single crystal by the physical vapor transport (PVT) method. As the crystal size increases, the influence of the crucible on the temperature distribution inside the growth chamber becomes greater. In order to optimize the field of temperature and study the specific effects of various parts of the crucible on the large size AlN single crystal growth system, this study carried out a series of numerical simulations of the temperature field of two crucibles of different materials and put forward the concept of a composite crucible, which combines different materials in the crucible parts. Four composite crucible models were established with different proportions and positions of tantalum carbide (TaC) parts and graphite parts in the crucible. Calculations reveal that different parts of the crucible have different effects on the internal temperature distribution. The axial temperature gradient at the crystal was mainly governed by the crucible wall, whereas the temperature gradient was determined by the integrated effect of the crucible lid and the crucible wall in the radial direction. One type of composite crucible was chosen to minimize the thermal stress in grown AlN crystal, which is applicable to the growth of large sized AlN crystals in the future; it can also be used to grow AlN single crystals at present as well.

Keywords: numerical simulation; thermal design; thermal stress; PVT growth; AlN single crystals

1. Introduction

As a third-generation semiconductor material, AlN has the advantages of a wide band gap, high resistivity, and high thermal conductivity [1]. Furthermore, both AlN and gallium nitride (GaN) crystals have a wurtzite structure, and their lattice and thermal expansion coefficients are quite close. Thus, compared with sapphire and silicon carbide (SiC), AlN is ideal for III-nitrides epitaxial growth and the ternary compounds [2]. However, at present, due to the lack of ideal III-nitride single crystal substrates, it has severely restricted the improvement of the lifetime and performance of III-nitride lasers and microelectronic devices. Therefore, it is necessary to research and develop the growth technology of AlN bulk single crystals to provide lattice-matched substrate materials for the epitaxial growth of devices and materials.

Due to the fact that the melting points of AlN are theoretically calculated to be as high as 2800 °C and the dissociation pressure to be 20 MPa, it is challenging to apply the melt Czochralski method to grow an AlN single crystal [3]. An AlN bulk single crystal generally grows at a high temperature via the PVT method [4]. Since the performing experiments are too time-consuming and expensive, numerical simulations allow preliminarily studies of the growth mechanism of an AlN bulk single crystal and the design of optimization schemes for the crystal growth process [5]. Some valuable simulation studies on the growth and material properties of AlN bulk single crystals have been conducted after Slack and Mcnelly [6] reported high purity AlN single crystal growth by the PVT method at an early stage. Dryburgh [7] estimated the maximum possible rate of substance transport and determined the possible rate-limiting steps by simple kinetic theory during sublimation. Inspired by this, Segal et al. [8] formulated the first one-dimensional model considering the convection and diffusion transport, as well as the kinetic constraints on the desorption and adsorption of N_2. After that, Liu [9] developed an elaborate 2D model containing both Stefan flow and thermal convection. Then, Bogdanov et al. [10] evaluated mass transport in the crucible to learn the species interchange effect upon the aluminum nitride growth rate about the environment and the crucible. The first global model was developed by Liu and Edgar [11] for simulating AlN sublimation growth including surface kinetics. To summarize the research mentioned earlier, Wu and Zhang [12] set out the model of diffusive transport, the model ruled by Al vapor, and the model ruled by N_2, and the operational terms that allow the implementation of different models were determined. Moreover, Cai et al. [13] accomplished the simulation by developing an internal integrated model to characterize conductive, radiative, and inductive heat transfer. Furthermore, Lee et al. [14] adopted a 3D numerical finite element modeling method to provide a detailed comparison of the distributions of residual interfacial thermal stress induced in AlN crystals deposited upon various substrates. Wolfson [15] investigated the growth rate dependence on the N_2 pressure. In the last decade, the numerical model has been constantly improved. Gao et al. [16] designed a compressible flow model, which is full-coupled to investigate the mass and sublimation transfer during the AlN crystal growth. Wang et al. [17,18] developed a completely 2D incompressible flow model and a 3D thermoelastic stress model to examine the effects of the crucible configuration on the mass distribution and transport and development of the total resolved shear stress of AlN crystal with numerical experiments.

The above research was all based on the small size AlN single crystal, which is below 50 mm. Since the growth process generates dislocations and defects, it is challenging to grow a larger-size AlN single crystal under the current experimental conditions. A numerical study can overcome this obstacle. The problem is solved and optimized to provide guidance and ideas for subsequent lab tests and commercial production of the AlN single crystal. The crucible is the main component used for crystal growth, and as the crystal size increases, the crucible has an increasing influence on the temperature distribution inside the cavity. Since crystals grown by the PVT method are very sensitive to temperature, and changes in temperature distribution can seriously affect the crystal growth quality, it is also necessary to study the effect of the crucible on the temperature distribution during the growth of large size AlN crystals.

Based on the above, the numerical simulations are conducted to accurately restore the temperature distributions of two crucibles with different materials, graphite crucible and TaC crucible, when growing a 90 mm AlN single crystal. Furthermore, the concept of the composite crucible combining two crucible materials is proposed, and the influence of different crucible parts on the temperature distribution is obtained by changing the proportion of the two crucible materials in the crucibles. Finally, one type of composite crucible that can reduce the internal thermal stress of the AlN crystal while maintaining a regular growth rate is carried out.

2. Simulation

The basic process of the PVT method for growing AlN crystal can be described as follows: In a high-temperature field, AlN charge powder decomposes and sublimates to produce the gaseous components aluminum vapor (A1) and N_2 vapor, driven by the negative axial temperature gradient; the gas phase components move to the AlN seed with relatively low temperature, and adsorb, migrate, crystallize, and desorb on the growth interface. This process continues. Then, the growth interface continues to move to the AlN powder region and grows into an AlN bulk crystal. Figure 1 presents a sketch of the growth cell of AlN crystal grown via the PVT method. In addition, the growth chamber is in a nitrogen atmosphere, it is heated by an induction coil (10 kHz), and the crucible needs to be covered by graphite felt to keep the temperature [19,20].

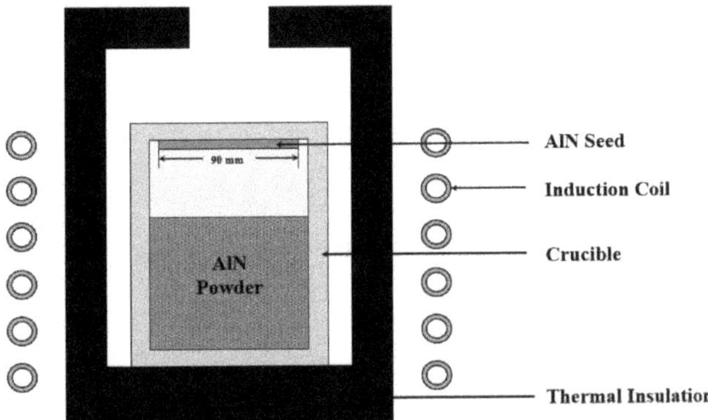

Figure 1. A sketch of the growth cell of AlN crystal.

Since the principal objective of this study deals with the optimization and simulation of the temperature field, according to [21], the gas convective heat transfer plays a small role in the temperature field. It can be negligible during the computation. Due to the symmetry of the crystal structure as well as the structure of the growth furnace system, the model can be simplified to a two-dimensional axisymmetric model in a cylindrical coordinate system. The models are analyzed by the finite element method (FEM). Assuming that the growth is in a steady-state equilibrium, the simulation steps can be summarized below: First, the temperature profile of the growth furnace is calculated with an established heat transfer model that considers heat conduction and heat radiation. Then, the first step temperature profile is taken as a boundary condition to calculate the intact model, including thermal stress.

The heat transfer governing equation is characterized as [22]:

$$\rho C_p \frac{\partial T}{\partial t} + \nabla (k \nabla T) = Q \tag{1}$$

$$\frac{Q_j}{\varepsilon_j} - \sum_{k=1}^{N} F_{j,k} \frac{1 - \varepsilon_k}{\varepsilon_k} Q_k = \sigma T_j^4 - \sum_{k=1}^{N} F_{j,k} \sigma T_k^4 \tag{2}$$

where ρC_p is the effective heat capacity, T is the Kelvin temperature, t is the time, k is the thermal conductivity, Q is the radiative heat flux, ε_j, ε_k is the emissivity, $F_{j,k}$ is the view factor, and σ is the Stefan Boltzmann constant.

The thermal physical and material properties of the system components are shown in Table 1 [23].

Table 1. Physical parameters of the materials in the growth chamber for simulation.

Material	Thermal Conductivity (W/m·K)	Density (kg/m³)	Heat Capacity (J/kg·K)
Graphite Crucible	115	1760	720
TaC Crucible	22	14,300	60.65
Insulation	0.5	170	2100
AlN powder	22.55	27, 0.34	11, 72.7
AlN seed	320	3250	1197

Since AlN crystal growth is in a nonequilibrium state, thermal stress is induced within the crystal throughout the growth process, which is the main driving force for the creation, slip, and proliferation of dislocations in the AlN crystal [24].

Assuming that the material is a linear elastic body, the governing equations of the thermal stress field are [25]:

$$\frac{1}{r}\frac{\partial}{\partial r}(r\sigma_{rr}) + \frac{\partial \tau_{rz}}{\partial z} - \frac{\sigma_{\varphi\varphi}}{r} = 0 \tag{3}$$

$$\frac{1}{r}\frac{\partial}{\partial r}(r\tau_{rz}) + \frac{\partial \sigma_{zz}}{\partial z} = 0 \tag{4}$$

where σ_{rr}, $\sigma_{\varphi\varphi}$ and σ_{zz} denote the normal stresses; τ_{rz} denotes the shear stress.

In addition, according to Hooke's law of thermoelastic solids, AlN crystal is a thermoelastic anisotropic body, and the stress–strain relation of AlN crystal can be taken as [26]:

$$\begin{pmatrix} \sigma_{rr} \\ \sigma_{\varphi\varphi} \\ \sigma_{zz} \\ \tau_{rz} \end{pmatrix} = \begin{pmatrix} c_{11} & c_{12} & c_{13} & 0 \\ c_{12} & c_{22} & c_{23} & 0 \\ c_{13} & c_{23} & c_{33} & 0 \\ 0 & 0 & 0 & c_{44} \end{pmatrix} \times \begin{pmatrix} \varepsilon_{rr} - \alpha_r\left(T - T_{ref}\right) \\ \varepsilon_{\varphi\varphi} - \alpha_\varphi\left(T - T_{ref}\right) \\ \varepsilon_{zz} - \alpha_z\left(T - T_{ref}\right) \\ \varepsilon_{rz} \end{pmatrix} \tag{5}$$

where c_{ij} represents the elastic constant, ε_{rr}, $\varepsilon_{\varphi\varphi}$, ε_{zz}, ε_{rz} represent the strain components, α_r, α_φ, α_z represent the thermal expansion coefficients, and due to the hexagonal structure, the thermal expansion coefficient of AlN crystal has only two independent components, $\alpha_r = \alpha_\varphi$. T_{ref} represents the reference temperature, which is chosen as the lowest temperature in the AlN seed.

The strain components can be described as [27]:

$$\varepsilon_{rr} = \frac{\partial u}{\partial r}$$

$$\varepsilon_{\varphi\phi} = \frac{u}{r}$$

$$\varepsilon_{zz} = \frac{\partial w}{\partial z} \tag{6}$$

$$\varepsilon_{rz} = \frac{\partial u}{\partial z} + \frac{\partial w}{\partial r}$$

where u and w represent the horizontal and vertical displacements, respectively.

3. Results and Discussions

3.1. Effect of Crucible Material on Temperature Distribution

Two growth systems were set up in simulation with a difference in crucible materials. According to the literature [28,29], the two most commonly used crucible materials for AlN growth by PVT are high-purity tungsten (W) metal and tantalum carbide, which are selected based on the fact that the two crucible materials do not interact with each other and do not have the same heat transfer capacity. Since W reacts with TaC at high temperatures to form WC, which affects the quality and lifetime of the composite crucible, graphite and TaC were chosen to meet both conditions and were easy to prepare. The temperature distributions of the two crucibles under the same heating conditions are displayed in Figure 2. In general, the temperature distributions of both crucibles were quite similar. The temperature distribution in the crystal growth region exhibited a trend of the lower high and upper low. Thus, a negative temperature gradient formed axially in the gas area between the AlN crystal and the powder source (the growth region). It was the major driving force of the AlN crystal growth. In contrast, at the bottom part, the temperature distribution in the graphite crucible was gentler than that in the TaC crucible. At the top part, it was the exact opposite condition. The influence of the crucible material on the temperature distribution was different in various parts. This difference was mainly related to the thermal conductivity of the crucible material itself. The higher the thermal conductivity inside the material, the higher the temperature gradient will be.

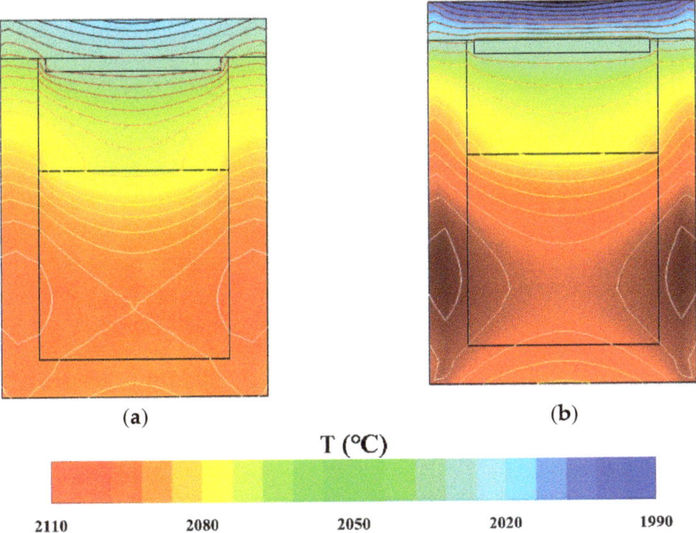

Figure 2. Distributions of temperature in (**a**) Graphite crucible and (**b**) TaC crucible.

During the simulation of AlN crystal growth, the influence of impurities was ignored. The crucible mainly contains two vapor species, Al and N_2. The crystal growth rate V_g is deduced from the growth kinetics, which is expressed as follows [30]:

$$V_g = k_g \frac{\exp\left(A - \frac{B}{T}\right)}{P^{1.5}T^{1.2}} \frac{\Delta T}{z} \tag{7}$$

where k_g is the growth rate coefficient obtained from the experiment, k_g = 407.539, A and B are the constants based on the thermodynamic data, A = 27.055, B = 75788, P is the internal pressure of the furnace, and ΔT and z are the difference of temperature and the length between the AlN powder and AlN seed, respectively.

As the gas convective is neglected in the calculation, the pressure P is regarded as a fixed value in this research, P = 300 Torr. According to Equation (8), only the temperature distribution dictates the growth rate. We set the center of the crystal surface as the temperature control point. Then the growth rate of the AlN crystal was mainly related to the axial gradient of temperature in the growth region.

Figure 3 illustrates the growth rates of AlN crystals grown in the two crucibles. The growth rate was linearly related to the reciprocal of the temperature. The profiles reveal that AlN crystal grew faster in the TaC crucibles than in the graphite crucibles. From Figure 2, the formation of a higher temperature gradient is evident on the longitudinal wall of the TaC crucible. So, the growth rate of the AlN crystal grown in it is higher than that of the graphite crucible.

For the purpose of increasing the diameter of the epitaxial growth crystals, it is essential to establish a reasonable temperature distribution of the as grown AlN crystal in the radial direction as well. An excessively large radial temperature gradient causes anisotropic thermoelastic stress of the crystal, resulting in the formation and propagation of various defects, such as low-angle grain boundaries (LAGB), basal plane dislocations (BPD), and so on. The multiple defects formulated during growth are the dominant limitations to AlN single crystal growth. Therefore, to successfully grow a 90 mm aluminum nitride single crystal, the thermal stress generated within it should be minimized.

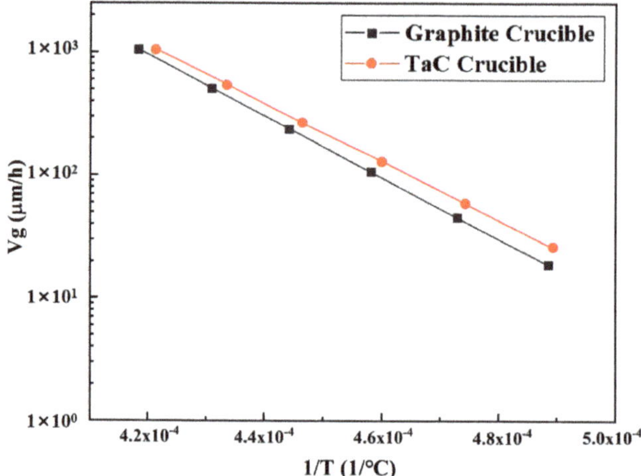

Figure 3. Growth rates of the AlN crystal grown in different crucibles.

Current theory believes that the resolved shear stress (RSS) due to the inhomogeneous radial temperature distribution is the main driving force for creating, slipping, and proliferation of the dislocations in the grown AlN single crystal. As a reference, the Von Mises stress (VMS) can be applied to evaluate the level of the stress inside the crystal [31].

$$\sigma_{Mises} = \sqrt{\frac{1}{2}\left[\left(\sigma_{zz} - \sigma_{rr}\right)^2 + \left(\sigma_{zz} - \sigma_{\varphi\varphi}\right)^2 + \left(\sigma_{\varphi\varphi} - \sigma_{rr}\right)^2 + 6\tau_{rz}\right]} \tag{8}$$

When the VMS stress of the AlN crystal excels the critical resolved shear stress (CRSS), dislocations or even cracks occur. Therefore, reducing the VMS stress is an effective method to reduce the risk of dislocation defect formation minimizing the density of dislocations as well as the macro cracks in AlN crystal.

It is assumed that the crystal surface attached to the crucible lid is rigid, while the other two surfaces of the crystal are not in contact with the crucible wall and stress-free. The boundary conditions for thermal stress calculations can be expressed as:

$$\begin{aligned}
u = 0, v = 0 \quad &at\ z = 0 \\
u = 0, \tfrac{\partial v}{\partial r} = 0 \quad &at\ r = 0 \\
\sigma \cdot n = 0 \quad &at\ the\ two\ free\ moving\ surfaces
\end{aligned} \tag{9}$$

The Von Mises stress distributions inside the grown AlN crystals in the two crucibles are shown in Figure 4. The maximum stress appeared at the crystal top edge on account of the rigid body. The minimum stress occurred at the center of the crystal surface attached to the crucible lid, where the temperature was the lowest, and the stress distributions near the crystal surfaces tended to be flat. The stress level of the AlN crystal grown in the graphite crucible was relatively high. As for the TaC crucible, since the internal radial temperature distribution is more uniform, the influence of the rigid body constraint was reduced. So, the stress of the fixed AlN crystal grown inside it is smaller. In addition, the high temperature at the free surface of the crystal led to higher stress there.

From the above calculations, the high thermal resistance of the TaC crucible allows for more uniform temperature distribution in the radial direction, which results in lower thermal stress.

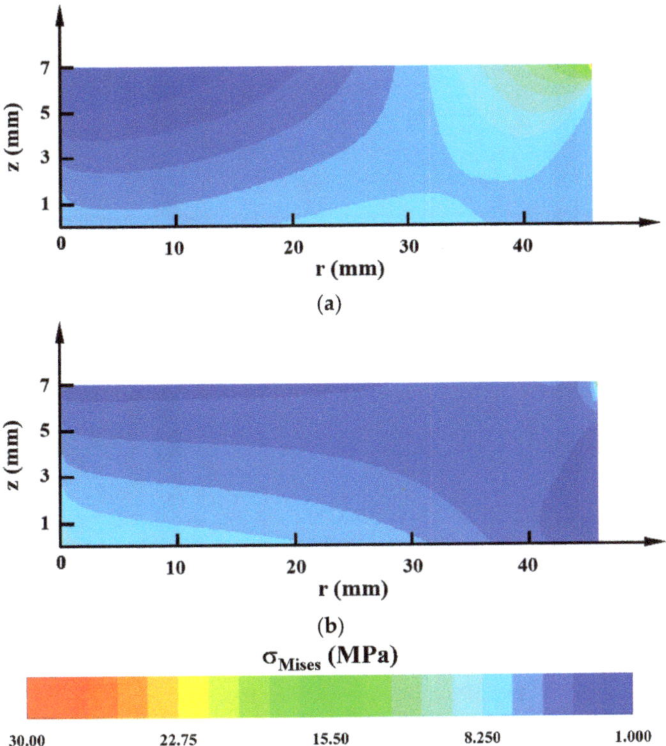

Figure 4. Distributions of thermal stress inside the AlN crystals grown in (**a**) Graphite crucible and (**b**) TaC crucible.

The difficulty in growing a large-sized AlN single crystal lies in the huge thermal stresses generated during the lateral growth of the crystal. As a result, the crystal defects and dislocations increase, which leads to failure growth. Judging by the calculation of this study, when a material with low thermal conductivity is used as a crucible, the crystal growth rate is high and the growth quality is high, so there is a greater potential to grow large size crystals.

3.2. Effect of Crucible Parts on Temperature Distribution

During the calculation, it was found that the various crucible parts had different effects on the distribution of temperature inside the crucible. Therefore, as shown in Figure 5, the above two materials (graphite and TaC) were combined into composite crucibles to determine the impact of the different crucible components on the temperature field. The main difference between these four crucibles is the different proportions of TaC and graphite forming the crucible. It can be seen as replacements of parts of the graphite crucible by TaC, and the replacements change from top to bottom. The details are explained in the description of Figure 5. As this study targets optimizing the growth conditions for large sized AlN single crystal, the dimensions follow the previously designed growth chambers for growing 90 mm AlN single crystal.

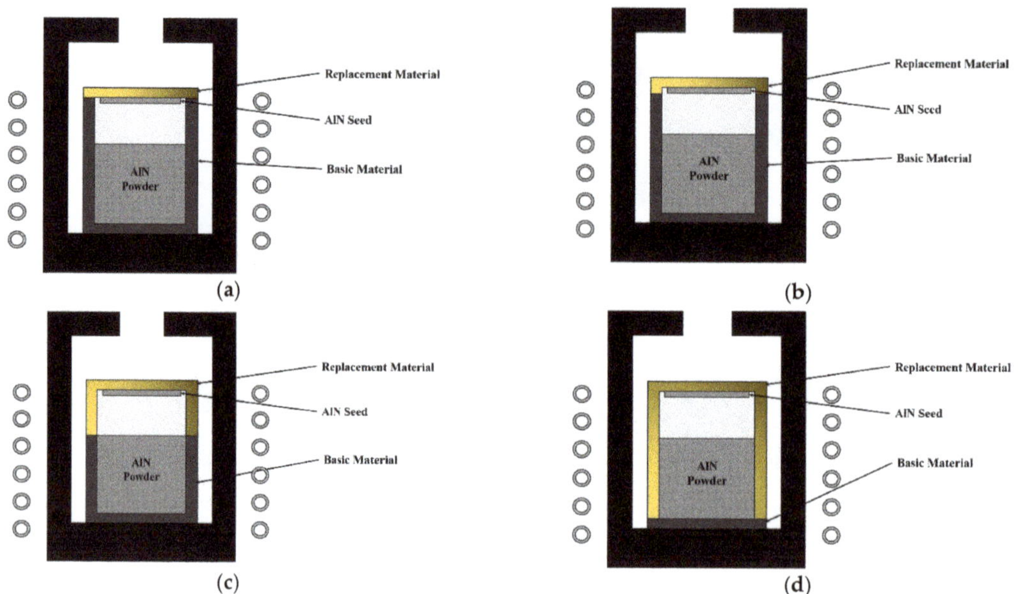

Figure 5. Schematic diagrams of the composite crucibles (**a**) Composite crucible A, (**b**) Composite crucible B, (**c**) Composite crucible C, (**d**) Composite crucible D.

Figure 6 shows the growth rates of the AlN crystals grown in four composite crucibles. Compared with the monomaterial crucibles, the growth rates of the four composite crucibles were higher, of which composite crucible C had the highest growth rate and A the lowest. As mentioned above, the crystal growth rate in this study was mainly affected by the axial temperature gradient. From a partial view, when the crucible parts were of varying thermal conductivity, the stronger thermally conductive part had a higher heat flux. Therefore, the growth rates of AlN crystals in the composite crucibles were higher than in the graphite crucible. On the whole, the combination of two materials with different thermal conductivity generated additional thermal resistance on the contact surface and increased the temperature difference, so that the growth rates were higher than that of the pure TaC crucible also.

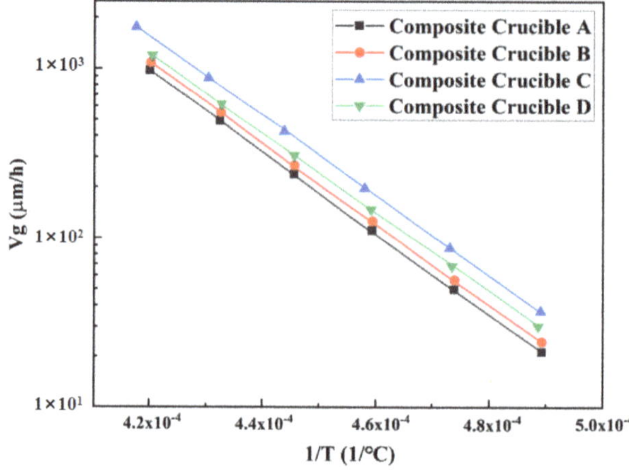

Figure 6. Comparison of the growth rates of the AlN crystals grown in four composite crucibles.

Because the longitudinal length of TaC in the composite crucible A and B occupied a very low proportion of the entire crucible, the axial temperature gradients in those two crucibles were lower. However, the gas area of the crucible is the main region for crystal growth, so variations in crucible material in this part significantly affected either the axial temperature distribution or the growth rate. Therefore, the growth rate of crucible C was the largest among the four crucibles. As for the composite crucible D, only the material of the crucible bottom is graphite, which has higher thermal conductivity, so its heat transfer effect was worse than that of C, resulting in a relatively low temperature gradient in the axial direction.

Figure 7 shows the radial temperature distributions of the horizontal free surfaces in the four composite crucibles. The radial temperature gradient turned from positive to negative as the proportion of TaC in the crucibles increased. However, when the TaC crucible part exceeded the AlN powder region, the negative temperature distribution tended to become flat again. It implies that the temperature distribution in the radial direction was not only dependent on the crucible lid, but also on the axial part of the crucible. The radial temperature profile on the surface of crystal grown in the composite crucible B was affected most, which was even flatter than that of the TaC crucible. It indicates that the crystal and its vicinity are the main components that govern the temperature distribution of the AlN crystal in the radial direction.

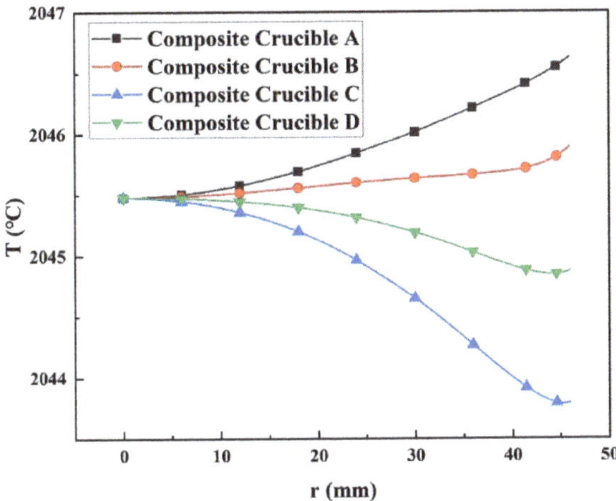

Figure 7. Radial temperature distributions of the AlN crystal horizontal free surface grown in four composite crucibles.

As Figure 8 shows, the AlN crystal grown in the composite crucible A had a thermal stress distribution resembling that of the W crucible but with a lower stress level. The thermal stress distributions of the crystals in the other three composite crucibles were similar to that of the TaC crucible. Among them, only the thermal stress level of the crystal in the composite crucible B was lower than that of the TaC crucible. Moreover, it is worth noting that the positive and negative radial temperature distributions had a remarkable impact on the level of thermal stress inside the crystals, and the negative radial temperature distribution produced a much higher minimum thermal stress. This was mainly attributed to a shift in the location where the minimum temperature occured. The implication is that a negative temperature gradient of the radial direction should be avoided to minimize the thermal stress generated in the as grown AlN single crystal.

The above discussion reveals that different crucible parts seriously impact the temperature distribution within the growth chamber. The crucible wall significantly influences the axial temperature distribution, while the joint action of the crucible lid and the crucible wall affects the radial temperature distribution. Since the bottom of the crucible is far from the growth region, it has less influence on the temperature distribution within the growth region. Moreover, for the current as well as the future growth of AlN single crystal, the composite crucible formed by the crucible parts covering the crystal region with low thermal conductivity materials and the rest of the high thermal conductivity crucible part has excellent potential for the growth of larger size and higher quality AlN single crystal given the extremely low internal thermal stress in the crystal.

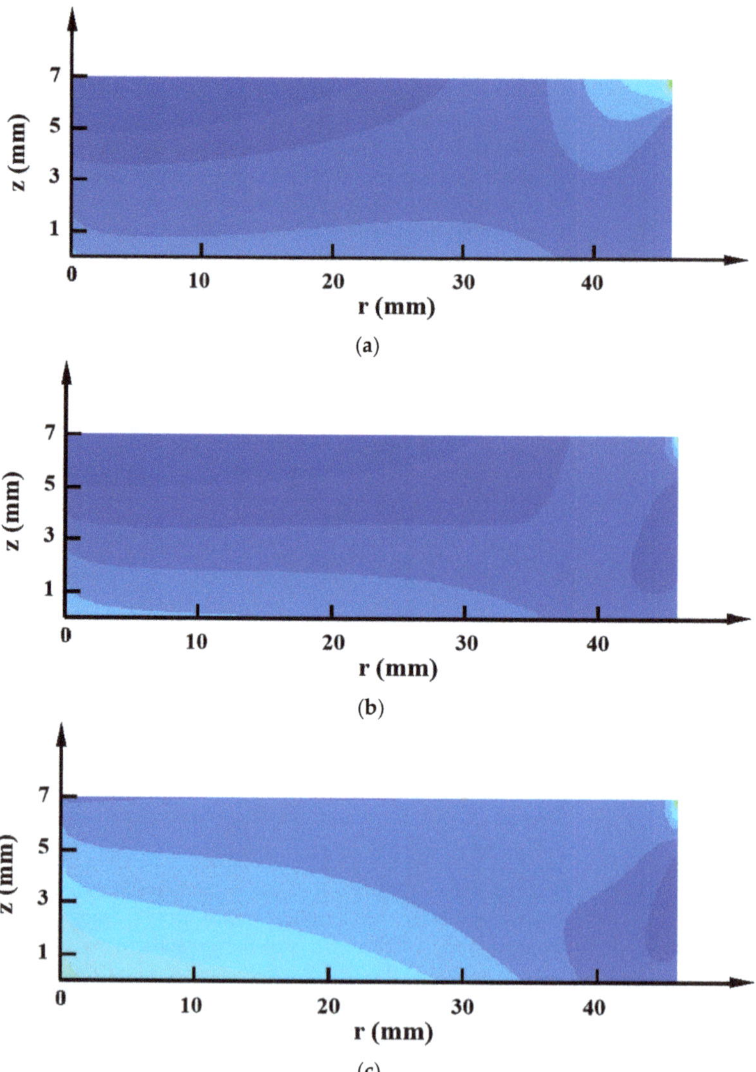

Figure 8. *Cont.*

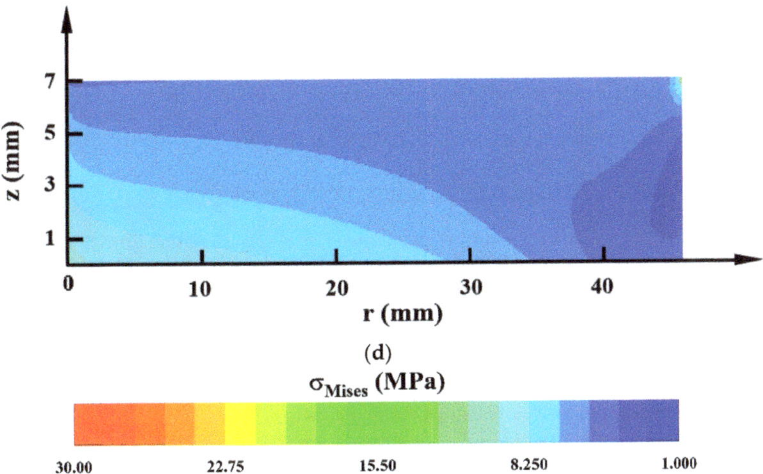

Figure 8. Distributions of thermal stress inside the AlN crystal grown in (**a**) Composite crucible A, (**b**) Composite crucible B, (**c**) Composite crucible C, and (**d**) Composite crucible D.

4. Conclusions

Concerning the problem of growing an AlN single crystal with a larger size and higher quality in the future, this paper used a numerical simulation to compare the two most suitable materials for growing 90 mm AlN single crystal concerning thermophysical properties and analyzed the influence of different crucible materials on the temperature field in the AlN crystal growth cell via the PVT method. The results showed that materials with weak thermal conductivity were more suitable as crucibles for growing large size and high quality AlN crystals because of the rapid crystal growth rate as well as the lower thermal stress level of the grown crystal. Moreover, the concept of composite crucible combining graphite and TaC to examine the effect of various crucible parts on the temperature distribution of the AlN crystal growth cell was put forward. According to the result, the temperature distribution in the axial direction was strongly influenced by the crucible wall. In contrast, the temperature distribution in the radial direction was mainly influenced by the joint action of the crucible lid and the crucible wall. The composite crucible formed by the TaC part covering the crystal region and the remaining part graphite minimized the thermal stress in the AlN crystal while ensuring a high growth rate. This study also suggests that a negative radial temperature gradient is not conducive to reducing the thermal stress inside the crystal. It provides a new direction for optimizing the temperature field for growing large-size AlN single crystal in the future.

Author Contributions: Conceptualization, Y.Y. and B.G.; methodology, B.G.; software, B.L.; validation, Y.Y., B.G. and S.L.; formal analysis, Y.Y.; investigation, B.S.; resources, Y.Y.; data curation, X.T.; writing—original draft preparation, Y.Y.; writing—review and editing, Y.Y.; visualization, P.H.; supervision, B.G.; project administration, B.G.; funding acquisition, B.G. All authors have read and agreed to the published version of the manuscript.

Funding: This research received no external funding.

Institutional Review Board Statement: Not applicable.

Informed Consent Statement: Not applicable.

Data Availability Statement: Not applicable.

Acknowledgments: All the authors listed have approved the manuscript that is enclosed.

Conflicts of Interest: The authors declare no conflict of interest.

References

1. Rojo, J.C.; Schowalter, L.J.; Gaska, R. Growth and characterization of epitaxial layers on aluminum nitride substrates prepared from bulk, single crystals. *J. Cryst. Growth* **2002**, *240*, 508–512. [CrossRef]
2. Hirayama, H.; Yatabe, T.; Noguchi, N. 231–261 nm AlGaN deep-ultraviolet light-emitting diodes fabricated on AlN multilayer buffers grown by ammonia pulse-flow method on sapphire. *Appl. Phys. Lett.* **2007**, *91*, 071901. [CrossRef]
3. Bockowski, M. Growth and doping of GaN and AlN single crystals under high nitrogen pressure. *Cryst. Res. Technol. J. Exp. Ind. Crystallogr.* **2001**, *36*, 771–787. [CrossRef]
4. Zhuang, D.; Herro, Z.G.; Schlesser, R.; Sitar, Z. Seeded growth of AlN single crystals by physical vapor transport. *J. Cryst. Growth* **2006**, *287*, 372–375. [CrossRef]
5. Klein, O.; Philip, P. Transient temperature phenomena during sublimation growth of silicon carbide single crystals. *J. Cryst. Growth* **2003**, *249*, 514–522. [CrossRef]
6. Slack, G.A.; McNelly, T.F. Growth of high purity AlN crystals. *J. Cryst. Growth* **1976**, *34*, 263–279. [CrossRef]
7. Dryburgh, P.M. The estimation of maximum growth rate for aluminium nitride crystals grown by direct sublimation. *J. Cryst. Growth* **1992**, *125*, 65–68. [CrossRef]
8. Segal, A.S.; Karpov, S.Y.; Makarov, Y.N. On mechanisms of sublimation growth of AlN bulk crystals. *J. Cryst. Growth* **2000**, *211*, 68–72. [CrossRef]
9. Liu, L.; Edgar, J.H. Transport effects in the sublimation growth of aluminum nitride. *J. Cryst. Growth* **2000**, *220*, 243–253. [CrossRef]
10. Bogdanov, M.V.; Karpov, S.Y.; Kulik, A.V. Experimental and Theoretical Analysis of Heat and Mass Transport in the System for AlN Bulk Crystal Growth. *MRS Online Proc. Libr. (OPL)* **2002**, *743*, 1545–1548. [CrossRef]
11. Liu, L.; Edgar, J.H. A global growth rate model for aluminum nitride sublimation. *J. Electrochem. Soc.* **2001**, *149*, G12. [CrossRef]
12. Wu, B.; Zhang, H. Isotropic and anisotropic growth models for the sublimation vapour transport process. *Model. Simul. Mater. Sci. Eng.* **2005**, *13*, 861. [CrossRef]
13. Cai, D.; Zheng, L.; Zhang, H.; Zhuang, D.; Herro, Z.; Schlesser, R.; Sitar, Z. Effect of thermal environment evolution on AlN bulk sublimation crystal growth. *J. Cryst. Growth* **2007**, *306*, 39–46. [CrossRef]
14. Lee, R.G.; Idesman, A.; Nyakiti, L. Modeling of the effects of different substrate materials on the residual thermal stresses in the aluminum nitride crystal grown by sublimation. *J. Appl. Phys.* **2009**, *105*, 033521. [CrossRef]
15. Wolfson, A.A. Dependence of the growth rate of an AlN layer on nitrogen pressure in a reactor for sublimation growth of AlN crystals. *Semicond.* **2010**, *44*, 1383–1385. [CrossRef]
16. Gao, B. The Impact of Pressure and Temperature on Growth Rate and Layer Uniformity in the Sublimation Growth of AlN Crystals. *J. Cryst. Growth* **2012**, *338*, 69–74. [CrossRef]
17. Wang, Q.K. Influence of Crucible Shape on Mass Transport in AlN Crystal Growth by Physical Vapor Transport Process. *J. Cryst. Growth* **2019**, *515*, 21–25. [CrossRef]
18. Wang, Q.K. Optimization of Total Resolved Shear Stress in AlN Single Crystals Homoepitaxially Grown by Physical Vapor Transport Method. *J. Cryst. Growth* **2019**, *519*, 14–19. [CrossRef]
19. Dupret, F.; Nicodeme, P.; Ryckmans, Y.; Wouters, P.; Crochet, M. Global modelling of heat transfer in crystal growth furnaces. *Int. J. Heat Mass Transf.* **1990**, *33*, 1849–1871. [CrossRef]
20. Chen, Q.S.; Zhang, H.; Prasad, V.; Balkas, C.M.; Yushin, N.K. Modeling of heat transfer and kinetics of physical vapor transport growth of silicon carbide crystals. *J. Heat Transf.* **2001**, *123*, 1098–1109. [CrossRef]
21. Wang, Z.; Deng, X.; Cao, K.; Wang, J.; Wu, L. Hotzone design and optimization for 2-in. AlN PVT growth process through global heat transfer modeling and simulations. *J. Cryst. Growth* **2017**, *474*, 76–80. [CrossRef]
22. Li, H.; Liu, X.; Feng, Y.; Wei, H.; Yang, S. Numerical study of radial temperature distribution in the AlN sublimation growth system. *Cryst. Res. Technol.* **2013**, *48*, 321–327. [CrossRef]
23. Toth, L. (Ed.) *Transition Metal Carbides and Nitrides*; Elsevier: Amsterdam, The Netherlands, 1971; Volume 1, pp. 145–148.
24. Miyazaki, N. Thermal Stress Analysis of Silicon Bulk Single Crystal during Czochralski Growth. *J. Cryst. Growth* **1992**, *125*, 102–111. [CrossRef]
25. Jordan, A.S.; Caruso, R.; VonNeida, A.R. A comparative study of thermal stress induced dislocation generation in pulled GaAs, InP, and Si crystals. *J. Appl. Phys.* **1981**, *52*, 3331–3336. [CrossRef]
26. Kamitani, K. The Elastic Constants of Silicon Carbide: ABrillouin-Scattering Study of, 4.H.; 6HSiCSingle Crystals. *J. Appl. Phys.* **1997**, *82*, 3152–3154. [CrossRef]
27. Ma, R.; Zhang, H.; Ha, S.; Skowronski, M. Integrated process modeling and experimental validation of silicon carbide sublimation growth. *J. Cryst. Growth* **2003**, *252*, 523–537. [CrossRef]
28. Liu, B.; Edgar, J.H.; Gu, Z. The durability of various crucible materials for aluminum nitride crystal growth by sublimation. *Mater. Res. Soc. Internet J. Nitride Semicond. Res.* **2004**, *9*, 6. [CrossRef]
29. Schlesser, R.; Dalmau, R.; Zhuang, D. Crucible materials for growth of aluminum nitride crystals. *J. Cryst. Growth* **2005**, *281*, 75–80. [CrossRef]
30. Noveski, V. Mass Transfer in AlNCrystal Growth at High Temperatures. *J. Cryst. Growth* **2004**, *264*, 369–378. [CrossRef]
31. Selder, M. Global Modeling of the SiC Sublimation Growth Process: Prediction of Thermoelastic Stress Control of Growth Conditions. *J. Cryst. Growth* **2001**, *226*, 501–510. [CrossRef]

MDPI

St. Alban-Anlage 66

4052 Basel

Switzerland

www.mdpi.com

Materials Editorial Office

E-mail: materials@mdpi.com

www.mdpi.com/journal/materials